The Measurement of Environmental and Resource Values

Theory and Methods

SECOND EDITION

A. Myrick Freeman III

RESOURCES FOR THE FUTURE
WASHINGTON, DC

Printed in the United States of America

An RFF Press book
Published by Resources for the Future
1616 P Street, NW, Washington, DC 20036–1400
www.rff.org

Library of Congress Cataloging-in-Publication Data

Freeman, A. Myrick, 1936–
The measurement of environmental and resource values : theory and methods / A. Myrick Freeman III — 2nd ed.
p. cm.
Includes bibliographical references and index.
ISBN 1-891853-63-5 (hardcover: alk. paper) — ISBN 1-891853-62-7 (pbk. : alk. paper)
1. Environmental policy—Cost effectiveness. 2. Environmental policy—Evaluation. I. Title.
HC79.E5F713 2003
333.7—dc21 2002155435

f e d c b a

The paper in this book meets the guidelines for permanence and durability of the Committee on Production Guidelines for Book Longevity of the Council on Library Resources.

The text of this book was designed and typeset by Betsy Kulamer in ITC New Baskerville. It was copyedited by Paula Berard. The cover was designed by Marek Antoniak.

ISBN 1–891853–63–5 (hardcover) and ISBN 1–891853–62–7 (paper)

About Resources for the Future and RFF Press

Resources for the Future (RFF) improves environmental and natural resource policymaking worldwide through independent social science research of the highest caliber.

Founded in 1952, RFF pioneered the application of economics as a tool to develop more effective policy about the use and conservation of natural resources. Its scholars continue to employ social science methods to analyze critical issues concerning pollution control, energy policy, land and water use, hazardous waste, climate change, biodiversity, and the environmental challenges of developing countries.

RFF Press supports the mission of RFF by publishing book-length works that present a broad range of approaches to the study of natural resources and the environment. Its authors and editors include RFF staff, researchers from the larger academic and policy communities, and journalists. Audiences for RFF publications include all of the participants in the policymaking process—scholars, the media, advocacy groups, nongovernmental organizations, professionals in business and government, and the general public.

Resources for the Future

Dedicated to the Memory of
Allen V. Kneese

Contents

Figures and Tables

Foreword

Young environmental economists may find it hard to imagine a situation where a clear understanding of the fundamentals of random utility, hedonic property and wage, averting behavior, and discrete response contingent valuation models was not an accepted part of the intellectual baseline. Today serious discussion of benefit measurement accepts these "details" as given and can focus on the specific attributes of each new challenge. Indeed, now the idea that economists could spend years and countless pages debating the limitations in Marshallian consumer surplus and its relationship to Hicksian measures probably seems silly. Look back a little further, not yet 50 years, and the comparison is even more surprising. One of the seminal contributions to benefit–cost analysis in 1958 concluded that public goods could *not* be valued in monetary terms. The author warned that attempts to do it would discredit benefit–cost analyses. How were these changes in beliefs as well as other common understandings of the main principles and practices of nonmarket valuation so widely recognized and accepted? The answer is simple—over this time, we had "Freeman" to start from.

Almost a quarter century ago, in 1979, Freeman's first book, *The Benefits of Environmental Improvement: Theory and Practice,* changed the intellectual landscape for nonmarket valuation. His goal was to "…describe the techniques for estimating various forms of benefits, show how they were related to the underlying welfare theory and discuss some of the pitfalls and problems in the empirical implementation…".* He sought to help practitioners and academics estimate how differences in nonmarket goods affected peo-

*A. Myrick Freeman. 1979. *The Benefits of Environmental Improvement: Theory and Practice.* Johns Hopkins University Press for Resources for the Future: Baltimore, MD, p. xiii.

ple's choices by making use of linkages between nonmarket environmental resources and private goods. These different decisions offered the keys to recovering measures for the implied economic values.

The Benefits of Environmental Improvement clarified the fundamentals and outlined the research for the U.S. Environmental Protection Agency (EPA) and the profession. *And both groups listened!* Among other things, Freeman called for:

- advancing our understanding of the framework used to measure values for small risks to life,
- an effort to examine regulatory choices to assess the implied value for changes in risk associated with the costs and outcomes of the policies,
- systematic study of recreation demand models and water quality,
- large scale recreation surveys, and
- surveys to evaluate stated preference methods.

We need only look at the literature in the eighties to see the responses. It is little wonder that, by the late 1980s, when Rick began work on a revision, the challenge was even greater. Rick's agenda for next steps and the profession's response to it created a daunting task. As a result, we didn't get a revision to *Benefits.* Freeman's new work "tamed the explosion" of new theory and applications. Once again, Rick captured the attention of the profession. Double the size of its influential predecessor, *The Measurement of Environmental and Resource Values: Theory and Methods* became *the* standard text for graduate courses in environmental economics that deal with the methodology for nonmarket valuation. Citations confirm this judgement, with about 500 to these books by November 2002. However, a citation count greatly understates the book's impact in providing the access point for serious conversations about nonmarket valuation.

This second edition of *Measurement* continues this remarkable tradition. The volume of applications continues to grow. Moreover, the literature has stretched outside the economic issues associated with consistent definition and measurement of welfare concepts. Now the issues include econometric and psychological questions that are relevant to the implementation of nonmarket valuation methods.* Rick's new edition remains true to his

* Another indication of Freeman's contribution in establishing the recognized economic infrastructure for analysis can be found in the new books designed as complements to develop the recent econometric methods used in implementing the theory he has explained. The two most recent volumes developed in this spirit are by Joseph A. Herriges and Catherine L. Kling (editors, 1999, *Valuing Recreation and the Environment,* Cheltenham, U.K., Edward Elgar) and by Timothy C. Haab and Kenneth E. McConnell (2002, *Valuing Environmental and Natural Resources,* Cheltenham, U.K., Edward Elgar). The most recent of these offers a great companion explaining the econometric issues associated with implementation of benefit estimation methods.

objective—explaining the economic dimensions of nonmarket valuation. His revision targets several of the important advances since 1993 and integrates them within the basic framework of *Measurement.* As a result, its accessibility is maintained, and its relevance to current practice enhanced.

Resources for the Future recently celebrated its 50th anniversary. For almost half of these fifty years, RFF has sustained its commitment to enhancing professional understanding of the principles and the practice of nonmarket valuation through support to assure we would have the Freeman trilogy. In my view, they could not have picked a better intellectual investment. As I have argued, the intellectual consequences of this decision have been dramatic. Its policy impact has also been substantial. The existence of the Environmental Economics Advisory Committee, a standing committee of EPA's Science Advisory Board; the strategies used in responding to Presidential Executive Orders for benefit–cost analysis; and the framework used to describe the information to be developed for the new Regulatory Flexibility Act of 2001 can all be traced to the enhanced understanding of the importance of nonmarket valuation for policy that has been provided through Rick's books.

Great teachers change the way their students think. Unfortunately, the scope of their direct impact is limited by classroom capacity. A few teachers write about their subject in a way that expands this scope of influence. Very few change a generation of scholars. Freeman has! With this edition he assures the sustained access he gave us in 1979 will continue for a new generation of environmental economists.

V. KERRY SMITH
Raleigh, North Carolina

Preface

I have been very gratified by the reception that the first edition of this book has received. And I appreciate the many requests that I undertake a revision of the book to reflect recent developments in the field of nonmarket valuation. I hope that this new edition will be viewed as an adequate response to these requests.

The objectives of this edition are essentially the same as those of the first edition and of its predecessor, *The Benefits of Environmental Improvement: Theory and Practice* (1979). These objectives are, first, to provide an introduction and overview of the principal methods and techniques of resource valuation to professional economists and graduate students who are not directly engaged in the field and, second, to give practitioners in the field an up-to-date reference on recent developments in the theory and methods underlying the practice of resource valuation. While I have tried to be comprehensive in my coverage of topics, this book is not a "how-to" manual. That kind of book would have to deal with a host of econometric and related technical issues that are outside of my area of expertise. For details of the empirical implementation of the methods and techniques described in this book, readers will have to go to the sources cited herein

What Is New

The chapter on stated preference methods has been completely rewritten and given a new title. In the chapter on longevity and health, I have added materials on such topics as quality-adjusted life years, estimating and using the value of a statistical life in policy analysis, and valuing health effects in children. And in a concluding chapter, I have added discussions of benefits transfer, combining stated preference and revealed preference data, and valuing ecosystem services.

Much of the new work in nonmarket valuation involves more sophisticated model specifications and econometric estimation methods. This is especially true of the discrete choice methods involved in analyzing many forms of stated preference data and in random utility modeling of recreation demand. Since econometrics is not my strength, I have not been able to do this work justice. But I have provided references to the relevant literature. The new book by Haab and McConnell* also should be a very useful complement to this one.

Acknowledgements

I want to thank RFF for providing financial and intellectual support for this project. Being at RFF while writing the first edition gave me the opportunity to collaborate with Maureen Cropper and Winston Harrington on two projects, the results of which found their way into Chapters 9 and 10 of this book. I also am indebted to my other colleagues at RFF during that time, for support, assistance, and stimulation, especially Adam Finkel, Carol Jones, Allen Kneese, Ray Kopp, Alan Krupnick, and Paul Portney. In addition, I want to thank the publications professionals at RFF Press, especially Paula Berard and Betsy Kulamer, for an outstanding job in editing and producing this edition of the book.

Richard Bishop, Glenn Blomquist, Catherine Kling, Raymond Palmquist, Andrew Plantinga, and Stephen Swallow read portions of the first edition and made suggestions for this revision. Two of my colleagues at Bowdoin, David Vail and Guillermo Herrera, also read and commented on portions of the new edition. I am very grateful to all of them for their help.

Finally, I want to thank the publishers of the *Southern Economic Journal* and the *Journal of Risk and Uncertainty* for permission to adapt materials which first appeared in their journals.†

A. MYRICK FREEMAN III
Georgetown, Maine

*Timothy C. Haab and Kenneth E. McConnell. 2002. *Valuing Environmental and Natural Resources: The Econometrics of Non-Market Valuation.* Cheltenham, U.K.: Edward Elgar.

†These publications are: "Measuring Welfare Values of Productivity Changes," with Winston Harrington, *Southern Economic Journal* 56(4), April 1990; "Welfare Measurement and the Benefit-Cost Analysis of Projects Affecting Risks," *Southern Economic Journal* 58(1), July, 1991; and "Indirect Methods for Valuing Changes in Environmental Risks with Non-expected Utility Preferences," *Journal of Risk and Uncertainty* 4(2), April, 1991.

The Measurement of Environmental and Resource Values

CHAPTER 1

Resource Evaluation and Public Policy

Estimates of the economic values of environmental and resource services can be a useful part of the information base supporting environmental and resource management decisions. The importance of this premise is illustrated by several current environmental and resource policy issues, all of which involve questions of economic values and trade-offs. Consider these items:

- Achieving the air and water pollution control objectives established by Congress requires massive expenditures on the part of both the public and private sectors. Is this diversion of resources from the production of other goods and services making us better off?
- Economists since A.C. Pigou (1929) have advocated placing taxes on emissions of air and water pollutants based on the damages they cause. How are these damages to be evaluated in monetary terms? And are the gains from moving to pollution taxes greater than the costs of estimating the relevant marginal damages?
- The development of new reserves of petroleum and minerals increasingly intrudes on wild and natural areas that provide other environmental and resource services. Areas that might be affected include the Arctic National Wildlife Refuge, with its fragile habitat for caribou and other species, and the outer continental shelf, where commercial and recreational fisheries may be threatened by petroleum exploration and production. Are restrictions on development in ecologically sensitive areas worth the costs they impose on society in reduced availability of and higher prices for energy and minerals?
- The development and management of large river systems such as the Columbia River basin involve choosing among alternative combinations

of hydroelectric power, water supply, and commercial and recreational fishing. There are also proposals to remove existing dams from many rivers. Are the ecological and recreational benefits of removing a dam greater than the costs in reduced power generation and water storage? And is it worthwhile to curb water withdrawals for irrigation or reduce discharges for power production to protect populations of salmon and other migratory fish?
- The commercial exploitation of some natural resource systems may be proceeding at unsustainable rates. Examples include some tropical forests and many of the world's fisheries. Shifting to sustainable rates of harvest may involve substantial short-term costs in forgone incomes. Are the long-term gains from achieving sustainable rates of harvest greater than or less than the short-term costs?
- The scientific consensus is that substantial reductions in the emissions of greenhouse gases will be required to slow or reverse the warming of the global climate. What degree of emissions reduction can be justified by the benefits of slowing or preventing global warming?
- Many people are now advocating that the U.S. system of national income accounts be substantially expanded to include measures of the values of nonmarket environmental services and deductions for the costs of environmental degradation and resource depletion. How are these values and costs to be measured?

This book is about how economics can contribute to answering questions such as these by providing measures of the economic values of the services of environmental and natural resource systems. I begin by introducing the idea of the natural environment as a set of assets or a kind of natural capital (Prugh 1999).

The Assets of Nature

Natural resources, such as forests and commercially exploitable fisheries, and environmental attributes, such as air quality, are valuable assets in that they yield flows of services to people. Public policies and the actions of individuals and firms can lead to changes in the flows of these services, thereby creating benefits and costs. Because of externalities and the common-property and public-good characteristics of at least some of these services, market forces can be relied on neither to guide them to their most highly valued uses nor to reveal prices that reflect their true social values. Externalities arise when a real variable (not a price) chosen by one economic agent enters the utility or production function of other economic agents and there is no requirement to or incentive for the first agent to take the effect on others into account when making choices.

An example of externalities is the level of emissions of smoke chosen by an electricity-generating plant when that smoke causes ill health to people downwind of the plant. A public good is nonexcludable, that is, once the good has been provided to one individual, others cannot be prevented from making use of the good, and nondepletable, that is, one person's use does not diminish the use that others can make of the good. The externalities and public-good character of many environmental services are responsible for the failure of the market system to allocate and price resource and environmental services correctly and create the need for economic measures of values to guide policymaking.

Benefit–cost analysis as the basis for making decisions about water resource investments came into its own more than 50 years ago (see Eckstein 1961, especially p. 47). Since the 1950s when the techniques of conventional benefit–cost analysis were being developed and refined, there have been significant changes in the nature of the problems being dealt with and the analytical tools that have become available. V. Kerry Smith, calling attention to these changes in his keynote lecture at Resources for the Future's 35th anniversary celebration in 1987, went on to say that

> This expansion of applications has far-reaching implications for the techniques used and for the treatment of measures of the benefits and costs. Consequently, it has led me to argue for the use of a broader term, *resource evaluation,* to describe more adequately the amendments and expansions to benefit–cost methods in evaluating today's environmental and natural resource issues. (Smith 1988, 2)

One of the changes noted by Smith is the expanding range of resource and environmental management problems being subjected to economic analysis. As Smith pointed out, benefit–cost analysis was first developed to assess the net economic values of public works projects, especially water resource developments, that withdrew productive factor inputs (land, labor, capital, and materials) from the economy to produce tangible outputs (for example, water, hydroelectric power, transportation). Many of the outputs had market counterparts, so estimation of monetary values was relatively straightforward. For example, the savings in the monetary costs of repairing flood damages was taken to be a measure of the benefits of controlling floods. In contrast, today the effects of many public actions are much more subtle and wide-ranging. This is true for both the favorable effects (benefits) and unfavorable effects (costs and damages). What were once considered unquantifiable and perhaps relatively unimportant intangibles such as improved recreation and visual amenities are now recognized as significant sources of value and are thought to be susceptible to economic measurement. Consequences that were once unrecognized (for example, small changes in the risk of cancer) or were thought to lie outside the realm of economic analysis (say, loss of biodiversity and the preservation of endangered species and unique ecological systems) are often central issues in the analysis of policy choices today.

Another change is that the distinction between natural resources and the environment that has prevailed in the discipline of economics for so long is probably no longer meaningful. The objects of analysis for natural resource economists have typically been such resources as the forest, the ore body, and the fish species that produced a flow of commodities to the economy such as wood, metal, and fish sticks. The environment has been viewed as the medium through which the externalities associated with air, noise, and water pollution have flowed and, sometimes, as the source of amenities. Increasingly this distinction appears to be artificial, as we recognize both the variety of service flows provided by natural resources and the importance of a variety of externalities. This recognition is apparently what Smith had in mind when he suggested the need "to model both natural and environmental resources as assets" (1988, 3) that yield a variety of valuable services. Freeman and others had earlier suggested that we "view the environment as an asset or a kind of nonreproducible capital good that produces a stream of various services for man. Services are tangible (such as flows of water or minerals), or functional (such as the removal, dispersion, storage, and degradation of wastes or residuals), or intangible (such as a scenic view)" (1973, 20). Ecologists are now also adopting this perspective as they refer to "natural capital" and the values of ecosystem services (Daily 1997; Daily et al. 1997; Daily et al. 2000; Prugh 1999).

As this change in perspective is adopted, it will be necessary to take a more expansive view of natural and environmental resources as complex systems with multiple outputs and joint products. The natural resource–environment complex can be viewed as producing four kinds of service flows to the economy. First, as in the conventional view of resource economics, the resource–environment system serves as a source of material inputs to the economy such as fossil fuels, wood products, minerals, water, and fish. Second, some components of the resource–environment system provide life-support services for people in the form of a breathable atmosphere and a livable climatic regime. Changes in the flows of some of these life-support services can be measured in terms of changes in the health status and life expectancies of affected populations. Third, the resource–environment system provides a wide variety of amenity services, including opportunities for recreation, wildlife observation, the pleasures of scenic views, and perhaps even services that are not related to any direct use of the environment (sometimes called nonuse or existence values). And fourth, this system disperses, transforms, and stores the residuals that are generated as by-products of economic activity. This is usually referred to as the waste receptor service of the environment (Kneese et al. 1970; Freeman et al. 1973). The resource–environment system also serves as a repository of genetic information that helps to determine the stability and resilience of the system in the face of anthropogenic and other shocks.

Many of the services provided by natural resource–environment systems can be characterized as direct services because their benefits accrue directly to people, for example materials flows and life-support services. Other environmental services could be better described as indirect services in the sense that they support other biological and ecological production processes that yield value to people. Examples include recycling of nutrients, decomposition of organic materials, generation and renewal of soil fertility, pollination of crops and natural vegetation, and biological control of agricultural and other pests.

A forest such as a unit in the U.S. national forest system is an example of a resource–environment system that provides a wide range of services, from materials such as wood and fiber to amenities such as scenic vistas, hiking, and wildlife observation, and from the regulation of stream flow and control of erosion to the absorption of atmospheric carbon dioxide. And because trees are known to emit nonmethane hydrocarbons, at least in some circumstances forests may contribute to the impairment of life-support services (Chameides et al. 1988). Some of the service flows are joint products, that is, pairs of services that can be increased or decreased together. But, generally speaking, increasing the flow of one type of service must be accompanied by a decrease in the flow of some other service, other things being equal. In other words, this system is characterized by scarcity and trade-offs and requires a multipurpose approach to its management (Bowes and Krutilla 1989).

The economic value of a resource–environment system as an asset is the sum of the discounted present values of the flows of all of the services. Because many of these service flows are not bought or sold in markets and therefore do not have market prices, the economic value of a natural asset may be quite different from its market value. For example, an acre of wetland might trade in the market for land on the basis of its value for commercial or residential development; but this value could be quite different from the value of its services as wildlife habitat and as a means of controlling floods and recharging groundwater aquifers.

The benefit of any public policy that increases the flow of one type of service is the increase in the present value of that service. However, the policy may have costs in the form of decreases in the flows of other services. Similarly, what is termed the damage due to pollution or some other human intervention is the reduction in the value of the flow of services it causes. All of these changes in resource flows, whether benefits, costs, or damages, have their counterparts in changes in the value of the resource–environment system as an asset. Some attention must therefore be devoted to the theories of asset pricing and the role of time and discounting in calculating changes in environmental and resource values. These topics will be taken up in Chapter 7.

Some of the service flows of resource–environment systems are linked directly or indirectly to markets and hence are responsive to market forces. However, many service flows are not properly regulated by markets because of externalities, their public-good characteristics of nonexcludability and nondepletability, and other factors. A decentralized market system is unlikely to lead to the optimal pattern of service flows. Hence there is a potential role for public policy in the management of resource–environment systems and a need for information on the values of the service flows.

The Economic Concept of Value

The term "value" can have several different meanings. For example, economists and ecologists use the term in two different ways in discussions of environmental services and ecosystems. Ecologists typically use the term to mean "that which is desirable or worthy of esteem for its own sake; thing or quality having intrinsic worth" (*Webster's New World Dictionary* 1988). Economists use the term in a sense more akin to "a fair or proper equivalent in money, commodities, etc." (*Webster's* again), where "equivalent in money" represents the sum of money that would have an equivalent effect on the welfare or utilities of individuals.

These two different uses of the word correspond to a distinction made by philosophers between *intrinsic* value and *instrumental* value. According to philosophers, something has *intrinsic* value "if it is valuable *in* and *for* itself—if its value is not derived from its utility, but is independent of any use or function it may have in relation to something or someone else .... an intrinsically valuable entity is said to be an 'end-in-itself,' not just a 'means' to another's ends [emphasis in original]" (Callicott 1989, 131). In contrast, something has *instrumental* value if it is valued as a means to some other end or purpose. In this view, the value of something lies in its contribution to some other goal (Costanza and Folke 1997, 49).

Some people have argued that nature has intrinsic value for various reasons, including its "harmony" or its natural balance. However, from the perspective of the "new ecology," which emphasizes disturbance and change in ecosystems (for example, Botkin 1990), this justification of an intrinsic value in nature is very problematic. A conservation biologist might argue that the part of nature consisting of the variety of organisms and their interactions and especially their genetic diversity has intrinsic value. However, this view does not endow any particular manifestation of nature with more or less intrinsic value than some alternative manifestation.

Nature's value is preserved as long as diversity in the broad sense is preserved. Although the concept of intrinsic value as applied to the environment is attractive in many respects, it does not provide a basis for dealing with the kinds of environmental management questions that I identified in the first section of this chapter. In contrast, the concept of instrumental value and in particular the economic form of instrumental value is well suited to helping answer these questions.

To assess the instrumental value of nature, we need to specify a goal and to identify the contributions that specific components of nature make toward the furtherance of that goal. Economics is the study of how societies organize themselves to provide for the sustenance and well-being of their members. Thus, in economics, the goal is increased human well-being. The economic theory of value is based on the ability of things to satisfy human needs and wants or to increase the well-being or utility of individuals. The economic value of something is a measure of its contribution to human well-being. The economic value of resource–environment systems resides in the contributions that the ecosystem functions and services make to human well-being.

The economic concept of value employed here has its foundation in neoclassical welfare economics. The basic premises of welfare economics are that the purpose of economic activity is to increase the well-being of the individuals who make up the society, and that each individual is the best judge of how well off he or she is in a given situation. Each individual's welfare depends not only on that individual's consumption of private goods and of goods and services produced by the government, but also on the quantities and qualities each receives of nonmarket goods and service flows from the resource–environment system, for example, health, visual amenities, and opportunities for outdoor recreation. The basis for deriving measures of the economic value of changes in resource–environment systems is the effects of the changes on human welfare. The anthropocentric focus of economic valuation does not preclude a concern for the survival and well-being of other species. Individuals can value the survival of other species not only because of the uses people make of them (for food and recreation, for example), but also because of an altruistic or ethical concern. The latter can be the source of existence or nonuse values, an economic value discussed in Chapter 5.

If society wishes to make the most (in terms of individuals' well-being) of its endowment of all resources, it should compare the values of what its members receive from any environmental change or use of a resource (that is, the benefits) with the values of what its members give up by taking resources and factor inputs from other uses (that is, the costs). A society that is concerned with the economic well-being of its citizens should make changes in environmental and resource allocations only if what is gained by the change is worth more in terms of individuals' welfare than what is given up by diverting resources and inputs from other uses.

The standard economic theory for measuring changes in individuals' well-being was developed to interpret changes in the prices and quantities of goods purchased in markets. This theory has been extended in the past 30 years or so to public goods and other nonmarket services such as environmental quality and health. The theory is based on the assumption that people have well-defined preferences among alternative bundles of goods, where bundles consist of various quantities of both market and nonmarket goods. The theory also assumes that people know their preferences and that these preferences have the property of substitutability among the market and nonmarket goods making up the bundles. By *substitutability*, economists mean that if the quantity of one element in an individual's bundle is reduced, it is possible to increase the quantity of some other element so as to leave the individual no worse off because of the change. In other words, the increase in the quantity of the second element substitutes for the decrease in the first element. The property of substitutability is at the core of the economist's concept of value because substitutability establishes trade-off ratios between pairs of goods that matter to people.

Given the central role of substitutability in the definition and measurement of economic values, it is important to consider the evidence supporting the assumption of substitutability. This assumption is the basis of most of the models of individual choice that are used to analyze and predict a wide variety of economic behavior both in and outside markets. These models include those of consumer demand and response to changes in prices, savings, and supply of labor. They also include models of a variety of individuals' behaviors related to environmental and health considerations, including participation in outdoor recreation activities, choices among jobs with varying degrees of risk of fatal accident, and choices of where to live and work when houses and urban centers offer different packages of amenities and pollution. The successful development and application of these models would not be possible if substitutability was not a common feature of individuals' preferences. However, some researchers have found evidence of lexicographic preferences. See, for example, Common et al. 1997 and references therein.

The trade-offs that people make as they choose less of one good and substitute more of some other good reveal something about the values people place on these goods. If one of the goods has a monetary value, the revealed values are monetary values. The money price of a market good is just a special case of a trade-off ratio because the money given up to purchase one unit of one element of the bundle is a proxy for the quantities of one or more of the other elements in the bundle that had to be reduced to make the purchase. However, even when money prices are not available, the trade-off ratios can be interpreted as expressions of economic values. In fact, there is a growing literature exploring such trade-off ratios. For example, Viscusi and colleagues (1991) asked respondents about their willingness to trade off the risk of contracting chronic bronchitis against the risk of death in an automobile accident.

Value measures based on substitutability can be expressed either in terms of willingness to pay (WTP) or willingness to accept compensation (WTA). WTP and WTA measures can be defined in terms of any good that the individual is willing to substitute for the good being valued. In the following discussion, I use money as the numeraire (unit in which prices are measured) in which trade-off ratios are expressed, but WTP and WTA could be measured in terms of any other good that mattered to the individual. The choice of a numeraire for measuring WTP or WTA is irrelevant in terms of its effect on how any one individual ranks alternative outcomes. But as Brekke (1997) has shown, the choice of a numeraire can affect the rankings of outcomes based on aggregation of welfare measures across individuals.

WTP is the maximum sum of money the individual would be willing to pay rather than do without an increase in some good such as an environmental amenity. This sum is the amount of money that would make the individual indifferent between the options of paying for and having the improvement and forgoing the improvement while keeping the money to spend on other things. WTA is the minimum sum of money the individual would require to voluntarily forgo an improvement that otherwise would be experienced; it is the amount that would make a person indifferent between the options of having the improvement and forgoing the improvement while getting extra money. Both value measures are based on the assumption of substitutability in preferences, but they adopt different reference points for levels of well-being. WTP takes as its reference point the absence of the improvement, whereas WTA takes the presence of the improvement as the base level of welfare or utility. In principle, WTP and WTA need not be exactly equal. WTP is constrained by the individual's income, but there is no upper limit on what a person could require as compensation for forgoing the improvement. Differences between WTP and WTA measures and the question of which measure is appropriate under various circumstances are discussed in more detail in Chapter 3.

Economic Values in Public Policy

There may be potential for substantial gains in economic welfare through better resource management and the judicious use of the principles of resource valuation in some cases, such as those involving the issues described at the beginning of this chapter. If the objective of management is to maximize the net economic values associated with the use of environmental and natural resources, then benefit–cost analysis becomes, in effect, a set of rules for optimum management and a set of definitions and procedures for measuring benefits and costs. Once the objective of maximum net economic value or economic efficiency has been accepted, policy becomes an almost mechanical (but not necessarily easy) process of working out estimates of marginal benefit and marginal cost curves and seeking their point of intersection.

However, most current resource and environmental policy is not based solely or even primarily on the efficiency criterion. One reason is that when many of the basic policy objectives were established, analysts were not able to provide the kind of information about values that would be required to implement the efficiency objective. But it is also true that decisionmakers may have other objectives besides economic efficiency. For example, decisionmakers may be concerned with equity considerations, intergenerational effects, the sustainability of resource systems, or social risk aversion. Thus it is not particularly useful to advocate benefit–cost analysis as a simple decision rule. Rather, as Arrow and colleagues (1996) argued, it should be considered as a framework and a set of procedures to help organize available information. Viewed in this light, benefit–cost analysis does not dictate choices; nor does it replace the ultimate authority and responsibility of decisionmakers. It is simply a tool for organizing and expressing certain kinds of information on the range of alternative courses of action. The usefulness of value estimates must be assessed in the context of this framework for arraying information.

Some people may distrust economists' efforts to extend economic measurements to such things as human health and safety, ecology, and aesthetics, and to reduce as many variables as possible to commensurate monetary measures. Some skepticism about the economist's penchant for monetary measurement is no doubt healthy, but it should not be overdone. Some people argue that some things, like human health and safety or the preservation of endangered species, cannot be valued in terms of dollars or some other numeraire. However, the real world often creates situations where trade-offs between such things as reducing risks of death and other things of value cannot be avoided. Where individuals can choose for themselves among these trade-offs, their values can be inferred from the choices. Where government policies affecting health and safety are involved, these policy choices imply values. We really need to know how to approach the problem of making choices about such trade-offs and what information can be gathered to help in the problem of choice.

Consider a hypothetical and highly simplified case of an air pollutant. Assume that the following information is known with certainty. At present levels of emissions, the pollutant causes excess mortality of 10 deaths per year in the population at risk. Reducing emissions by 30% would cost $5 million and would reduce the excess mortality to 5 deaths per year. Reducing emissions by 60% would reduce excess mortality to 3 deaths per year but would cost $15 million.

This information can be displayed as follows:

Level of control (percentage)	*Cost of control (million dollars)*	*Excess mortality (deaths per year)*
0	0	10
30	5	5
60	15	3

The problem is clearly one of the trade-offs between lives and the value of resources used up in the process of controlling emissions. If the monetary value of saving lives were known, the right-hand column of the table could be converted to dollar measures of benefits, and the appropriate benefit–cost rules could be applied to determine the optimum level of emissions control. However, in the absence of some agreed-upon basis for making lifesaving and control costs commensurable in dollar terms, no simple decision rule can be applied to determine the correct choice.

Choices of this sort are made in the political realm by decisionmakers such as the administrator of the U.S. Environmental Protection Agency (EPA). Whatever the choice, there is an implicit value of lifesaving that is consistent with that choice and can be said to have been revealed by that choice. In this example, if the decisionmaker chooses the 30% control level, the value of lifesaving is revealed to be at least $1 million per death avoided. The 30% control level "buys" 5 lives saved at a cost of $5 million. The choice further reveals that the value of lifesaving to the decisionmaker is less than $5 million, because the decisionmaker declined the opportunity to "purchase" the additional 2 lives saved that the additional $10 million of control costs would make possible. If the 60% control level had been chosen, this would have revealed a value of lifesaving of at least $5 million.

In this example with only three data points, the implicit value can only be determined within some range. If control costs and mortality as functions of the level of control were continuous relationships and known with certainty, the choice of a control level would imply a precise value of life. If it is assumed that the control level was established to equate marginal benefits and marginal costs, and marginal costs are known, the marginal benefit or value can be inferred.

A number of recent studies have shed light on the implied value of lifesaving by estimating the costs per death avoided for various regulatory policies. See, for example, Cropper and coauthors 1992; van Houtven and Cropper 1996; Hamilton and Viscusi 1999; and Tengs and coauthors 1995.

In this example, choice revealed value, rather than value determining the choice. But either way, the problem of valuation cannot be avoided. It can be hidden. However, I would argue that in a democratic society, the more open decisionmakers are about the problems of making choices and the values involved and the more information they have about the implications of their choices, the better their choices are likely to be. Estimates of values in monetary terms are one such source of information.

Most resource and environmental management problems have structures similar to the one just discussed. For example, in the preceding table the first column could be rate of harvest in a forest, the second column the net economic value of the harvest, and the third column the probability of survival of an endangered species requiring old-growth forest habitat. For an example, see Montgomery et al. 1994. Reducing harvest levels increases the probability of survival of the species, but at a cost. The rate of harvest actually chosen implies something about the value to the decisionmakers of increasing the probability of survival. Of course, the second column need not be measured in dollars. The trade-off could be between having more timber wolves and fewer deer. The fact that there is no monetary measure in that example does not make it any less an economic problem. Whatever choice is made about a population level for wolves implies something about the relative values placed on wolves and deer by the person making the decision. This example is another manifestation of the fundamental economic fact of scarcity, that is, that more of one thing means less of something else that people value.

Because policy choices about resources and environmental quality are made in a political context and are likely to involve comparisons and trade-offs among variables for which there is no agreement about commensurate values, monetary benefit and cost data will not always be the determining factors in decisionmaking. However, benefit and cost estimates are an important type of information. Their usefulness lies in the fact that they use easily understood and accepted rules to reduce complex clusters of effects and phenomena to single-valued commensurate magnitudes, that is, to dollars. The value of the benefit–cost framework lies in its ability to organize and simplify certain types of information into commensurate measures (Arrow et al. 1996).

Classifications of Values

In this section I describe some ways of classifying the types of environmental and resource service flows for which value measures might be desired. Any classification system contains a certain element of arbitrariness. The usefulness of any particular classification depends on how well it illuminates important similarities and differences among types of service flows. Which similarities and differences are important depends upon the particular questions being examined.

One basis for classification is the type of resource or environmental media. Environmental effects are often classified according to whether they stem from changes in air quality, water quality, land quality, and so forth. The current legal and administrative division of responsibilities for environmental management and pollution control is consistent with this basis for classification. However, this division is becoming less and less relevant as cross-media effects are becoming better understood. For example, controlling emissions of nitrogen oxides from coal-burning power plants might be part of a cost-effective strategy for improving water quality in estuaries because of the impact of nitrate deposition on nutrient levels in these waters. Controls on land use are a key part of the strategy for reducing nonpoint-source water pollution.

A second type of classification is based on the economic channel through which human well-being is affected. Environmental and resource service flows can be classified according to whether they convey their effects through the market system (as changes in incomes to producers and changes in the availability of and prices for marketed goods and services to consumers) or through changes in the availability of goods and services not normally purchased through markets, for example, health, environmental amenities such as visibility, and opportunities for outdoor recreation. This book is about the methods and techniques for measuring the values of these latter, nonmarket, services. However, many of the policies for managing environmental and resource systems will affect the flows of both market and nonmarket goods and services. So policy assessments will need to make use of market as well as nonmarket valuation methods.

A third way of classifying environmental and resource service flows is according to whether they affect humans directly, indirectly through their impact on other living organisms, or indirectly through inanimate systems. Direct effects on humans include the morbidity and mortality effects associated with air and water pollutants, hazardous wastes, pesticide residues, and the like, and the nonhealth effects of pollutants manifesting themselves as odors, reduced visibility, and reduced visual attractiveness of outdoor settings.

Effects on humans involving biological mechanisms and other organisms include those on the economic productivity of both managed and natural ecosystems, such as agricultural croplands, commercial forests, and commercial fisheries. Market valuation methods are used to value these effects. Nonmarket direct service flows also affect people, such as recreational uses of ecosystems for hunting, fishing, and nature observation. There are also impacts on indirect or intermediate ecosystem services such as pollination, decomposition, biological pest control, and nutrient recycling.

Effects acting through nonliving systems include damages to materials and structures and increases in cleaning and repair costs at commercial activities, which would be measured by market valuation techniques; damages to materials and structures and increases in cleaning and repair costs for households, which would be measured by nonmarket valuation techniques; and effects on weather and climate, which would be measured by either market or nonmarket valuation techniques, depending on the nature of the activity affected.

Finally, we can distinguish between those services that individuals value because they make use of them in some way (*use values*) and those that they value independent of any kind of observable use. These have been called *nonuse values,* or more recently, *passive use values.* Questions of defining *use* and use versus nonuse values will be taken up in Chapter 5.

Ex Post and Ex Ante Analysis of Values

The decisionmaker who is trying to allocate scarce resources and is faced with a number of competing goals needs ex ante analyses of the effects of alternative policy actions to guide decisionmaking. *Ex ante analysis* involves the prediction of the physical and economic consequences of policies on the basis of a model of the physical and economic processes involved. It involves visualizing two alternative states of the world, one with the policy in question and one without, and then comparing these alternative futures in terms of some established criterion, such as net economic efficiency. *Ex post analysis* of a policy involves measuring the actual consequences of the policy by comparing the observed state with a hypothetical alternative, the state of the world without the policy. Ex post analysis, in effect, treats the policy as a controlled laboratory experiment, except that the control is hypothetical rather than real. Natural resource damage assessment is an example of ex post evaluation, in that the damaged state is observed and must be compared with a hypothetical or counterfactual alternative in which there was no pollution event but all other factors are assumed to have remained unchanged.

Ex post and ex ante analyses are not competitive alternatives; rather they should be viewed as complementary techniques for improving our knowledge. An ex post analysis of a policy can be viewed as a check on the validity of the ex ante analysis. The ex ante analysis is a prediction of what will happen; the ex post analysis is a check of what actually did happen.

It is particularly important that the economic analysis of environmental and resource policies include ex post analysis. Our knowledge of the physical and economic systems on which present ex ante analyses are based is extremely limited. We need not only to develop more comprehensive models of the physical, biological, and economic aspects of the system but also to devote more effort to verifying these models through ex post comparisons of the predictions with observed results.

Although ex ante analyses of environmental, health, and safety regulations have become quite common in the United States (Hahn 2000), ex post analyses of environmental and resource policies are quite rare. An early study of the realized benefits and costs of U.S. Army Corps of Engineers' water resource development projects was done by Haveman (1972). The most notable and controversial recent study was U.S. EPA's retrospective analysis of the benefits and costs of the Clean Air Act (U.S. EPA 1997).*

The ex post verification of the analytical models used in resource valuation is not simply a comparison of actual results with predictions. Ex ante models are based upon some view of the future with projections of economic magnitudes such as population levels, real income, and price levels. Care must be taken in ex post analysis to sort out the effects of unforeseen developments such as war or uncontrolled inflation on the variables in question. For example, if the failure of income levels to rise on the projected path results in a shortfall of recreation benefits at a particular site, this is not a failure of the analytical model so much as a reflection of our inability to perceive the future. The real benefit of ex post analysis is in making the most of the opportunity to improve on the analytical models used.

Preview

This book reviews and summarizes the basic theory of economic welfare measurement and presents resource evaluation and benefit measurement techniques that are consistent with this underlying theory. In the next chapter, I provide an overview of valuation and welfare measurement methods and discuss the relationship between the economic methods of valuation and the physical and biological relationships that define the resource and environmental systems being valued.

Chapters 3 and 4 constitute the theoretical core of the book. Chapter 3 lays out the basic premises and value judgements that underlie the economic concept of benefits and presents the basic theory of the measurement of economic welfare changes. Chapter 4 introduces the basic methods and models for deriving welfare and value measures from the revealed choices of individuals and from observed changes in market prices.

*For some of the controversy generated by this report, see Crandall 1997; Lutter and Belzer 2000; and Brown et al. 2001.

Chapter 5 introduces the concept of nonuse or existence value and shows that the indirect methods of measurement outlined in Chapter 4 are not likely to be capable of measuring this type of value. Chapter 6 describes the stated preference methods for calculating welfare measures, which are based on responses to hypothetical questions rather than on observed prices and quantities. They include WTP surveys and bidding games, direct referendum questions, and questions about how individuals would rank alternative bundles of environmental and private goods and how their behavior would change under alternative hypothetical circumstances.

Chapter 7 takes up the question of valuation across time and the role of discounting in welfare measurement. Chapter 8 extends the theory of value and welfare change to a situation of risk where people are uncertain about what the actual state of the world will be. Public policies toward the management of environmental resources can affect either the probabilities of alternative outcomes or the magnitudes of environmental services in alternative states of the world. Thus this chapter describes the application of the basic theory of welfare change to evaluating environmentally induced changes in risks.

The remaining chapters describe the application of the various models and methods for welfare measurement to specific situations such as measuring the values of environmental changes affecting producers' costs and productivity (Chapter 9), environmentally induced changes in longevity and health (Chapter 10), applications of the hedonic price model to housing prices and wage rates (Chapters 11 and 12), and the valuation of resources that support recreation activities (Chapter 13). The final chapter contains sections on benefits transfer, combining stated preference and revealed preference data, and the valuation of ecosystem services and then offers some conclusions.

Mathematical Notation

In this book I use the following conventions regarding mathematical notation:

- Vectors are represented by boldface uppercase letters; lowercase letters with or without subscripts represent values for individual variables in these vectors, for example, $\boldsymbol{X} = x_1, \ldots, x_i, \ldots, x_n$.
- The subscript letters $i, j, k, m,$ and n and subscripted numbers index elements of vectors.
- The meanings of other subscripted letters are specified when they are first used.
- Superscripted letters are used to index such things as utility functions and production functions to specific individuals and businesses. For

example, $u^j(\mathbf{X})$ gives individual j's utility as a function of that individual's consumption of goods $x_1, \ldots, x_i, \ldots, x_n$.

- Primes and superscripted numerals represent specific values for variables. For example, M^0 represents the initial value of the variable M. Similarly, $\Delta x = x'' - x'$ means the change in x from x' to x''.
- Uppercase letters represent variables expressed as quantities of money. For example, M represents income, and CV is the compensating variation measure of welfare change.

References

Arrow, Kenneth J., Maureen L. Cropper, et al. 1996. Is There a Role for Benefit–Cost Analysis in Environmental, Health, and Safety Regulation? *Science* 272: 221–222.

Botkin, Daniel. 1990. *Discordant Harmonies: A New Ecology for the Twenty-First Century*. New York: Oxford University Press.

Bowes, Michael D., and John V. Krutilla. 1989. *Multiple-Use Management: The Economics of Public Forestlands*. Washington, DC: Resources for the Future.

Brekke, Kjell Arne. 1997. The numeraire matters in cost–benefit analysis. *Journal of Public Economics* 64: 117–123.

Brown, Gardner M., Trudy A. Cameron, et al. 2001. Misrepresenting ACCACA. *Regulation* 24(4): 4–5.

Callicott, J. Baird. 1989. *In Defense of the Land Ethic: Essays in Environmental Philosophy*. Albany, NY: State University of New York Press.

Chameides, W.L., R.W. Lindsay, J. Richardson, and C.S. King. 1988. The Role of Biogenic Hydrocarbons in Urban Photochemical Smog: Atlanta as a Case Study. *Science* 241: 1473–1475.

Common, M., I. Reid, and R. Blamey. 1997. Do Existence Values for Cost-Benefit Analysis Exist? *Environmental and Resource Economics* 9(2): 225-238.

Costanza, Robert, and Carl Folke. 1997. Valuing Ecosystem Services with Efficiency, Fairness and Sustainability as Goals. In *Nature's Services: Societal Dependence on Natural Ecosystems*, edited by Gretchen C. Daily. Washington, DC: Island Press.

Crandall, Robert W. 1997. The Costly Pursuit of the Impossible. *The Brookings Review* Summer: 41–47.

Cropper, Maureen L., William N. Evans, Stephen J. Beradi, Maria M. Ducla-Soares, and Paul R. Portney. 1992. The Determinants of Pesticide Regulation: A Statistical Analysis of EPA Decision Making. *Journal of Political Economy* 100(1): 175–189.

Daily, Gretchen C. (ed.). 1997. *Nature's Services: Societal Dependence on Natural Ecosystems*. Washington, DC: Island Press.

Daily, Gretchen C., S.E. Alexander, et al. 1997. Ecosystem Services: Benefits Supplied to Human Societies by Natural Ecosystems. *Issues in Ecology*. 2(Spring). http://www.esa.org/sbi/issues2.pdf (accessed October 9, 2002).

Daily, Gretchen C., T. Söderqvist, et al. 2000. The Value of Nature and the Nature of Value. *Science* 289: 395–396.

Eckstein, Otto. 1961. *Water-Resource Development: The Economics of Project Evaluation.* Cambridge, MA: Harvard University Press.

Freeman, A. Myrick, III, Robert H. Haveman, and Allen V. Kneese. 1973. *The Economics of Environmental Policy.* New York: John Wiley.

Hahn, Robert W. 2000. *Reviving Regulatory Reform: A Global Perspective.* Washington, DC: AEI–Brookings Joint Center for Regulatory Studies.

Hamilton, James T., and W. Kip Viscusi. 1999. How Costly Is 'Clean'? An Analysis of the Benefits and Costs of Superfund Remediations. *Journal of Policy Analysis and Management* 18(1): 2–27.

Haveman, Robert H. 1972. *The Economic Performance of Public Investments.* Baltimore, MD: The Johns Hopkins University Press for Resources for the Future.

Kneese, Allen V., Robert U. Ayres, and Ralph C. d'Arge. 1970. *Economics and the Environment: A Materials Balance Approach.* Baltimore, MD: The Johns Hopkins University Press for Resources for the Future.

Lutter, Randall, and Richard B. Belzer. 2000. EPA Pats Itself on the Back. *Regulation* 23(3): 23–28.

Montgomery, Claire A., G. M. Brown Jr., and Darius M. Adams. 1994. The Marginal Cost of Species Preservation: The Northern Spotted Owl. *Journal of Environmental Economics and Management* 26(2): 111–128.

Pigou, Arthur C. 1929. *The Economics of Welfare.* London: Macmillan.

Prugh, Thomas. 1999. *Natural Capital and Human Economic Survival.* 2nd ed. Boca Raton, FL: Lewis Publishers.

Smith, V. Kerry. 1988. Resource Evaluation at the Crossroads. *Resources* 90: 2–6.

Tengs, Tammy O., E. Winer, et al. 1995. Five-Hundred Life-Saving Interventions and Their Cost-Effectiveness. *Risk Analysis* 15: 369–390.

U.S. EPA (Environmental Protection Agency). 1997. *The Benefits and Costs of the Clean Air Act—1970 to 1990.* Washington, DC: U.S. EPA.

van Houtven, George, and Maureen L. Cropper. 1996. When Is a Life Too Costly To Save? The Evidence from U.S. Environmental Regulations. *Journal of Environmental Economics and Management* 30(3): 344–368.

Viscusi, W. Kip, Wesley A. Magat, and Joel Huber. 1991. Pricing Environmental Health Risks: Survey Assessments of Risk–Risk and Risk–Dollar Trade-offs for Chronic Bronchitis. *Journal of Environmental Economics and Management* 21(1): 322–351.

Webster's New World Dictionary, Third College Edition. 1988. New York: Simon & Schuster.

CHAPTER

2

Measuring Values, Benefits, and Costs: An Overview

In the first section of this chapter, I present a simple general equilibrium model of an economy with nonmarket environmental and resource service flows to show how the values of these flows emerge as shadow prices from the solution to a welfare maximization problem. In the second section, I describe in general terms the major types of methods for estimating values and benefits. In the third section, I discuss some theoretical frameworks used to model revealed preference models. In the fourth section, I present a simple model to illustrate the relationships between physical and biological changes in environmental and resource systems and the changes in well-being and values realized by the people affected by these changes. These models help to make clear the kinds of data required to carry out resource evaluations and to show the roles of economists and physical and biological scientists in the evaluation process. The fifth section describes the noneconomic foundations of resource valuation. I conclude this chapter with discussions of the nature of cost in welfare theory and the appropriate treatment of uncertainties in the measurement of values.

Resource Values as Shadow Prices

Economic values can only be defined in terms of some underlying criterion that identifies what is to be considered good. As discussed in Chapter 1, in neoclassical welfare economics, good is defined in terms of the well-being of individuals. Here, I assume that an individual's well-being can be represented by an ordinal utility function. This assumption is only a first step in defining "good" because it does not deal with interpersonal comparisons. Specifically, it does not answer the question of whether it is good when one individual's utility increases while another individual's utility decreases.

The concept of *Pareto optimality* is the route I choose for dealing with this set of problems. An allocation of resources, goods, and services in an economy is Pareto optimal if there is no feasible reallocation that can increase any one person's utility without decreasing someone else's utility. Of course, there is an infinite number of Pareto optimum allocations for an economy, each with a different distribution of utilities across individuals. To rank the allocations it is necessary to have a social welfare function that aggregates the utilities of the individuals, perhaps by assigning social welfare weights. If such a social welfare function exists, Pareto optimality is a necessary but not a sufficient condition for maximizing that function. Hence, despite its limitations, Pareto optimality has some usefulness in evaluating economic outcomes.

Pareto optimality is the solution to a constrained maximization problem in which some of the constraints are the exogenously determined environmental and resource service flows. The shadow prices on these constraints are the economic values of these service flows. In the general equilibrium model I use here, it is clear that the values assigned to the environmental and resource service flows are not fixed parameters but are determined by their roles in enhancing individuals' well-being and arise from their scarcity or limited availability. This model is a variation of the general equilibrium externalities model presented by Baumol and Oates (1975, 1988).

In this model the subscript j indexes a vector of m commodities ($j = 1, \ldots, j, \ldots, m$), i indexes the n individuals in the economy, and k indexes the h firms. Each of the commodities is divisible, with well-defined and enforceable property rights. The variable names for these commodities reflect their role in economic activity. $\boldsymbol{X}$ represents the vector of commodities used as consumer goods by individuals. Specifically, let x_{ji} = the amount of commodity j consumed by individual i ($j = 1, \ldots, m$) ($i = 1, \ldots, n$).

$\boldsymbol{Y}$ represents the vector of commodities being produced by firms, where negative values indicate commodities being used as resource inputs to production. Specifically, y_{jk} = the amount of commodity j produced (used) by firm k ($j = 1, \ldots, m$) ($k = 1, \ldots, h$).

There is also an endowment of commodities represented by $\boldsymbol{S}$: s_j = the total of endowment of commodity j available to the economy ($j = 1, \ldots, m$).

For each commodity, the initial endowment plus the quantities produced by firms must just equal the sum of the quantities used as inputs by firms and consumed by individuals. This requirement is expressed in a set of production–consumption constraints:

$$s_j + \sum_{k=1}^{h} y_{jk} - \sum_{i=1}^{n} x_{ji} = 0 \quad (j = 1, \ldots, m) \tag{2-1}$$

In addition, let there be a resource service flow r that is nondivisible, so that it is, in effect, a public good. The endowment of r is determined exogenously. For simplicity, I assume that r does not vary across space. This means that r takes the same value for all individuals. Finally, let d be some environmental quality parameter, say the concentration of a pollutant at a specific place. The level of d will depend on the discharges of pollutants by firms and will determine the quantities of pollution experienced by each individual. These relationships will be specified below.

Individuals' preferences can be represented by utility functions:

$$u^i = u^i(x_{ji}, r, d_i)\,(i = 1, \ldots, n) \quad (j = 1, \ldots, m) \tag{2-2}$$

Let $\partial u^i/\partial x_{ji} \geq 0$, $\partial u^i/\partial r \geq 0$, and $\partial u^i/\partial d_i < 0$, where d_i is the concentration of the pollutant to which the ith individual is exposed.

The production side of the economy can be represented by a set of production functions for multiproduct firms:

$$f^k(y_{1k}, \ldots, y_{jk}, \ldots, y_{mk}, r, d_k) = 0 \quad (k = 1, \ldots, h) \tag{2-3}$$

where y_{jk} is the production of good j by firm k, r is the level of the nonmarket resource service flow, and d_k is the firm's contribution to the environmental quality parameter d as represented by, for example, emissions of a pollutant.

The variable d is a function of the emissions of all firms. For simplicity, let $d = \Sigma^k d_k$. Also, assume that $d_i = \alpha_i \times d$, where α_i relates each individual's exposure to the pollutant to the aggregate level of pollution, d. For a perfectly mixed unavoidable pollutant, $\alpha_i = 1$ for all i. It would also be possible to represent averting behavior by making α_i a choice variable, perhaps a function of some x_{ji}.

The Pareto optimum conditions for this economy can be found by determining the conditions that maximize each individual's utility, subject to the production function and resource constraints and the constraint that all other individuals' utilities are held constant at some level representing the status quo. This procedure makes it clear that any Pareto optimum allocation is only as good as our judgement about the associated distribution of utilities across individuals. For individual one, the problem is to choose the values for x_{j1} and d_1 that

$$\max L_1 = u^1(\cdot) + \sum_{i=2}^{n} \lambda_i \left[u^{i*} - u^i(\cdot) \right] - \sum_{k=1}^{h} \mu_k f^k(\cdot) + \sum_{j=1}^{m} \rho_j \left(s_j + \sum_{k=1}^{h} y_{jk} - \sum_{i=1}^{n} x_{ji} \right) + \gamma (r^* - r) \quad (2\text{-}4)$$

where r^* is the exogenously determined quantity of r. Assuming an interior solution, the first-order conditions for a Pareto optimum are the following:

$$\lambda_i \left(\frac{\partial u^i}{\partial x_{ji}} \right) - \rho_j = 0 \quad (i = 1, \ldots, n)(j = 1, \ldots, m) \quad (2\text{-}5)$$

$$-\mu_k \left(\frac{\partial f^k}{\partial y_{jk}} \right) + \rho_j = 0 \quad (k = 1, \ldots, h)(j = 1, \ldots, m) \quad (2\text{-}6)$$

$$\sum_{i=1}^{n} \lambda_i \left(\frac{\partial u^i}{\partial r} \right) - \sum_{k=1}^{h} \mu_k \left(\frac{\partial f^k}{\partial r} \right) - \gamma = 0 \quad (i = 1, \ldots, n)(k = 1, \ldots, h) \quad (2\text{-}7)$$

$$\sum_{i=1}^{n} \alpha^i \lambda_i \left(\frac{\partial u^i}{\partial d_i} \right) + \mu_k \left(\frac{\partial f^k}{\partial d_k} \right) = 0 \quad (i = 1, \ldots, n)(k = 1, \ldots, h) \quad (2\text{-}8)$$

where $\partial u_i / \partial d_i$, μ_k, and $\partial f^k / \partial d_k < 0$.

Conditions 2-5 and 2-6 generate the marginal rates of substitution and of transformation for the marketed goods. They can be interpreted as saying either that each individual's marginal valuation of each good must equal the marginal cost to firms of its production, or that each individual's marginal rate of substitution between any pair of goods must equal each firm's marginal rate of transformation between that pair of goods.

In condition 2-7 the Lagrangian term γ is the shadow value for r^*, that is, it gives the increase in the objective function for a marginal increase in the constrained resource service r. Condition 2-7 also shows that the shadow price depends on the utility and production functions. Specifically, the shadow price is equal to the sum of the marginal values attached to r by all individuals and producers in the constrained solution.

The first term in condition 2-8 is the value to individuals of reducing firm k's emissions by one unit. Condition 2-8 says, in effect, that Pareto optimality requires that the aggregate value of reducing the emissions of firm k must be just equal to the marginal cost of that reduction (the second term in the equation). The second term in condition 2-8 can also be interpreted as the marginal value to the kth firm of being able to emit one unit to the environment, that is, it is the marginal value of the waste receptor services of the environment.

Conditions 2-7 and 2-8 give the marginal values for changes in r and d_k. They also imply the existence of aggregate marginal willingness-to-pay curves that give the marginal willingness to pay (WTP) for the service flows as functions of the quantities of the flows being supplied. As is shown in Chapter 3, WTP can be taken to be equal to the area under such a marginal WTP curve. Value estimation, then, involves determining directly or indirectly the shapes of these marginal WTP curves for environmental services.

If the services of the environment could be purchased in a perfectly functioning market, there would be observable demand functions for them, making estimation of the marginal WTP curves a fairly straightforward econometric problem. Also, then, environmental and resource management would not be an important public policy matter. However, environmental and resource service flows typically have characteristics such as nonexcludability and nonrivalry in consumption, which make it difficult or impossible for markets for these services to function well. Often individuals are not free to vary independently the level of the services they consume. The public-good character of environmental services then leads to market failure. Without a market, there are no price and quantity data from which the demand relationships can be estimated.

To sum up, the economic values of r and of reducing d emerge as part of the solution to the welfare maximization problem. These values depend on context in the sense that changes in preferences (equation 2-2), production technology (equation 2-3), or the resource endowments and constraints will affect these values. This also means that if the economy is not at a Pareto optimum allocation as defined by conditions 2-5–2-8, for example, because of an additional constraint, then the shadow prices or values attached to r and d will be different.

Methods for Measuring Values

The principal distinction among methods for valuing changes in environmental goods is based on the source of the data (Mitchell and Carson 1989, 74–87). The data can come either from observations of people acting in real-world settings where people must live with the consequences of their choices or from people's responses to hypothetical questions of the form, what would you do if …? or, how much would you be willing to pay for …? In the first edition of this book (Freeman 1993), I distinguished between "observed" methods and "hypothetical" methods. But it has become more common now to refer to these as *revealed preference* methods and *stated preference* methods.

Revealed preference methods are based on actual behavior reflecting utility maximization subject to constraints. One type of revealed preference method is based on observed choices in a referendum setting. If an individual is offered a fixed quantity of a good at a given price on a take-it-or-leave-it or a yes–no basis (as in a referendum), observation of the choice reveals only whether the value of the offered good to the individual was greater than or less than the offering price. A family of discrete choice and random utility models has been developed for the purpose of deriving value measures from these choices.

In most instances the environmental service does not have an offering price, but sometimes its quantity affects the choices people make about other things, such as quantities of market goods. In these cases, the value of the environmental service must be inferred through the application of some model of the relationship between market goods and the environmental service. Most such models are based on the assumption of some kind of substitute or complementary relationship between the environmental service and marketed goods and services. Examples of these models include the household production model (which includes models of household spending on cleaning and on repair of materials damaged by air pollution), the travel cost demand model for visits to a recreation site, and the hedonic property value and hedonic wage models. Revealed preference methods involve a kind of detective work in which clues about the values individuals place on environmental services are pieced together from the evidence that people leave behind as they respond to prices and other economic signals. The basic properties of these models are discussed in Chapter 4.

The principal difference between revealed preference and stated preference methods is that the latter draw their data from people's responses to hypothetical questions rather than from observations of real-world choices. The earliest techniques for estimating values using stated preference methods involved asking people directly about the values they place on environmental services by creating, in effect, a hypothetical market. For example, people could be asked what value they place on a specified change in environmental services, or how much of an environmental service they would "purchase" at a given price. Stated preference questions of this type simply ask people what value they place on a specified change in an environmental amenity or the maximum amount they would be willing to pay to have an event occur. The responses, if truthful, are direct expressions of value and would be interpreted as measures of compensating variation. The term *contingent valuation method* is conventionally used to refer to approaches based on this type of questioning.

While a variety of elicitation formats is possible, the most popular today is the *referendum format,* which asks for a yes or no answer to the question, would you be willing to pay $X for ...? In their simplest form, referendum-type questions reveal only an upper bound (for a no) or a lower bound (for a yes) on the relevant welfare measure. Questions of this sort are often referred to as referendum questions because of the analogy with voting on such things as bond issues. Discrete choice methods applied to a large sample of individual responses can be used to estimate WTP functions or indirect utility functions from data on responses and on the characteristics of the people in the sample.

In a second form of stated preference question, respondents are given a set of hypothetical alternatives, each depicting a different situation with respect to some environmental amenity and other characteristics and are asked to rank the alternatives in order of preference or to pick the most preferred alternative. Several names have been applied to variations of this approach, including *contingent ranking* and *choice experiments.* The rankings or choices can then be analyzed to determine, in effect, the marginal rate of substitution between any other characteristic and the level of the environmental amenity. If one of the other characteristics is a monetary price, then it is possible to compute the respondent's willingness to pay for the good on the basis of the ranking of alternatives.

In the third type of stated preference question, known as *contingent activity* or *contingent behavior* questions, individuals are asked how they would change the level of some activity in response to a change in an environmental amenity. If the activity can be interpreted in the context of some behavioral model, such as an averting behavior model or a recreation travel cost demand model, the appropriate revealed preference method can be used to obtain a measure of WTP, as if the reported behavioral intentions were actual behaviors. McConnell (1986), for example, applied a recreation travel cost demand model to questions of the form, how often would you visit these beaches if they were free of PCBs? to estimate the damages (resource value lost) from the pollution of the waters of New Bedford Harbor, in Massachusetts, with polychlorinated biphenyls.

A fourth type of stated preference question asks respondents to rate a set of bundles on some scale. As in the case of contingent ranking, if one of the characteristics of a bundle is a money price, values can be inferred from the ratings. This approach is sometimes referred to as *conjoint analysis* or *attribute-based stated choice.* For a recent example, see Roe et al. 1996.

Some issues and problems in stated preference methods are specific to the particular form of the question being asked. For example, when people are asked how much they would be willing to pay for something, they might say "zero" because they reject the idea of having to pay for something they consider rightfully theirs. Other problems are generic to all methods based on hypothetical questions, for example, problems in scenario specification, sampling, and item nonresponse. The major questions regarding all stated preference methods concern the validity and reliability of the data, that is, whether the hypothetical nature of the questions asked inevitably leads to some kind of bias or results in so much "noise" that the data are not useful for drawing inferences. Further discussion of these questions is left to Chapter 6.

The Methodology of Revealed Preference Models

Because the revealed preference methods for measuring values use data on observed behavior, some theoretical framework must be developed to model this behavior and to relate the behavior to the desired monetary measures of value and welfare change. A key element in the theoretical framework is the model of the optimizing behavior of an economic agent (individual or firm) that relates the agent's choices to the relevant prices and constraints, including the level of environmental or resource quality q. If a behavioral relationship between observable choice variables and q can be specified and estimated, this relationship can be used to calculate the marginal rate of substitution between q and some observed choice variable with a money price tag, thereby revealing the marginal value of changes in q.

Welfare measurement in the case where changes in q affect individuals involves three steps. The first is to derive the expression for WTP as a function of the environmental variable, usually either from the indirect utility function or the expenditure function. This expression gives the compensating change in income that holds utility constant for the change in the environmental parameter. The second step is to develop a model of individual utility-maximizing behavior that relates the individual's choices to the relevant prices and constraints, including the level of environmental quality. The first-order conditions for optimization can then be derived. These first-order conditions involve equating measures of marginal value to price or equating some marginal rate of substitution or marginal rate of transformation to some price ratio. The third step is to examine the model to see whether the first-order conditions include a relationship between the desired marginal value for the environmental change and some observable variable. If they do, then the observable variable can be taken as a measure of the marginal change in welfare. Several types of models using this methodology are presented in Chapter 4.

Drawing inferences about the marginal values of environmental changes is relatively straightforward. However, policymakers will typically need estimates of values for nonmarginal changes. Deriving these estimates is considerably more difficult because what is needed is not the marginal values but the marginal value *functions.* I will have more to say about how these might be obtained in subsequent chapters.

The measures of value and welfare change derived from optimization models often produce results that are quite different from those of the naive models used in the early literature on value and benefit measurement. The early models were often based on what has come to be known as the *damage function approach.* The damage function approach involves estimating some physical relationship between a measure of environmental quality (or its converse, pollution) and some physical measure of damage or loss (such as number of workdays lost to sickness in the case of health, or percentage of crop lost in the case of effects of air pollution on agricultural productivity). Then some unit price is applied to the physical measure to convert it to monetary terms. In some studies, for example, lost wages and medical costs were used to determine the value of avoiding one day of illness induced by air pollution. Similarly, the market price of a crop was often used to determine the value of lost productivity. The benefit of a pollution control program would be estimated as the reduction in the damages calculated according to this approach.

I have characterized the damage function approach as naive because this approach implies a model of the world in which behavioral and market responses to changes in q are implicitly ruled out. Farmers can, in fact, adjust to changes in air pollution by changing cultivation practices, shifting to more resistant cultivars of the same crop, or even changing to entirely different crops that are less sensitive to pollution. Furthermore, the prices of agricultural crops may change because of changes in crop supplies. It may be that the changes in prices are of greater significance to human welfare than the changes in physical yields of crops. Similarly, people can choose defensive or mitigating activities in response to air pollution that affects health. These behavioral changes result in welfare consequences that have a monetary dimension, which should be taken into account in calculating values.

A Model of Environmental and Resource Values

Although measuring values involves the use of economic theory and technique, value measures must be based on other types of knowledge. For example, estimates of the value of a salt marsh in sustaining a marine fishery must be based on knowledge of the biological and ecological links between the marsh and the exploited fish species. Estimates of the health benefits from air pollution control must be based on scientific knowledge of the relationship between pollutant concentrations and human health. Estimates of the recreational fishing benefits stemming from water pollution control require knowledge of the relationships among pollutant levels, biological productivity, and anglers' activities (Freeman 1995). Lack of knowledge of these relationships may, in some instances, be a major barrier to empirical measurement of values. In this section, I lay out a very simple model for examining the relationship between the economic concept of value and the physical and biological dimensions of the resource system being valued. The model helps clarify the fact that economic valuation requires some knowledge of the underlying physical and biological relationships that determine the quantity and quality of environmental and resource service flows.

The economic values of the service flows from a resource–environment system can be viewed as the product of three sets of functional relationships, which we will discuss in the next few pages. The first relates some measure of environmental or resource quality to the human interventions that affect it. Let q represent a qualitative or quantitative measure of some environmental or resource attribute. Examples include the biomass of some species of fish of commercial or recreational interest, the stock of standing timber in a forest, or the concentration of some pollutant in the atmosphere. Two kinds of human intervention need to be specified. One involves the unregulated activities of the market economy (for example, the commercial exploitation of a fishery or the discharge of pollutants into the air); these will be left implicit in the relationships presented here. The other kind of intervention is government actions taken to prevent or ameliorate the adverse effects of unregulated market activities or to protect or enhance the value of market and nonmarket services provided by the environment. Let $\boldsymbol{S}$ represent the set of government interventions. For example, if q represents the population of waterfowl, $\boldsymbol{S}$ could be the stock of protected habitat and breeding grounds. Alternatively, $\boldsymbol{S}$ could represent a set of regulations designed to attain a stated ambient air quality standard, or it could be a management plan for a national forest. Let us represent the relationship between q and $\boldsymbol{S}$ as

$$q = q(\boldsymbol{S}) \tag{2-9}$$

As discussed below, this relationship could be quite complex in its spatial and temporal dimensions.

Where the government regulates private activities that influence q, the effect of a change in $\boldsymbol{S}$ on q can depend in complex ways on the responses of private decisionmakers to the public regulation. The most obvious example of this dependence is the question of compliance with pollution control regulations. For any given $\boldsymbol{S}$, q increases as the degree of compliance with the regulation increases. Some public regulations are only indirectly linked to the relevant environmental quality measure. An example is the automotive emissions standards set under the Clean Air Act, which regulate emissions of certain pollutants in grams per mile traveled by the automobile. In this case the effect of a change in $\mathbf{S}$ on q depends also on how automobile use is affected. To account for these complexities, it might be more appropriate to write q as a function of both $\mathbf{S}$ and some measure of private responses to the government regulation, $\mathbf{R}(\mathbf{S})$:

$$q = q[\mathbf{S}, \mathbf{R}(\mathbf{S})] \tag{2-9'}$$

The second set of functional relationships involves the human uses of the environment or resource and their dependence on q. Let $\mathbf{X}$ represent the levels of some set of activities involving use of the environment or resource. For example, $\mathbf{X}$ could be days of recreational activity on some body of water, tons of fish caught from some commercially exploited fishery, and, where human health depends on the level of environmental quality, some measure of health status. Typically, the level of $\mathbf{X}$ will depend not only on q but also on the inputs of labor, capital, and other materials and resources, including time; these will also depend on q. For example, if $\mathbf{X}$ is agricultural output and q is air pollution, farmers might adjust to changes in pollution by changing inputs of water, fertilizer, or labor. Let $\mathbf{Y}$ represent these other inputs into the production of environmental services or activities based on the resource. The second functional relationship can be written as

$$\mathbf{X} = \mathbf{X}[q, \mathbf{Y}(q)] \tag{2-10}$$

This expression ignores possible feedback from $\mathbf{X}$ to q such as when agricultural output is associated with increases in water pollution from pesticide or fertilizer runoff.

The third set of functional relationships gives the economic value of the uses of the environment. Let V represent the money value of the flow of services or activities based on the environment or resource. The relationship

$$V = V(\mathbf{X}) \tag{2-11}$$

embodies the value judgements society has adopted for economic welfare purposes. Here I assume that the value function is a simple aggregation of individuals' values; however, $V(\mathbf{X})$ could also incorporate social welfare weights that reflect some social equity goals. Alternatively, $V(\cdot)$ could incorporate concepts of environmental ethics or social norms. Also, this expression could incorporate nonuse values as in $V = V(\mathbf{X}, q)$.

By substitution of equations 2-9′ and 2-10 into equation 2-11, we have

$$V = V\{\mathbf{X}[\mathbf{S}, \mathbf{R}(\mathbf{S})], \mathbf{Y}[\mathbf{S}, \mathbf{R}(\mathbf{S})]\} \tag{2-12}$$

The marginal value of the change in $\boldsymbol{S}$ can be found by taking the total derivative of equation 2-12 to reflect the private adjustments of $\boldsymbol{R}$ and $\boldsymbol{Y}$ to the public intervention. The benefits of a nonmarginal change in policy intervention that increases q are given by

$$\begin{aligned} B \equiv \Delta V = V\{\boldsymbol{X}[\boldsymbol{S}^2, \boldsymbol{R}(\boldsymbol{S}^2)], \boldsymbol{Y}[\boldsymbol{S}^2, \boldsymbol{R}(\boldsymbol{S}^2)]\} \\ - V\{\boldsymbol{X}[\boldsymbol{S}^1, \boldsymbol{R}(\boldsymbol{S}^1)], \boldsymbol{Y}[\boldsymbol{S}^1, \boldsymbol{R}(\boldsymbol{S}^1)]\} \end{aligned} \qquad (2\text{-}13)$$

where $\boldsymbol{S}^1$ and $\boldsymbol{S}^2$ are the pre- and postpolicy levels of intervention, and $\boldsymbol{R}$ and $\boldsymbol{Y}$ are optimally adjusted to the change in $\boldsymbol{S}$.

The relationships represented in equation 2-9′ are largely noneconomic because they involve a variety of physical and biological processes. However, there is a social science or behavioral component to the private responses to $\boldsymbol{S}$. The relationship reflected in equation 2-11 is wholly within the realm of economics because it involves the theory of economic welfare and the use of economic data. The set of relationships reflected in equation 2-10 represents the interface between natural science and social science. Some aspects of these relationships, for example, how recreation use varies with changes in water quality, are primarily behavioral or social. Other aspects are almost wholly physical or biological, as in the effects of air pollution on human health and mortality. But even here, to the extent that people can "defend themselves" against the adverse effects of air pollution (say, by purchasing home air purifiers) or mitigate the symptoms of illness induced by air pollution, behavioral relationships are embedded in equation 2-10. The effect of an air pollutant on a particular type of vegetation is also a biological question; but if farmers alter crop patterns as a way of adapting to changes in air pollution, then the behavioral and biological aspects of the relationship must be considered together.

Figure 2-1 illustrates the three sets of relationships for the case of the benefits associated with an improvement in ambient water quality. A variety of substances is discharged into bodies of water. Reductions in the discharges can affect physical, chemical, and biological indicators of water quality such as dissolved oxygen, temperature, algae levels, and fish populations (stage 1). Changes in the indicators can be predicted with water-quality models. The resulting water quality, as measured by the indicators, can in turn affect human uses of the body of water (stage 2). These uses include withdrawal uses, for example, for industrial or municipal water supply, and irrigation or in-stream uses, for example, for fishery production or recreation. The major difficulty at stage 2 arises from the fact that only rarely is the level of use a simple function of a single water-quality indicator like dissolved oxygen. Rather, some uses (as in the case of commercial fisheries and recreation) depend in complex ways on the whole range

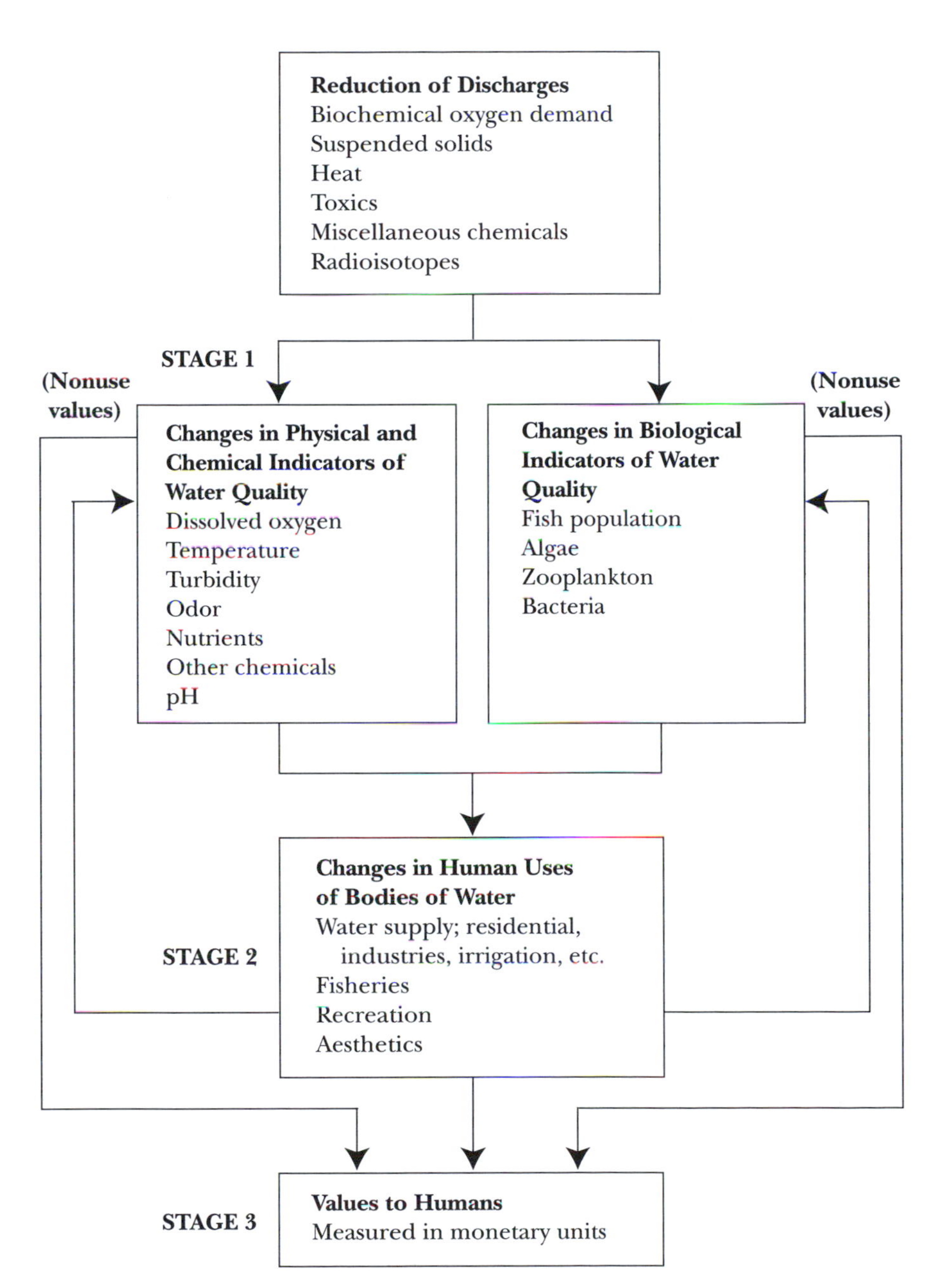

Figure 2-1. The production of benefits from improved ambient water quality

of physical, chemical, and biological water-quality indicators. The feedback loops from stage 2 to stage 1 reflect possible effects of changes in human uses on measures of water quality, for example, reductions in fish stocks due to overfishing. The figure also shows nonuse values that are independent of stage 2.

In our illustration, estimating water-quality benefits involves determining the monetary values that people place on such things as improved recreational opportunities, increases in fish production, and the availability of particular species of fish (stage 3). There is a well-developed theory of economic value relevant to the analysis of this stage. As discussed earlier, there are several approaches for estimating these values under different circumstances.

Figure 2-2 illustrates a similar application of the model to the abatement of air pollution. Understanding stage 1 is the task of atmospheric scientists who study photochemistry, atmospheric dispersion and transport, and the like. They must provide the air-quality models that relate changes in emissions to changes in the ambient concentrations of the substances of interest.

Understanding the effects of changes in air quality on agricultural productivity and materials damages is a technical task for botanists, plant physiologists, and materials engineers. However, economic models are also necessary because the effects and their valuation also depend on human responses, such as changes in crop patterns, materials substitution, and a variety of mitigating activities. These must be analyzed properly if an accurate picture of the effects of air pollution on human activity and welfare is to be obtained. Similarly, we must look to the biomedical sciences for information on the health effects of air pollution. A full understanding of the relationship between air pollution and the health status of the population, though, requires a comprehensive epidemiological analysis that controls for socioeconomic influences and other confounding variables, such as diet, lifestyle, and occupational exposures to harmful substances.

The Noneconomic Foundations of Resource Valuation

In a book that is about determining the economic values attached to the services affected by environmental change, the discussion cannot proceed far without acknowledging the importance of the relationship between environmental and resource quality and the uses of the environment. For example, the value of a recreation user-day at a lake is affected by fish populations and species distribution, algae levels, the number and type of bacteria present, temperature, smell, turbidity, and concentration of toxic substances. To further complicate matters, an increase in the magnitude of

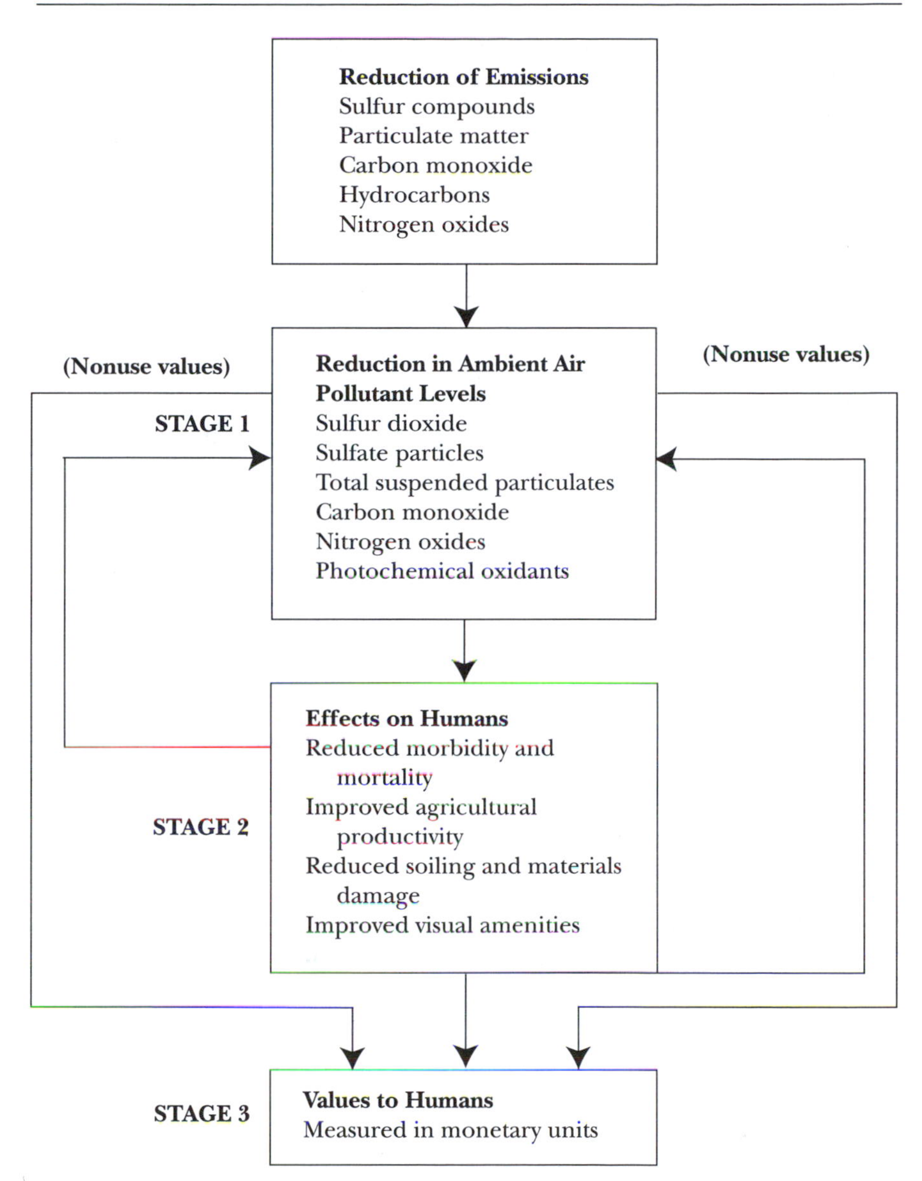

Figure 2-2. The production of benefits from improved ambient air quality

one characteristic may affect one use favorably while affecting an alternative use in a negative way. Higher water temperatures, for example, may make for better swimming while adversely affecting trout and salmon populations. Industrial discharges of acids may adversely affect recreation and fisheries while improving the value of water for industrial uses because of retarded algae growth.

The difficulties in tracing the effects of the discharge of a pollutant on the many parameters of environmental quality and, in turn, their effects on human uses of the environment substantially limit our ability to do careful benefit–cost analyses of environmental quality and improvements. This fact has not always been fully appreciated by many advocates of greater use of benefit–cost analysis in this field.

In 1968 Allen Kneese wrote

> I believe that our limited ability to evaluate the recreational losses associated with poor quality water, or conversely, the benefits of water improvement, is an extremely important barrier to rational water quality management .... The first [complexity] is the relationship between the level of various water quality parameters and the recreational attractiveness of the water resource. This relationship can be viewed as being composed of two linkages: a natural one and a human one. I think these are both about equally ill-understood. It is my impression ... that the biological sciences are almost never able to tell us specifically what difference a change in measured parameters of water quality will make in those biological characteristics of the water that contribute to its recreational value .... Perhaps the undeveloped state of forecasting is a result of the fact that biologists have seldom been confronted with the types of questions we would now like them to answer .... There is also a human linkage that is ill-understood. What quality characteristics of water do human beings find attractive for recreation? This is still largely an area of ignorance. (Kneese 1968, 180–181)

Although substantial progress has been made in the past 35 years or so since Kneese wrote this, there are still substantial gaps in our understanding of these linkages. What is true of water quality and recreation is also true of the other uses we make of bodies of water. It is also true of other dimensions of environmental quality and the uses we make of the environment.

Describing Resource and Environmental Quality

An analysis of the value of a resource or of the benefits of an environmental or resource policy change must begin with a description of the resource flow or some measure of environmental quality. This description requires choices about what attributes or characteristics of the resource–environment service are important. Suppose the question is, what will be the benefits of achieving the automotive emissions standards mandated by federal law? To answer this question we need to first determine which things that matter to people are adversely affected by automotive emissions, and then to trace out the links between emissions and those things that are valued. It is now understood that automobile emissions and the subsequent products of their photochemical reactions adversely affect human health, visibility, and agricultural productivity, among other things. So, to estimate values it is necessary to determine what specific measures of air quality are linked to these effects and how these measures of air quality are affected by the mandated standards. The measures used to characterize effects on visibility may be quite different from those relevant to effects on human health.

Two kinds of problems arise in this stage of the analysis. The first concerns the choice of parameters for describing the resource or environmental quality. The second involves determining the functional relationship between the policy instrument and the resource service flow or environmental quality measure.

Consider the case of water quality. A single effluent discharge can contain many substances that affect water quality—for example, oxygen-demanding organic wastes (biochemical oxygen demand), suspended solids, waste heat, and toxic chemicals. When these substances enter the waterway, they affect—in sometimes simple and sometimes complex ways—such measurable components of water quality as dissolved oxygen, temperature, and concentrations of chemicals. A nondegradable chemical substance, for example, will simply be diluted, and its concentration in the body of water will be a calculable fraction of its concentration in the effluent stream. In contrast, organic wastes affect water-quality parameters in a more complicated way. As organic wastes are degraded by bacteria, they reduce dissolved-oxygen levels to an extent and at a rate that depends on water temperature, wind, rates of river flow, and other physical and biological characteristics of the receiving water.

Some of the physical measures of water quality, such as turbidity and smell, affect human uses of the water directly. In addition, these and other physical parameters affect the stream ecology in complex and not always well-understood ways. The populations and species distributions of fish, algae, zooplankton, and bacteria may also be affected, and not necessarily in the same direction, by changes in the physical and chemical parameters of water quality.

Even providing a descriptive characterization of this first stage is a formidable task. Water quality cannot be represented by a single number on some scale, but rather is an *n*-dimensional vector of the relevant parameters. Which subsets of these parameters are most important in influencing the uses of a body of water (commercial fishing, boating, or swimming, for example) is still a major question for research.

Developing predictive models for these parameters is also a major research priority. The most commonly used water-quality models relate dissolved oxygen to the biochemical oxygen demand of discharges of organic wastes. However, dissolved-oxygen levels are only crudely related to the suitability of a body of water for fishery production or recreational use.

In the case of air quality, the choice of parameters is somewhat easier, but not without pitfalls. Only in the past 15 years or so have scientists turned attention from measurement of sulfur dioxide to its transformation products, sulfate particles. The latter measure of air quality is now known to be a more important air pollutant indicator than sulfur dioxide because of its effects on human health and ecology (U.S. EPA 1997 and references therein).

Temporal and Spatial Aggregation

Typically, measures of environmental quality vary over time and space. The dissolved-oxygen level at any point in a river rises and falls with changes in stream flow, discharge rates, water temperature, and the like, and is different at different locations. Air pollution readings vary with the time of day, day of the week, time of year, and from one place to another. One problem in empirical research on effects of pollution is how to define a variable or set of variables in a way that adequately reflects the temporal and spatial variations in environmental quality while still being manageable.

To put the problem in a concrete setting, consider an attempt to estimate the relationship between the level of an air pollutant and some health effect, say the occurrence of an asthma attack. The air pollution level at any particular point in an urban area is an instantaneous variable that fluctuates over time. The true exposure of an individual located at that point is measured by tracing the time path of that instantaneous variable over the relevant time. However, the individual may also move from one point to another in the urban space or outside the urban space. The published data on air pollution levels that are used to generate exposure variables for empirical research involve various approaches to summarizing the instantaneous time paths at the locations of air pollutant measurement devices. One common measure is the annual mean, either arithmetic or geometric. Averages are also struck over shorter periods, for example, a 24-hour average for particulates and 8-hour and 1-hour averages for other pollutants. Also, readings taken at one or two points in the space must be used to represent the space as a whole. Inevitably, summarizing involves loss of information. No one or two of these temporal and spatial summaries can completely represent the true exposure of any individual. Research on health effects is hampered by our inability to characterize accurately the exposure of individuals.

The Welfare Economics of Costs

So far, this chapter has focused attention on valuing the benefits of environmental changes. It is time now to turn attention to the costs of achieving these changes. It is a commonly held view both within and outside the economics profession that the costs of environmental regulations are relatively easy to measure, at least in comparison with the task of measuring environmental benefits. This optimistic view is consistent with what I would call a naive theory of cost, which takes the following form. Firms respond to pollution control regulations by purchasing and installing waste-treatment equipment and control systems that are, in effect, bolted on to the existing factory. The purchase and installation costs of this equipment, plus the added operating and maintenance costs it entails, are readily identified in the firm's accounts. These expenditures can be taken, at least as a first approximation, to be the social costs of complying with the regulation.

The naive theory fails to recognize the fundamental symmetry between benefits and costs as changes in the utilities of individuals. It also neglects several important realities concerning the ways that government regulations can affect people's welfare. The symmetry of benefits and costs stems from the fact that ultimately all costs take the form of utility losses to individuals in their dual roles as receivers of income and consumers of market and nonmarket goods and services. These losses have their monetary counterparts in compensating measures of welfare change, that is, willingness to pay to avoid the cost and willingness to accept compensation for bearing the cost. Because of this fundamental symmetry, proper measurement of costs involves the same kinds of problems as, and is likely to be as difficult as, the measurement of the benefits of environmental improvement. Once the symmetry is acknowledged, the whole economist's tool kit of revealed preference methods and stated preference methods of measuring welfare changes becomes available for the cost analyst.

The naive view about cost estimation is also implicitly based on the presumption that only firms cause pollution and that therefore only firms must incur costs in the process of moving to meet environmental quality objectives. However, this is not the case. Regulations, for example, may be placed on household activities, and these regulations might not require any identifiable expenditure of money. Suppose that commuters are required to form car pools as part of a program to meet air-quality standards for ozone. Household expenditures on commuting may, in fact, be lower as a consequence of the regulation. The cost of such regulation takes the form of increased commuting time and loss of convenience. These things may be difficult to quantify and value. Consider also a regulation that decreases the quality or stream of services from a consumer good. In the 1970s, for example, new cars complying with tailpipe emissions standards had lower performance and fuel efficiency than earlier models. The true cost of complying with these standards consisted of both the higher cost of purchasing and operating the car and the decreased value of the services from the car.

When an environmental regulation affects household activities or the availability of nonmarket goods and services, there may be opportunities for using revealed preference methods to draw inferences about the negative values or costs of these changes. For example, if the quality of a good is reduced as a consequence of a regulation, hedonic price models might be used to estimate the marginal implicit price of the relevant attribute. In the example of imposed car-pooling, an estimate of the shadow price of time could be used to derive the costs of the increased commuting time. Also, stated preference methods such as contingent valuation or contingent ranking might be used where a comprehensive set of regulations affects a wide range of activities.

Another factor neglected in the naive theory of costs is that market mechanisms are likely both to shift the burden of firms' expenditures and to change the magnitude of the burden so that the true costs are not accurately measured by summing the expenditures by firms. At least for nonmarginal changes, firms will raise prices and will experience decreases in quantity demanded. Price increases cause losses of consumer surplus, which are part of the social costs. There may also be losses in producers' surpluses and in the incomes of factors of production as well. Thus it is not correct simply to equate pollution control expenditures and social costs (Portney 1981).

As an extreme example, consider the case of a tax on the carbon content of fuels as a means of reducing the emissions of carbon dioxide (CO_2). Because there do not appear to be economically feasible technologies for controlling CO_2 emissions from combustion, the tax would work entirely through raising the prices of carbon-based fuels and reducing demands for them. Consumers would bear at least part of the costs of the revenues raised by the tax; the rest would come from decreases in resource rents. However, because these tax revenues represent a transfer to the government, they are not part of the social cost of reducing CO_2 emissions. The social cost of controlling CO_2 by taxation comes entirely in the equivalent of the dead-weight losses or welfare triangles associated with the tax.

Where the direct effect of a regulation is on firms, and where its effects are transmitted to individuals through changes in prices and incomes, the process of cost estimation must call on two types of models. The first is a model of the firm's production technology and costs. The second is a market model that can be used to calculate the changes in prices and incomes of the affected individuals. This model could in some circumstances be a partial equilibrium model. For broad social regulations, a general equilibrium model of the economy may be required (Hazilla and Kopp 1990).

In modeling the behavior of the firm, several questions must be considered. For example, can the technology be modeled as additively separable so that total cost is the sum of the costs of producing the marketed outputs and the cost of treating the waste? Or must the technology and costs be modeled as fully joint? The assumption of additive separability may be appropriate if, in fact, firms would respond to a regulation by purchasing add-on control or treatment equipment. However, other kinds of technological responses, such as input substitution and recycling, may have to be modeled in a joint production framework.

Once a model of firms' costs has been obtained, the next step is to embed that model in a model of the market economy so that the changes in all relevant quantities, prices, and incomes can be predicted. In the simplest case, where the regulation involves a marginal change in costs and where the economy is perfectly competitive, costs can be measured by the predicted change in expenditures on factor inputs. This is because the invariant factor prices are equal to the values of the marginal products of those inputs in other uses; and, in turn, those values of marginal products measure the opportunity losses to consumers associated with the marginal reallocation of inputs.

With nonmarginal changes, it can be expected that there will also be changes in product prices and perhaps factor prices. If a regulation affects only one industry, a partial equilibrium model may be appropriate. Rather than have costs be a function of q, among other things, a parameter reflecting the stringency of the regulation or the degree of pollution control required could be included as a shifter of the cost or production function.

Implicit in this approach are the assumptions that the regulation does not cause any shifts in either the output demand functions or the factor supply functions, either immediately or over time. However, if either of these assumptions is not valid, then a general equilibrium framework would be required. For example, if the regulation affected both the x and y industries, the regulation-induced change in the price of y could shift the demand function for x. Also, if y were an intermediate product and an input in the x industry, the increase in its price or the upward shift in its supply function would have a secondary impact on the cost of producing x. Hazilla and Kopp (1990) showed that estimating true social costs in a general equilibrium framework can lead to quite different results in comparison with the pollution control expenditure approach used by the U.S. Environmental Protection Agency.

More recent work has called attention to another, perhaps more important, general equilibrium effect. If an environmental regulation leads to an increase in the prices of goods and services, the result is a fall in real wages and, if labor supply elasticities are positive, a reduction in the quantity of labor supplied. Since the labor market is already distorted because of the presence of income and payroll taxes, the marginal social value of labor exceeds its marginal social cost by a substantial amount. Even a small decrease in the quantity of labor supplied can have a large net welfare cost.

This cost is in addition to the direct cost of the environmental regulation. The impact of regulatory costs on the labor market is known as the tax interaction effect. For a clear explanation of this effect and a discussion of its significance for the economic analysis of environmental and other regulatory policies, see Parry and Oates 2000.

Uncertainty

One problem in carrying out an analysis on environmental benefits or costs is that the values for some physical, technical, or economic parameters of the model may not be known with certainty. The state of the art in measurement is not sufficiently advanced to produce exact measures of value for many kinds of environmental and resource changes. This fact leads to the question, must policymakers wait for further research to produce exact measures before they can use value and benefit information to guide decisionmaking? If not, how should they interpret the ranges of values that current research has produced?

To counsel waiting for exact measures is equivalent to saying that in many cases value measures should never be used. The state of the art cannot be expected to advance to the point of producing exact values for all kinds of environmental change. This is because of the inherent uncertainty and imprecision in measurement techniques based on statistical inference and because of the fact that the true values held by individuals will vary with their circumstances (age, income, and so forth) and with the description of the specific changes being valued. So how are policymakers to proceed in the face of continued and inherent uncertainty about values?

The simplest approach is to base the calculations of benefits and costs on the expected values of the uncertain parameters and to base decisions on these expected values. However, decisionmakers will often want to know more about the magnitude of the uncertainties in the estimates of benefits and costs. Decisionmakers could be provided with the upper and lower bounds of the ranges of values along with the expected values. If the benefits of a policy calculated with the upper end of the range are less than the lower end of the range of estimated costs, the policy is unlikely to be justifiable on economic grounds. And if the benefits calculated with the lower end of the range exceed the upper end of the range of costs, the economic case for the policy is quite strong.

This simple-minded approach is a step in the right direction, but it can be criticized because it does not use all the relevant information contained in the estimates making up the range of values. Formally, the range reflects only the information contained in the two estimates yielding the highest and lowest values. It ignores information on the quality of these two estimates, and it ignores the information contained in the other estimates that yielded values within the range. There is a way to use the results of all the available estimates and to incorporate judgements about the quality of each of these estimates. This formal approach is based on viewing probabilities as statements about the degree of confidence held about the occurrence of some possible event. The approach involves assigning probabilities to all of the values produced by the available estimates, where a higher probability reflects a greater degree of confidence in that estimate. For example, the assignment of a probability of 1 to a particular estimate means that we can be certain that this study has produced the correct value. Once the probabilities have been assigned, various statistical manipulations can be performed. For example, the expected value of the parameter in question (the mean of its probability distribution) can be calculated and used for benefit–cost calculations. The variance of the distribution can be used to determine confidence intervals on the value to be used, thus preserving for policymakers information on the uncertainty about values. When there are multiple uncertainties, Monte Carlo methods can be used to draw from the assumed distributions to generate a probability distribution of outcomes.

Summary

Economists seek measures of values that are based on the preferences of individuals. When value measures are derived using models of behavior, these models should be internally consistent and based on accepted theories of preferences, choice, and economic interactions. Equally important is the need for a sound understanding of the underlying biological and physical processes by which environmental and resource service flows are generated. However, if empirical observations of individuals' choices are taken without benefit of an underlying theoretical model, researchers may be led to make faulty or erroneous interpretations of the data. An interesting example is the early studies of the land value–air pollution relationship. Researchers discovered that land values and air pollution levels were inversely related in urban areas, other things being equal. They then assumed that changes in welfare associated with reduced pollution would be accurately measured by the associated increases in land values as predicted by the regression equation relating land values at a point in time to air pollution. Subsequent research based on theoretical models of urban land markets has shown that this assumption is not true in general. The relationships among air pollution, land values, and measures of welfare change are discussed in Chapter 11.

References

Baumol, William J., and Wallace E. Oates. 1975. *The Theory of Environmental Policy.* Englewood Cliffs, NJ: Prentice-Hall.

———. 1988. *The Theory of Environmental Policy.* 2nd ed. Cambridge, U.K.: Cambridge University Press.

Freeman, A. Myrick, III. 1993. *The Measurement of Environmental and Resource Values: Theory and Methods.* Washington, DC: Resources for the Future.

———. 1995. The Benefits of Water Quality Improvements for Marine Recreation: A Review of the Empirical Evidence. *Marine Resource Economics* 10(4): 385–406.

Hazilla, Michael, and Raymond J. Kopp. 1990. The Social Cost of Environmental Quality Regulations: A General Equilibrium Analysis. *Journal of Political Economy* 98(4): 853–873.

Kneese, Allen V. 1968. Economics and the Quality of the Environment: Some Empirical Experiences. In *Social Sciences and the Environment,* edited by Morris Garnsey and James Gibbs. Boulder, CO: University of Colorado Press.

McConnell, Kenneth E. 1986. *The Damages to Recreational Activities from PCBs in New Bedford Harbor.* Report prepared by Industrial Economics, Inc., Cambridge, MA, for the Ocean Assessment Division, National Oceanic and Atmospheric Administration.

Mitchell, Robert Cameron, and Richard T. Carson. 1989. *Using Surveys To Value Public Goods: The Contingent Valuation Method.* Washington, DC: Resources for the Future.

Parry, Ian W.H., and Wallace E. Oates. 2000. Policy Analysis in the Presence of Distorting Taxes. *Journal of Policy Analysis and Management* 19(4): 603–613.

Portney, Paul R. 1981. The Macroeconomic Impacts of Federal Environmental Regulation. In *Environmental Regulation and the U. S. Economy,* edited by Henry M. Peskin, Paul R. Portney, and Allen V. Kneese. Washington, DC: Resources for the Future.

Roe, Brian, Kevin J. Boyle, and Mario F. Teisl. 1996. Using Conjoint Analysis To Derive Estimates of Compensating Variation. *Journal of Environmental Economics and Management* 31(2): 145–159.

U.S. EPA (U.S. Environmental Protection Agency). 1997. *The Benefits and Costs of the Clean Air Act—1970 to 1990.* Washington, DC: U.S. EPA.

CHAPTER

3

Defining and Measuring Welfare Changes: Basic Theory

The theory of the measurement of welfare change has been discussed by others both at the most rigorous levels of abstraction and in pragmatic, practical terms of application. See, for example, Boadway and Bruce 1984; Johansson 1987; and Just et al. 1982. Most of this discussion has focused on the welfare effects of changes in the prices people pay for the goods they consume. My major purpose in adding to this literature is to provide a definition and measurement of welfare effects of changes in quantities or qualities of environmental and resource service flows.

Changes in environmental quality can affect individuals' welfares through any of the following four channels: changes in the prices they pay for goods bought in markets; changes in the prices they receive for their factors of production; changes in the quantities or qualities of nonmarketed goods (for example, public goods such as air quality); and changes in the risks individuals face. The first three of these channels are the focus of this chapter. After a brief review of the theory of individual preferences and demand, I review the principles of welfare measurement for price changes. These principles are relevant because some forms of environmental change affect people only indirectly through their effects on prices. Also, a review of these principles will provide a good foundation for the treatment of quantity and quality changes that follows. The extension of these principles to the valuation of changes in risk, the fourth channel, raises some interesting questions, which will be left to Chapter 8.

The principles and measures developed in this chapter apply equally to decreases and increases in individuals' welfare. It is a basic principle of welfare economics that all costs ultimately take the form of reductions in the utility welfare of individuals. This principle applies equally to the costs of public policies (for example, investment in resource development and the regulation of private activities) and to the costs of private uses of the environment (for example, harvesting from a common property resource and using the waste receptor services of the environment). Hence, the welfare measures developed here provide a foundation for the analysis of both the benefits and the costs of environmental change.

In this chapter, I consider three sets of questions in some detail. The first set concerns how to define an acceptable monetary measure of changes in economic welfare for an individual. The answer to this question hinges in part on what the measure would be used for, that is, a welfare measure should answer the questions posed by policymakers. However, policymakers ask different kinds of questions. For example, suppose that policymakers wish to evaluate proposed policy changes in terms of an aggregate social welfare function that placed different weights on individuals' changes in utility depending on their positions in the income distribution (Bergson 1966). In that case, the welfare measure that answers the policymakers' question must be a money metric of utility changes. Alternatively, if policymakers wish to select policies on the basis of the potential Pareto improvement criterion, they will want measures of required compensation and willingness to make compensating payments. In the concluding section of this chapter, I return to the question of choosing from among the alternative measures described here.

The second set of questions concerns how changes in welfare would be measured both in theory and in practice. Theory suggests several alternative ways of calculating either exact or approximate welfare measures, using data on observed behavior of individuals, for example, their demand functions for market goods. These alternatives will be described and evaluated, especially from the practical perspective of implementation.

The third set of questions concerns how any measure of welfare changes for individuals might be used to make judgements about social policies affecting many individuals. For example, is it possible to speak of a measure of aggregate welfare for the society as a whole? If so, what significance can be attached to changes in such a measure? Measures of welfare change for an individual can be defined and analyzed without reference to the notions of efficiency and equity. In this sense the concept is objective; that is, one can define and measure a monetary equivalent of an individual's welfare change without being committed to any particular set of value judgements concerning aggregation across individuals or the role of such welfare measures in social choice. Of course, in a more fundamental sense focusing attention on changes in income and consumption to the exclusion of other aspects of well-being, such as social relationships and sense of self-worth, itself involves a kind of value judgement. However, this judgement is implicit in the division of labor between economists and others. It is in answering the third set of questions that value judgements about the relative deservingness of individuals, the meaning of efficiency, and the objectives of public policy come into play. Some of these issues are discussed in the section headed Aggregation and Social Welfare.

In the following section I review some of the basic terminology and theory involving individual preferences and demand. I then take up the standard case where utility depends only on the consumption of market goods. I examine the theory of measuring the welfare value of changes in the prices of these goods and the relationships among the Marshallian and Hicksian surplus measures of welfare change. There are two reasons for choosing this order of presentation. First, it parallels the historical evolution of the theory of welfare change. And second, it makes for an easier exposition of the basic principles. The section concludes with a review of methods for obtaining exact measures of and approximations to the desired Hicksian surpluses. In the third section I briefly describe how the theory of welfare value for changes in prices of goods can be applied to changes in factor prices. I then take up, in the fourth section, the case of the welfare effects of changes in the quantities of nonmarket goods where individuals are not free to adjust these quantities in response to price changes. In the final two sections, I review some of the issues involved in aggregating measures of individual welfare change for public policy decisionmaking and in selecting the appropriate welfare measure.

Individual Preferences and Demand

Before introducing the various possible welfare measures, it will be useful to review briefly the basic theory of individual preferences and the demand for goods as it relates to welfare theory. For alternative treatments of this and related topics, the reader may wish to consult other texts, such as Just et al. 1982; Boadway and Bruce 1984; Varian 1984; and Johansson 1987. This theory starts with the premises that individuals are their own best judge of their welfares and that inferences about welfare can be drawn for each individual by observing that individual's choices among alternative bundles of goods and services. If an individual prefers bundle *A* over bundle *B*, then bundle *A* must convey a higher level of welfare.

What things are to be included in the bundles (such as *A* and *B*) among which individuals are assumed to have preferences? There is little controversy over the inclusion of all the goods and services that can be bought or sold in markets: consumer goods, the services of household assets such as a house or a car, and consumer durables. Because time can be used in leisure activities or sold at some wage rate in the labor market, individuals must also have preferences among alternative uses of time such as reading, outdoor recreation, and working at some wage rate. And because government and the environment both provide a variety of services that enhance the welfares of individuals, these services should also be included in the bundles among which people have preferences. Environmental services include those provided by cleaner air, cleaner water, and scenic amenities.

If we assume that individuals can rank the alternative bundles according to their preferences, what properties will the resulting ordering of bundles display? For our purpose, two are important. The first is *nonsatiation*, or the "more-is-better" property. This means that a bundle with a larger quantity of an element will be preferred to a bundle with a smaller quantity of that element, other things being equal. Formally, if X' consists of $x_1', \ldots, x_i', \ldots, x_n'$, and X'' consists of $x_1', \ldots, x_i'', \ldots, x_n'$ and $x_i' > x_i''$, then this individual will prefer X' to X''.

The second property is *substitutability* among the components of bundles. This means that if the quantity of one element of a bundle, say x_i, is decreased, it is possible to increase the quantity of another element, say x_j, sufficiently to make the individual indifferent between the two bundles. More formally, suppose that X' consists of $x_1', \ldots, x_i', \ldots, x_j', \ldots, x_n'$; and X''consists of $x_1', \ldots, x_i'', \ldots, x_j', \ldots, x_n'$ with $x_i'' < x_i'$. Substitutability means that there is another bundle X^* consisting of $x_1', \ldots, x_i'', \ldots, x_j^*, \ldots, x_n'$ with $x_j^* > x_j'$, such that the individual is indifferent as to X' and X^*. In other words, X' and X^* lie on the same indifference surface.*

The property of substitutability is at the core of the economist's concept of value. This is because substitutability establishes trade-off ratios between pairs of goods that matter to people. In this formulation, the trade-off ratio is $(x_j^* - x_j')/(x_i' - x_i'')$ or $|\Delta x_j/\Delta x_i|$. In the limit for infinitesimally small changes, this reduces to $|dx_j/dx_i|$, which is the definition of the marginal rate of substitution between x_i and x_j or the slope of the two-dimensional indifference curve between these two elements. The money price of a market good is just a special case of a trade-off ratio because the money given up to purchase one unit of one element of the bundle is a proxy for the quantities of one or more of the other elements in the bundle that had to be reduced to make the purchase.

If the preference ordering has the properties described here, it can be represented by an ordinal preference function or utility function that assigns a number to each bundle as a function of the quantities of each element of the bundle. Specifically,

---

*Two other important properties are transitivity and quasiconcavity. If there are three bundles X', X'', and X^*, and the individual prefers X' over X'' and X'' over X^*, then transitivity is satisfied if the individual prefers X' over X^*. For more on the axiomatic description of these properties of preference ordering, see Boadway and Bruce 1984 or Varian 1984.

$$u = u(\boldsymbol{X}, \boldsymbol{Q}, \boldsymbol{T}) \tag{3-1}$$

where $\boldsymbol{X}$ is a vector of the quantities of market goods, $\boldsymbol{Q}$ is a vector of public goods and environmental and resource services whose quantities or qualities are fixed for the individual, and $\boldsymbol{T}$ is a vector of the times spent in various activities that yield utility to the individual. This utility function is assumed to be increasing in all of its arguments and unique up to a monotonic transformation. For purposes of mathematical modeling and analysis, it is convenient also to assume that this function is continuous, convex, and twice differentiable. This preference function is not the same thing as the cardinal utility function of the classical utilitarians. Because there is no unit of measurement for this ordinal utility, it is not possible to add or otherwise compare the utilities of different individuals.

To simplify the exposition and notation, let us now consider an individual whose utility is a function only of private goods that can be bought and sold in markets. Assume that tastes and preferences (that is, the utility function) are given and do not change. The individual faces a set of given prices for these goods and is assumed to choose the quantities of the goods so as to maximize his or her utility, given the constraints of prices and a fixed money income M. The maximization problem can be expressed as

$$\begin{aligned} &\text{maximize } u = u(\boldsymbol{X}) \\ &\text{subject to } \sum^{i} p_i \cdot x_i = M \end{aligned} \tag{3-2}$$

where $\boldsymbol{X}$ is the vector of quantities ($\boldsymbol{X} = x_1, \ldots, x_i, \ldots, x_n$). The solution to this problem leads to a set of ordinary or Marshallian demand functions

$$x_i = x_i(\boldsymbol{P}, M) \tag{3-3}$$

where $\boldsymbol{P}$ is the vector of prices ($\boldsymbol{P} = p_1, \ldots, p_i, \ldots, p_n$).

Substituting the expressions for x_i as functions of $\boldsymbol{P}$ and M into the direct utility function gives the indirect utility function, that is, utility as a function of prices and income, assuming optimal choices of goods:

$$u = v(\boldsymbol{P}, M) \tag{3-4}$$

According to Roy's Identity, the demand functions can also be expressed in terms of derivatives of the indirect utility function,

$$x_i(\boldsymbol{P}, M) = -\frac{(\partial v / \partial p_i)}{(\partial v / \partial M)} \tag{3-5}$$

The expenditure function represents a useful perspective on the problem of individual choice. It is derived by formulating the dual of the utility maximization problem. The individual is assumed to minimize total expenditure,

$$e = \sum^{i} p_i \cdot x_i \tag{3-6}$$

subject to a constraint on the level of utility attained,

$$u(\boldsymbol{X}) = u^0 \tag{3-7}$$

where u^0 is the maximum utility attained with the solution to the primal problem. Just as the solution to the utility maximization problem yields a set of ordinary demand curves conditional on prices and income, the solution of the expenditure minimization problem yields a set of functions giving optimal quantities for given prices and utility. These are Hicks-compensated demand functions that show the quantities consumed at various prices assuming that income is adjusted (compensated), so that utility is held constant at u^0. Substituting these demand functions into the expression for total expenditure yields the expenditure function. This expression gives the minimum dollar expenditure necessary to achieve a specified utility level, given market prices. In functional notation:

$$e = e(\boldsymbol{P}, u^0) \tag{3-8}$$

where e is the dollar expenditure and u^0 is the specified utility level. The compensated demand functions can also be found by differentiating the expenditure function with respect to each of the prices:

$$\frac{\partial e}{\partial p_i} = h_i = h_i\left(\boldsymbol{P}, u^0\right) \tag{3-9}$$

where h_i is the compensated demand for x_i.

Now consider the set of ordinary demand functions derived from the utility maximization problem. To determine the functional form and parameters of these demand functions, we need to know the underlying utility function, and this may not be directly observable. Suppose instead that we observe an individual's behavior and estimate the demand functions that describe the individual's responses to changes in prices and income. These functions should contain all the same information as the underlying preferences. This is assured if the demand functions satisfy the so-called *integrability conditions.* These conditions require that the Slutsky matrix of substitution terms,

$$\frac{\partial h_i(\boldsymbol{P},U)}{\partial p_j}=\frac{\partial x_i(\boldsymbol{P},M)}{\partial p_j}+\frac{\partial x_i(\boldsymbol{P},M)}{\partial M}x_j \qquad (3\text{-}10)$$

be symmetric and negative semidefinite (Hurwicz and Uzawa 1971; Silberberg 1978; Varian 1984). If these conditions are satisfied, the system of demand functions can be integrated to yield the expenditure function, which in turn can be used to derive the indirect and direct utility functions. If the integrability conditions are not satisfied, the implication is that the observed demand functions are not consistent with the maximization of a well-behaved utility function. As I explain below, if the integrability conditions are satisfied, it may be possible to use empirically derived descriptions of demand behavior to obtain a complete description of the underlying preferences, as well as exact measures of welfare change for a wide range of postulated changes in economic circumstances.

Welfare Measures for Changes in Prices

An Overview

To introduce the alternative welfare measures, we consider first the simplest case of only two goods and the welfare gain associated with a nonmarginal decrease in the price of one of these goods. Five alternative measures of this welfare change have been identified in the literature. The first is the change in ordinary consumer's surplus, a concept whose origin can be traced back through Alfred Marshall (1920) to Dupuit. Currie and others (1971) provide a useful discussion of the history and evolution of the concept of consumer's surplus and of Dupuit's original contribution. As Marshall explained it,

> [The individual] derives from a purchase a surplus of satisfaction. The excess of the price which he would be willing to pay rather than go without the thing, over that which he actually does pay is the economic measure of this surplus of satisfaction. It may be called consumer's surplus. (1920, 124)

Ordinary consumer's surplus is measured by the area under a Marshallian ordinary demand curve but above the horizontal price line. As we will see, the consumer's surplus measure cannot be defined in terms of the underlying utility function. And, in general, it will be different from the other four measures defined below.

The other four measures of welfare change are theoretical refinements of the ordinary consumer's surplus (Hicks 1943). Each can be defined in

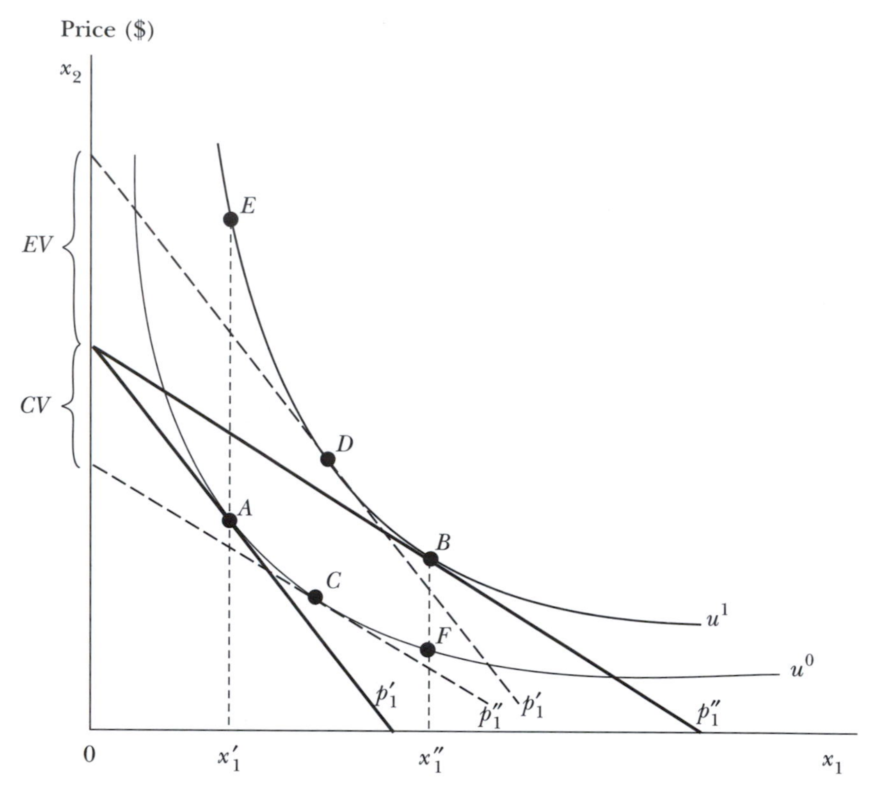

Figure 3-1. Four measures of the welfare gain from a price decrease

terms of the underlying individual preference mapping. Figure 3-1 shows two indifference curves for an individual. Assume that an environmental improvement reduces the cost of producing x_1, so that its price drops from p_1' to p_1''. In response to the price reduction, the individual shifts from the consumption bundle marked *A* at utility level u^0 to consumption bundle *B* at utility level u^1. What is the welfare benefit of the price reduction to this individual? Four alternative welfare measures can be defined in terms of good x_2, which is taken to be the numeraire. The units of x_2 are chosen so that the price of x_2 is equal to one. Thus, x_2 can be taken to represent income. In the many-good case, x_2 is a composite good that can be treated as an index of the consumption levels of all other goods except x_1. The aggregation of all other goods into a composite good for graphical representation is valid so long as the prices of all of the goods are assumed to move in the same proportion, that is, that there are no changes in the relative prices of components of the composite good bundle. This assumption can be maintained because we are analyzing only the consequences of the change in the price of x_1.

The four remaining measures of welfare changes are

1. Compensating Variation (*CV*). This measure asks what compensating payment (that is, an offsetting change in income) is necessary to make the individual indifferent between the original situation (*A* in Figure 3-1) and the new price set. Given the new price set with consumption point *B*, the individual's income could be reduced by the amount of *CV* and that person would still be as well off at point *C* as at point *A* with the original price set and income. The measure *CV* is often interpreted as the maximum amount that the individual would be willing to pay for the opportunity to consume at the new price set. However, for a price increase, *CV* measures what must be paid to the individual to make that person indifferent to the price change. For price decreases the *CV* cannot be greater than the individual's income, but for a price increase, the *CV* could exceed income.

2. Equivalent Variation (*EV*). This measure asks what change in income (given the original prices) would lead to the same utility change as the change in the price of x_1. As shown in Figure 3-1, given the original prices, the individual could reach utility level u^1 at point *D* with an income increase equal to *EV*. *EV* is the income change equivalent to the welfare gain due to the price change. The *EV* measure has also been described as the minimum lump-sum payment the individual would have to receive to induce that person to voluntarily forgo the opportunity to purchase at the new price set. For a price increase, *EV* is the maximum amount the individual would be willing to pay to avoid the change in prices.

   Note that both the *EV* and *CV* measures allow the individual to adjust the quantities consumed of both goods in response to both changes in relative prices and income levels. The remaining two measures are defined so as to place restrictions on the individual's adjustment of the consumption bundles.

3. Compensating Surplus (*CS*). This measure asks what compensating payment will make the individual indifferent as to the original situation and the opportunity to purchase the new quantity x_1'' of the good whose price has changed. The *CS* measure is the vertical distance between the indifference curves at the new quantity x_1''. This is the distance from *B* to *F* in Figure 3-1. This measure is closely related to the *CV* measure; the only difference is the restriction on adjusting the purchases of x_1 in response to the compensating change in income.

4. Equivalent Surplus (*ES*). This measure asks what change in income is required, given the old prices and consumption level of x_1, to make the individual as well off as that person would be with the new price set and

consumption point *B*. In Figure 3-1, the *ES* measure is the vertical distance between the two indifference curves, holding the consumption of good x_1 at the original level, that is, the vertical distance from *A* to *E*. The *ES* measure is closely related to the *EV* measure; the only difference is in the restriction on the adjustment of the consumption of x_1 in the former case. The *ES* measure is larger than the *EV* measure for price decreases because it must include an income equivalent to the individual's welfare loss stemming from the inability to adjust the consumption of x_1 so as to equate the marginal rate of substitution with the price ratio.

Each of these concepts measures something different and has a different meaning. The next subsection is devoted to a comparison and evaluation of the compensating and equivalent variations and their relationship to the ordinary consumer's surplus. The compensating and equivalent surplus measures for price changes do not answer very useful questions because they both arbitrarily restrict the individual to consuming a specific quantity of the good whose price has changed. Hence, they will not be considered further.

A Closer Look at the Welfare Measures

In this section I first present the basic welfare measure for a marginal change in one price. Then I give more rigorous derivations of the consumer's surplus, compensating variation, and equivalent variation measures of welfare change for the case of changes in price. For more detailed treatment of these topics, see Silberberg 1972 and 1978; Just et al. 1982, Appendix B; Varian 1984; and Johansson 1987. For a marginal change in, say p_1, the basic welfare measure is the change in expenditure necessary to hold utility constant. Using equation 3-9 from the previous section, we have

$$w_{p_1} = \frac{\partial e(\boldsymbol{P}, u)}{\partial p_1} = h_1\left(\boldsymbol{P}, u^0\right) \tag{3-11}$$

where w_{p_1}, is the marginal welfare measure. This result also follows from the indirect utility function and Roy's Identity:

$$w_{p_1} = x_1 = -\frac{(\partial v / \partial p_1)}{(\partial v / \partial M)} \tag{3-12}$$

or

$$\frac{dM}{dp_1} = x_1 \tag{3-12'}$$

In equation 3-12, the marginal utility of the price change is converted to monetary units by dividing by the marginal utility of income. Equation 3-12′ says that the change in income required to hold utility constant is equal to the change in price multiplied by the quantity of the good being purchased.

Marshallian Consumer's Surplus. In Figure 3-2, panel A shows one individual's preference mapping in the simple two-good case. Suppose that the price of good x_1 falls from p_1' to p_1''. The individual responds by moving from the original equilibrium at point *A* to point *B* on the new budget line. In panel B of Figure 3-2, these equilibrium positions are plotted in the price and quantity plane. Points *A* and *B* are on the ordinary demand curve holding the price of good x_2 and income constant. Because the Marshallian surplus associated with the consumption of a good at a given price is the area under the demand curve, the change in surplus for a change in the good's price is the geometric area $p_1'ABp_1''$ in panel B of Figure 3-2. In mathematical form,

$$S = \int_{p_1''}^{p_1'} x_1(\boldsymbol{P}, M)\, dp_1 \tag{3-13}$$

where *S* is the change in surplus.

The condition under which *S* can be interpreted as an indicator of utility change can be seen by employing Roy's Identity:

$$x_1(\boldsymbol{P}, M) = -\frac{[\partial v(\boldsymbol{P}, M)/\partial p_1]}{[\partial v(\boldsymbol{P}, M)/\partial M]} \tag{3-14}$$

and substituting this into equation 3-13 to obtain

$$S = -\int_{p_1''}^{p_1'} \frac{(\partial v/\partial p_1)}{(\partial v/\partial M)}\, dp_1 \tag{3-15}$$

If the marginal utility of income is constant over the range of the price change, this can be written as (see Varian 1984, 208–209)

$$S = \frac{[v(p_1'', p_2, M) - v(p_1', p_2, M)]}{(\partial v/\partial M)} \tag{3-16}$$

This expression shows that the Marshallian surplus can be interpreted as the utility change converted to monetary units by a weighting factor, the

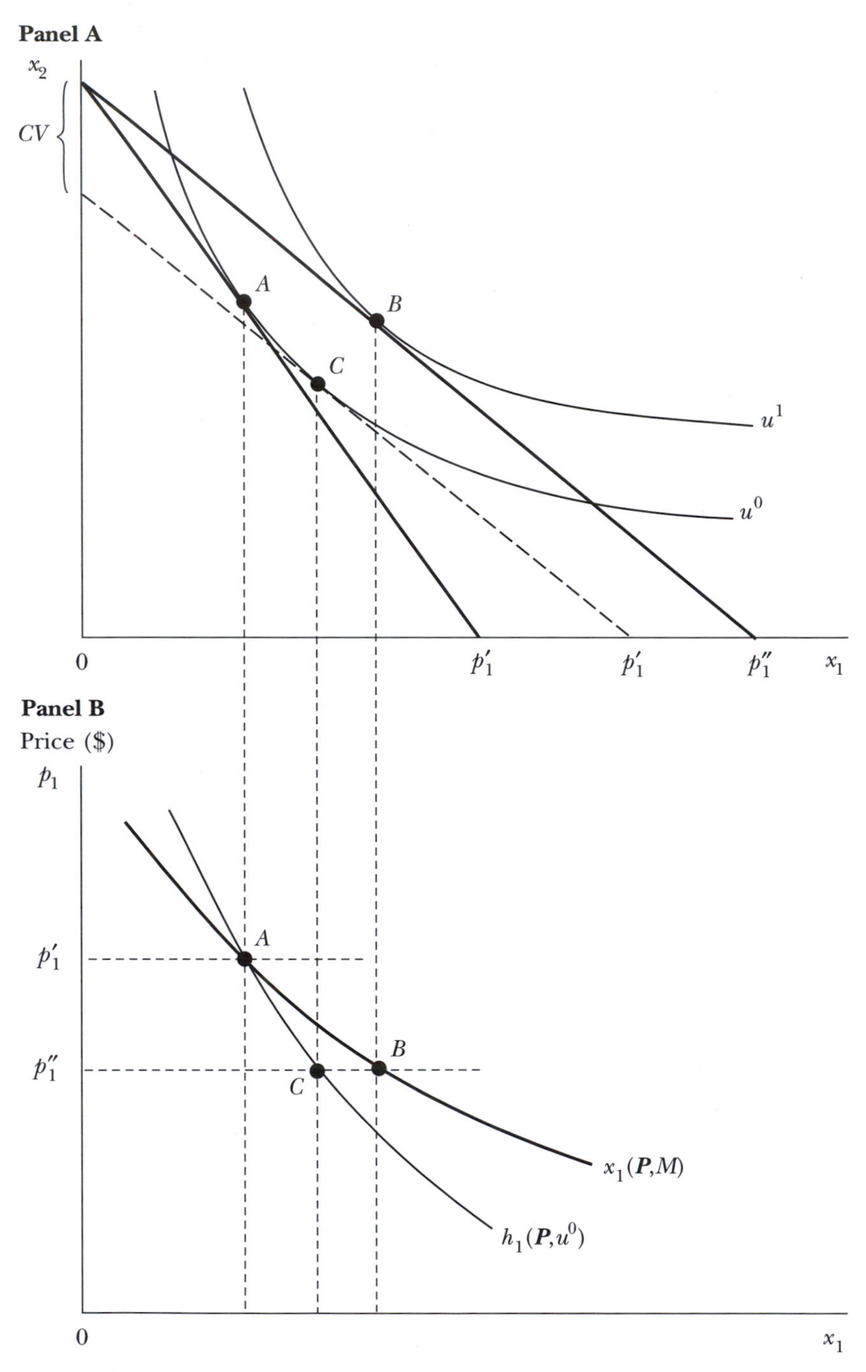

Figure 3-2. The compensating variation and the Hicks-compensated demand

marginal utility of income. If the marginal utility of income is constant, then S can be said to be proportional to the change in utility for any price change. However, the constancy of the marginal utility of income as any one price changes is a restrictive condition. The marginal utility of income cannot simultaneously be invariant with respect to income and to changes in all of the prices (Samuelson 1942; Johansson 1987, Chapter 4).

Alternatively, as Silberberg explains (1978, 350–361), the integral of equation 3-15 can be viewed as the sum of a series of small steps from an initial price and income vector of (p_1', p_2', M) to (p_1'', p_2', M) following a path on which p_2 and M are held constant. However, equation 3-15 can be integrated over other paths involving changes away from the initial values for p_2 and/or M as long as the terminal point is (p_1'', p_2', M). In general the other paths will not lead to the same solution value for the integral. In other words, the integral will not in general be path-independent.

A similar problem arises when the Marshallian surplus measure is generalized to simultaneous changes in all prices. In this case, S is defined as a line integral. This integral will be independent of the path of integration (that is, the order in which prices and/or incomes are assumed to change) only if the income elasticities of demand for all goods are equal. The income elasticities of all goods can be equal to each other only if they are all equal to one, in other words, if preferences are homothetic. Finally, if the prices of only a subset of all goods change, a unique S exists if the marginal utility of income is constant with respect to only those prices that are changed. (See Just and others [1982] for more details.)

Compensating Variation. Suppose now that as the price of good x_1 is decreased, income is taken away from the individual so that he or she remains at the initial utility level and indifference curve u^0. Given the price change and the compensating income change, the individual would be in equilibrium at point C in panel A of Figure 3-2. Point C is also plotted in panel B of Figure 3-2. Points A and C are on the Hicks-compensated demand curve, a demand curve that reflects only the substitution effect of the change in relative prices. The income effect of the price change has been eliminated by the device of compensating withdrawals of money income. Because x_1 is a normal good by assumption, that is, it has an income elasticity greater than zero, the Hicks-compensated demand curve is less price-elastic than the ordinary demand curve. The difference between the Hicks-compensated and the ordinary demand functions is one of the main considerations in the comparison of *EV*, *CV*, and consumer's surplus measures of welfare change.

Panel A of Figure 3-2 shows the compensating variation measure of the welfare change associated with the price decrease, that is, the reduction in income needed to hold the individual on the original indifference curve. In terms of the indirect utility function, *CV* is the solution to

$$v(\boldsymbol{P}', M) = v(\boldsymbol{P}'', M - CV) = u^0 \tag{3-17}$$

The *CV* can also be defined in terms of the expenditure function. It is the difference between the expenditures required to sustain utility level u^0, at the two price sets:

$$\begin{aligned} CV &= e\left(p_1', p_2, u^0\right) - e\left(p_1'', p_2, u^0\right) > 0 \\ &= M - e\left(p_1'', p_2, u^0\right) \end{aligned} \tag{3-18}$$

Because *CV* is defined as the difference between two levels of expenditure, it can also be written as the integral of the marginal welfare measure (equation 3-11) over the relevant range. Specifically,

$$CV = \int_{p_1''}^{p_1'} \frac{\partial e\left(\boldsymbol{P}, u^0\right)}{\partial p_1} dp_1 = \int_{p_1''}^{p_1'} h_1\left(\boldsymbol{P}, u^0\right) dp_1 \tag{3-19}$$

Because spending *M* at the new price set yields a higher level of utility, we can also write

$$M = e(p_1'', p_2, u^1) \tag{3-20}$$

and by substitution

$$CV = e(p_1'', p_2, u^1) - e(p_1'', p_2, u^0) > 0 \tag{3-21}$$

In other words, although the *CV* is defined in terms of u^0, it also measures the amount of money required to raise utility from u^0 to u^1 at the new set of prices.

The *CV* is equal to the area to the left of the Hicks-compensated demand curve between the two prices, that is, the area $p_1'ACp_1''$ in Panel B of Figure 3-2. The partial derivative of the expenditure function with respect to p_1 gives the change in expenditure (income) necessary to keep the individual on u^0 for small changes in p_1. As shown above, this derivative gives the Hicks-compensated demand curve, that is, it gives the optimal quantity for x_1, holding utility constant. For finite changes, the integral of this derivative is the area to the left of the Hicks-compensated demand curve, that is, the *CV*. In other words,

$$CV = \int_{p_1''}^{p_1'} h_1\left(\boldsymbol{P}, u^0\right) dp_1 \tag{3-22}$$

Unlike the Marshallian measure of surplus given by equation 3-13, this measure does not rely on any assumption about the constancy of the marginal utility of income. This is true because this measure integrates along a constant utility indifference curve at u^0. In the many-good case, when several prices change, the *CV* of the price changes taken together is the integral of the set of compensated demand functions evaluated by taking each price change successively. The order in which the price changes are evaluated is irrelevant. This follows from the symmetry of the cross-price substitution terms, that is, $\partial x_i/\partial p_j = \partial x_j/\partial p_i$ along the indifference curve.

Equivalent Variation. The equivalent variation can also be derived through the expenditure function. Panel A of Figure 3-3 shows the same preference mapping and price change for an individual. With a price decrease, the *EV* is defined as the additional expenditure (income) necessary to reach utility level u^1, given the initial set of prices. In terms of the indirect utility function, *EV* is the solution to

$$v(\boldsymbol{P}', M + EV) = v(\boldsymbol{P}'', M) = u^1 \tag{3-23}$$

In Figure 3-3, the *EV* is the additional expenditure necessary to sustain point *C′* over point *A* at the initial prices, or

$$\begin{aligned} EV &= e\left(p_1', p_2, u^1\right) - e\left(p_1', p_2, u^0\right) > 0 \\ &= e\left(p_1', p_2, u^1\right) - M \end{aligned} \tag{3-24}$$

Because the expenditure levels are the same at point *A* and point *B*, that is, $e(p_1', p_2, u^0) = e(p_1'', p_2, u^1)$, this can also be written as

$$EV = e(p_1', p_2, u^1) - e(p_1'', p_2, u^1) \tag{3-25}$$

In other words, although the *EV* is defined in terms of the monetary equivalent of a change from u^0 to u^1, it can also be measured by the change in expenditure associated with price changes given utility level u^1.

The *EV* can also be written as the integral of the marginal value measure (equation 3-11):

$$EV = \int_{p_1''}^{p_1'} \frac{\partial e\left(\boldsymbol{P}, u^1\right)}{\partial p_1} dp_1 \tag{3-26}$$

The price derivative of the expenditure function (this time holding utility constant at u^1) generates another Hicks-compensated demand curve

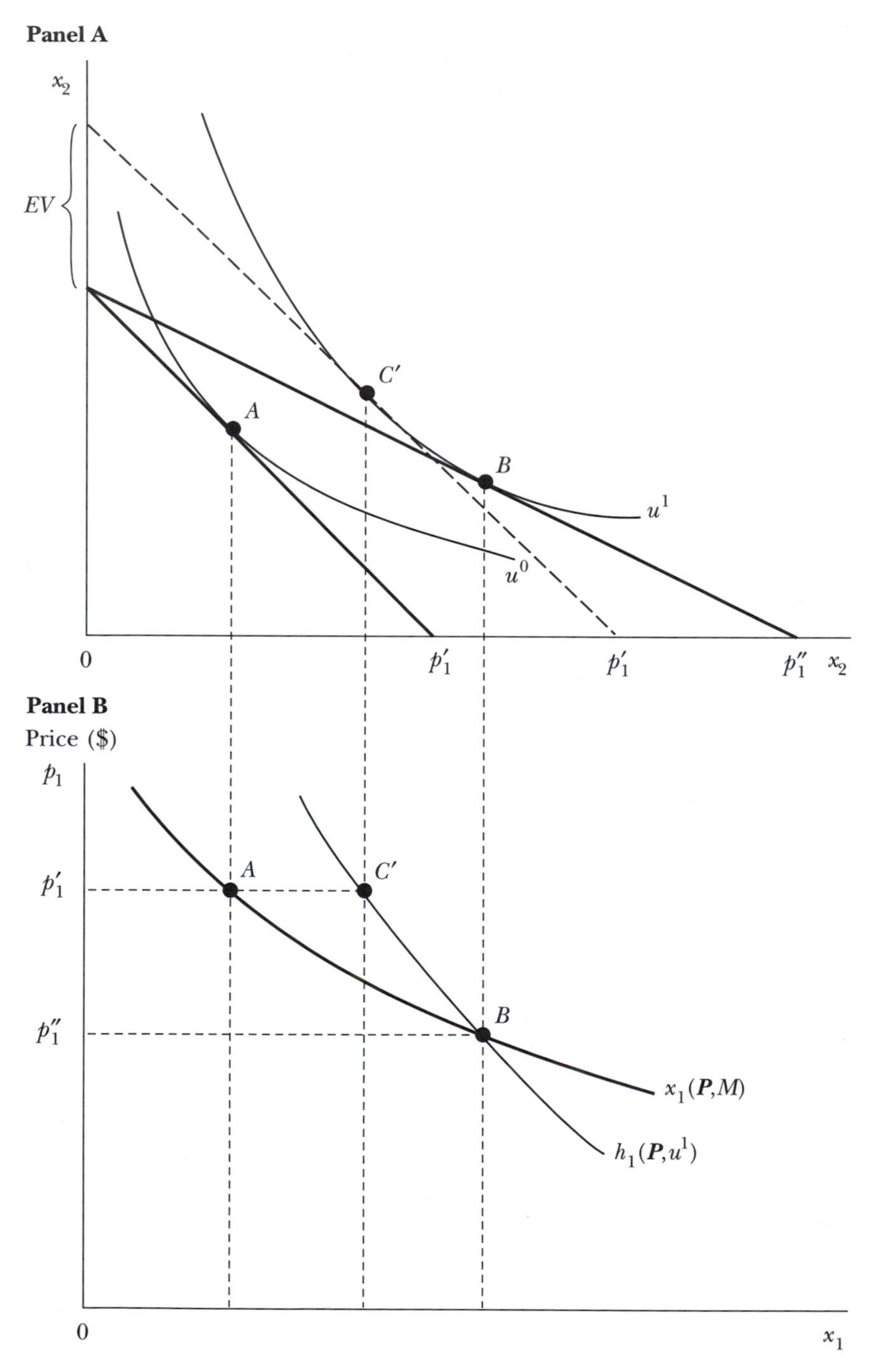

Figure 3-3. The equivalent variation and the Hicks-compensated demand

through point B in panel B of Figure 3-3. The area to the left of this Hicks-compensated demand curve between the two prices (area $p_1'C'Bp_1''$) is the equivalent variation welfare measure. In other words,

$$EV = \int_{p_1''}^{p_1'} h_1\left(\mathbf{P}, u^1\right) dp_1 \tag{3-27}$$

As in the case of the *CV*, this measure does not require any assumption about the constancy of the marginal utility of income. The measure for multiple price changes is path-independent.

All of this discussion has been in terms of the welfare gain due to a price decrease. The derivation of the welfare cost of a price increase can be worked out in a symmetrical fashion. In general, for any price change, the *CV* welfare measure is the area to the left of the Hicks-compensated demand curve that passes through the initial position. The *EV* measure of the welfare change is the area to the left of the Hicks-compensated demand curve that passes through the final position.

A Comparison of the Three Measures. Although the Marshallian consumer's surplus has some intuitive appeal as a welfare indicator, it does not measure any of the theoretical definitions of welfare change developed here. It is neither an index of utility change, except under special conditions, nor a measure of gain or loss that can be used in a potential compensation test. The Marshallian surplus lies between the *CV* and the *EV*. This fact opens the question of whether it can be a useful approximation to either of these other measures, a question that I take up below in the subsection "Consumer's Surplus without Apology."

In contrast, the *CV* and the *EV* represent welfare-relevant measures. The *EV* is the monetary equivalent of a price change. It can be interpreted as an index of utility in the sense that it imputes the same monetary value to all changes from an initial position that result in the same final utility level. This is because it evaluates all changes from an initial position at the same set of prices. *EV* is an ordinal utility index (Morey 1984). For example, suppose a change from initial position A to position B has an *EV* of $10, while a change from A to C has an *EV* of $20. It cannot be inferred that the second change conveys twice as much extra utility as the first change.

The *CV* cannot be interpreted as an index of utility. Rather, it measures the offsetting income change necessary to "prevent" a utility change. As Silberberg put it, "the [*EV*] imputes a dollar evaluation to a change in utility levels for a particular path of price changes, while the [*CV*] derives dollar values necessary to hold utility constant when prices change" (1972, 948).

The two measures *EV* and *CV* will be the same if the income elasticity of demand for good x_1 is zero. In this case, the ordinary and Hicks-compensated demand curves are identical. With a positive income elasticity, the *EV* exceeds the *CV* for price decreases, but the *CV* exceeds the *EV* when price increases are considered. The difference between points *C* and *B* in Figure 3-2 and between points *A* and *C′* in Figure 3-3 is one of income level. If the income elasticity of demand for x_1 were zero, the income differences would have no effect on the purchase of x_1. The *CV* and the *EV* would be exactly equal; they could both be measured by the area under the ordinary demand curve. The higher the income elasticity of demand for x_1, the larger is the difference between the *EV* and the *CV*, and the larger is the difference between either measure and the ordinary consumer's surplus.

There is a symmetry between the *CV* and the *EV* measures. This symmetry can be seen by comparing Figures 3-2 and 3-3 and by comparing equation 3-21 with equation 3-24 and equation 3-18 with equation 3-25. For simplicity, let I represent the initial price set (with p_1') and let II represent the second price set (with p_1''). The *CV* for moving from I to II with u^0 as the reference utility level is exactly equal to the *EV* of moving from II to I with u^1 as the reference utility level. The *CV* is a welfare measure for the move from *A* to *B* via point *C*; the *EV* starts at point *B* and measures the reduction in income necessary to get to point *A* and therefore u^0 via point *C′*. Similarly, the *EV* for the move from I to II is just equal to the *CV* starting at II and u^1 and moving to I.

This symmetry relates to the interpretation of *CV* and *EV* as measures of willingness to pay (WTP) and willingness to accept compensation (WTA). The *CV* is sometimes described as the maximum willingness to pay for the right to purchase the good at the new price level, that is, the lump-sum payment that the individual would be willing to make that would just exhaust the potential for welfare gain from the new price. This description is accurate only for a price decrease. For a price increase, the *CV* defines the minimum payment to the individual sufficient to prevent a utility decrease—in other words, it defines a WTA measure. Similarly, the *EV* defines a WTA measure for a price decrease, that is, the sum of money the individual would require to voluntarily forgo a proposed price decrease. However, for a proposed price increase, the *EV* is a WTP measure, that is, the maximum sum of money that could be taken away from the individual, yielding a loss of utility equivalent to that caused by the price change. Whatever the direction of the price change, the *CV* takes the initial utility as the reference point.

These two measures can also be interpreted in terms of the implied rights and obligations associated with alternative price sets. The *CV* carries with it implicitly the presumption that the individual has no right to make purchases at a new set of lower prices but does have a right to the original price set in the case of price increases. In contrast, the *EV* contains the presumption that the individual has a right to (an obligation to accept) the new lower (higher) price set and must be compensated (make a payment)

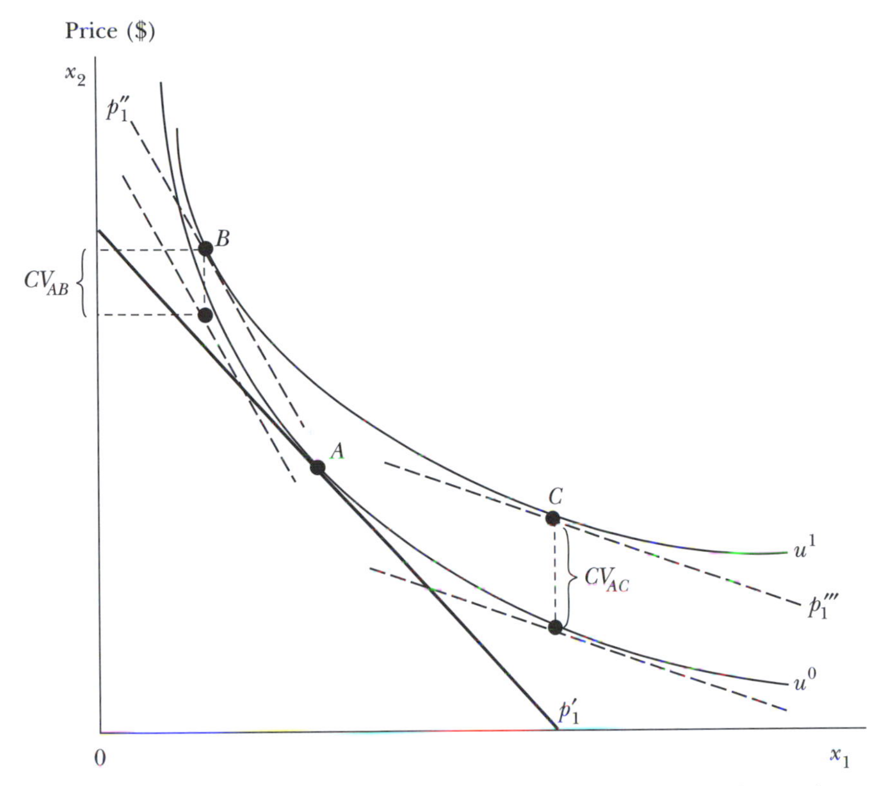

Figure 3-4. The compensating variation incorrectly ranks two alternative policies

if the new price set is not to be attained. Based on this interpretation of the two measures, some economists have argued that the choice between them is basically an ethical one, that is, one that depends on a value judgement as to which underlying distribution of property rights is more equitable (Krutilla 1967; Mishan 1976). All of this can be summarized as follows:

Welfare measure	*Price increase*	*Price decrease*
EV—Implied property right in the change	WTP to avoid	WTA to forgo
CV—Implied property right in the status quo	WTA to accept	WTP to obtain

For two alternative price changes, the welfare measures should be the same if both changes place the individual on the same higher indifference curve. However, if the two price changes place the individual on different indifference curves, the welfare measure should correctly indicate the preference ranking of the two alternatives. The *EV* measure always provides a consistent ranking in this sense, but the *CV* measure does not.

Figure 3-4 illustrates why this is the case. It shows an individual in equilibrium at point *A*, given prices and income. Suppose that one policy proposal would increase the price of x_1 and decrease the price of x_2 simultaneously. The individual would achieve a new equilibrium at point *B*. The *CV* measure of the welfare change is shown as CV_{AB}. The second policy alternative would decrease the price of x_1 while increasing the price of x_2. This would lead to a new consumer's equilibrium at point *C*. Point *C* has been drawn on the same indifference curve as point *B*. Therefore, the measure of welfare change should be the same for the two policy alternatives. However, as can be seen by inspection, the *CV* for the second policy, CV_{AC}, is larger. The *CV* measure would indicate a preference for the second policy while the individual is in fact indifferent between the two policies. The *EV* gives the same welfare measure for the two policy alternatives. This is because the *EV* measure bases its comparison on a point on the indifference curve passing through the new equilibrium, but with the old prices. If two policies are on the same new indifference curve, the *EV* measure picks the same point for measuring the welfare effects for both policies.

If the question being asked by policymakers is, does the proposed change pass the Kaldor potential compensation test? then *CV* is the measure to use. The *Kaldor potential compensation test* is one form of potential Pareto improvement test that asks whether it is possible for the winners to fully compensate all of the losers from the proposed policy change and still leave someone better off. For each person, the *CV* gives the compensating income change required to maintain that person at his or her initial utility level. If the sum of what could be collected from all gainers exceeds the sum of the required compensations for losers, the proposal passes this form of the potential Pareto improvement test. The fact that the *CV* cannot rank consistently two or more policy changes is no obstacle to its use in this manner. This is because the potential Pareto criterion itself provides no basis for ranking two or more proposed policy changes. If two proposed changes both pass the Kaldor potential compensation test, the potential Pareto improvement criterion provides no basis for choosing between them.

On the other hand, if the question being asked by policymakers is, does the policy pass the Hicks version of the potential compensation test? then *EV is* the appropriate measure. The Hicks test asks whether it is possible for the losers to bribe the gainers to obtain their consent to forgo the proposed policy change. The potential gainers would accept a bribe only if it were large enough to raise their utility by the same amount as the proposed project would have. The offered bribe would have to be as large as each individual's *EV* measure of welfare gain. The maximum bribe that would be offered by the potential losers would be their *EV* measure of loss. Thus if the sum of the *EV* of all gainers exceeded the sum of the *EVs* of all losers, the proposal would pass the Hicks form of the potential compensation test. Also, because the Hicks form of the compensation test is based on the *EV* measure, it will consistently rank two or more policy changes, provided that society is indifferent as to the distribution of gains and losses across individuals.

Measurement

Simply put, the problem posed for applied welfare economics is that the desired welfare measures, the *CV* or the *EV*, are based on the unobservable Hicks-compensated demand functions, whereas the one measure based on the observed Marshallian demand functions is flawed as a welfare indicator. The typical practice had been to use the Marshallian surplus anyway and to offer such justifications as "income effects are likely to be small"; "with only one price change, path dependence is not an issue"; and "it is the only measure we have and it is better than nothing." Robert Willig (1976), in a widely cited article, provided a justification for using the Marshallian surplus by examining the magnitude of the differences between *S* and *CV* or *EV* under different conditions. He argued that "in most applications the error of approximation will be very small. In fact the error will often be overshadowed by the errors involved in estimating the demand curve" (1976, 589). More recently, several authors have developed methods for direct calculation of the *CV* and *EV* from information contained in the ordinary demand function, either through a Taylor's series approximation (McKenzie and Pearce 1982; McKenzie 1983) or, as exact measures, through integration to obtain the indirect utility function and the expenditure function (see, for example, Hausman 1981). In this subsection I review these recent developments in applied welfare analysis and assess their significance for valuing environmental services.

"Consumer's Surplus without Apology." Willig (1976) offered rigorous derivations of expressions relating *CV*, *S*, and *EV*. These expressions provide a way of calculating the magnitude of the differences among the three measures for given prices, quantities, and income. The differences among the three measures depend on the income elasticity of demand for the good in question and consumer's surplus as a percentage of income. The differences among the measures appear to be small and almost trivial for most realistic cases. The differences are probably smaller than the errors in the estimation of the parameters of demand functions by econometric methods.

Willig's bounds for the approximation errors are based on the fact that the differences between *S* and *CV* or *EV* arise from an income effect on the quantity demanded; the size of that effect depends on the change in real income brought about by the price change and on the income elasticity of demand for the good. This can be shown in a nonrigorous way for the case

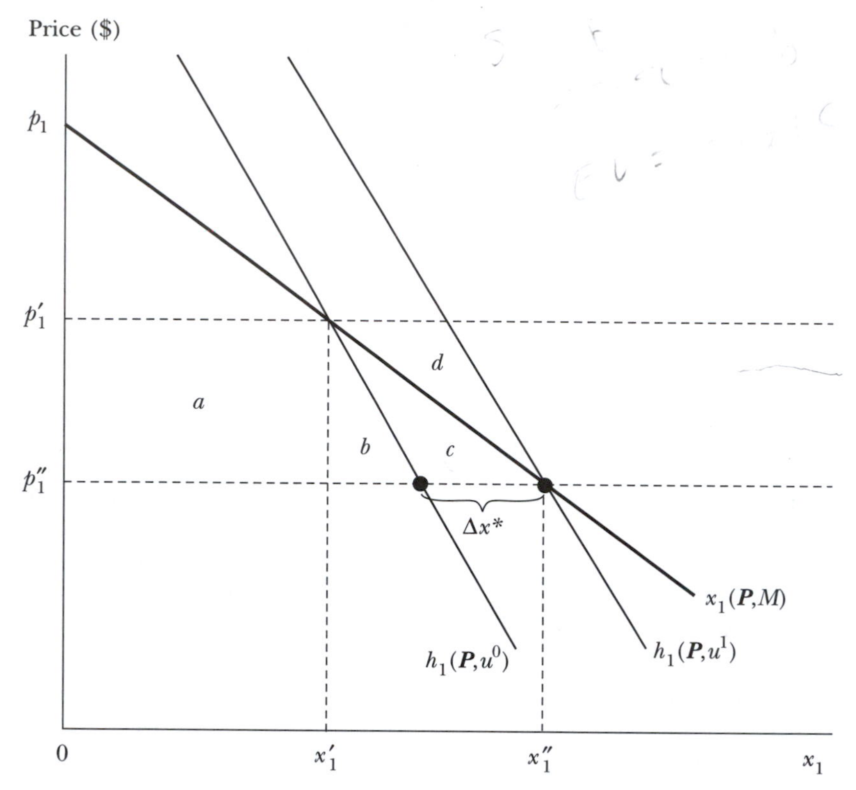

Figure 3-5. Deriving the Willig bounds for *S* as an approximation to *CV*

of one price change with the help of Figure 3-5. Although this exposition applies to the case of only one price change, the Willig expressions can be generalized to accommodate multiple price changes (Willig 1979), provided that a specific path of integration is chosen. In Figure 3-5 the ordinary and compensated demand curves are assumed to be linear. Let S represent the area $a + b + c$. So

$$CV = a + b = S - c \tag{3-28}$$

and

$$EV = a + b + c + d = S + d \tag{3-29}$$

The errors in using S to approximate CV and EV are equal to the areas c and d, respectively. For a price change from p_1' to p_1'', the factors influencing the size of the approximation error can be seen by examining the determinants of the area c:

$$\begin{aligned} CV - S &= -c \\ &= -\tfrac{1}{2}\Delta p \cdot \Delta x^* \end{aligned} \tag{3-30}$$

where Δx^* is the income effect on the quantity demanded of x associated with reducing income sufficiently to hold utility at u^0. Let ΔM^* represent this income change. By definition, ΔM^* is CV. The definition of income elasticity of demand is

$$E_M = \frac{\Delta x}{\Delta M} \cdot \frac{M}{x} \tag{3-31}$$

Solving this expression for Δx^* gives

$$\Delta x^* = E_M \cdot x \cdot \frac{\Delta M^*}{M} \tag{3-32}$$

Substituting this into equation 3-30, we obtain

$$CV - S = \frac{\Delta p \cdot x \cdot E_M \cdot CV}{2M} \tag{3-33}$$

In general, for small changes in p, $\Delta p \cdot x \simeq -S$. This is strictly true for the linear demand curve when x is evaluated at the midpoint between x' and x''. Finally, dividing both sides by CV to express the error in percentage terms gives

$$\frac{CV - S}{CV} \simeq -\frac{E_M}{2} \cdot \frac{S}{M} \tag{3-34}$$

This is similar to the Willig expression for the approximation error. The principal difference is that it expresses the error as a percentage of CV, whereas Willig's term makes the error a percentage of S. Willig's term says that the error is proportional to the income elasticity of demand and consumer's surplus as a percentage of income. A similar line of reasoning can be used to derive the relationship between EV and S.

Willig's analysis is more rigorous than this because it takes into account the possibility that for finite changes in price and quantity, the income elasticity of demand may vary over the range of the price change. He derives rules of thumb for calculating the maximum error in using S as an approximation for EV or CV. The rules of thumb are applicable if the following conditions are met:

$$\frac{S}{M} \cdot \frac{\underline{E}_M}{2} \leq 0.05$$
$$\frac{S}{M} \cdot \frac{\bar{E}_M}{2} \leq 0.05 \qquad (3\text{-}35)$$

and

$$\frac{S}{M} \leq 0.9 \qquad (3\text{-}36)$$

where $\underline{E}_M$ and $\bar{E}_M$ are the smallest and largest values, respectively, of the income elasticity of demand for the good in the region under consideration.

Given these conditions, the rule of thumb for *CV* is

$$\frac{S}{M} \cdot \frac{\underline{E}_M}{2} \leq \left|\frac{CV - S}{S}\right| \leq \frac{S}{M} \cdot \frac{\bar{E}_M}{2} \qquad (3\text{-}37)$$

and the rule of thumb for *EV* is

$$\frac{S}{M} \cdot \frac{\underline{E}_M}{2} \leq \left|\frac{S - EV}{S}\right| \leq \frac{S}{M} \cdot \frac{\bar{E}_M}{2} \qquad (3\text{-}38)$$

The first thing to note is the conditions under which these rules of thumb are valid. Consider equation 3-36 first. The change in consumer's surplus as a percentage of income depends on the size of the price change, the price elasticity of demand, and expenditure on this good as a percentage of total income. The smaller the price change and the smaller the proportion of income spent on the good, the smaller is *S/M*. It can readily be shown that

$$\frac{S}{M} \leq \left|\frac{\Delta p}{p}\right| \cdot \frac{p \cdot x}{M} \qquad (3\text{-}39)$$

From a given initial situation, *S* is largest when the demand curve is perfectly inelastic. Then $S = |x \cdot \Delta p|$ and equation 3-39 holds as an equality. With more elastic demand, $S < |x \cdot \Delta p|$ and the condition follows. It shows that, for example, for a good absorbing 50% of total income and for a 100% price change, *S/M* cannot exceed 0.5, whereas for a 10% price change for a good absorbing 10% of income, *S/M* will be less than 0.1.

Thus condition 3-36 is likely to be satisfied except for very large price increases for goods with low price elasticities that also absorb a large proportion of the total budget.

As for the first condition, the smaller that consumer's surplus is as a percentage of income, and the smaller that the income elasticity of demand is, the more likely condition 3-35 is to be satisfied. For example, if consumer's surplus is 5% of income, the income elasticity of demand can be as high as 2.0 and still satisfy condition 3-35. If S/M just barely satisfies condition 3-36, the income elasticity cannot exceed 0.11 to satisfy condition 3-35.

Assuming that conditions 3-35 and 3-36 hold, let us turn to the rules of thumb. First, according to condition 3-35, the maximum error involved in using S as an approximation for either CV or EV is 5%. Second, the smaller the change in income elasticity over the range being considered, the more precise are equations 3-37 and 3-38 as statements of the error involved in using S rather than CV or EV. If the income elasticity of demand does not change over the range being considered, the left-hand and right-hand sides of equations 3-37 and 3-38 are equal to each other and the errors are zero, as discussed above. Finally, as the income elasticity of demand for the good decreases, the differences among ordinary consumer's surplus, CV, and EV decrease, disappearing as E_M goes to zero.

Willig's analysis has been interpreted as providing a justification for using consumer's surplus as an approximation of the CV or the EV. However, there are two reasons why one should be cautious about adopting the Willig approach to welfare measurement. The first has to do with limitations on the applicability of the Willig conditions to some kinds of problems of welfare measurement, including some of specific interest to environmental and resource economists. The second arises because of the recent development of new methods for obtaining exact measures of CV and EV from the same information that is required to use the Willig approximation.

The Willig conditions for valid approximation were developed for changes in S as a result of changes in the price of some market good. Many environmental and resource policy issues require information on the total value of some environmental service as a measure of what would be lost if the resource were destroyed or diverted to some other use. For example, the economic cost of damming a river that provides whitewater canoeing and trout fishing would be measured by the total areas under the Hicks-compensated demand curves for these activities. This is equivalent to measuring the change in consumer's surplus for a price increase from the present price to the vertical intercept of the Hicks-compensated demand curve. Bockstael and McConnell (1980) point out that for the linear demand function, the income elasticity of demand goes to infinity as the price approaches the vertical intercept. Thus the approximation error can-

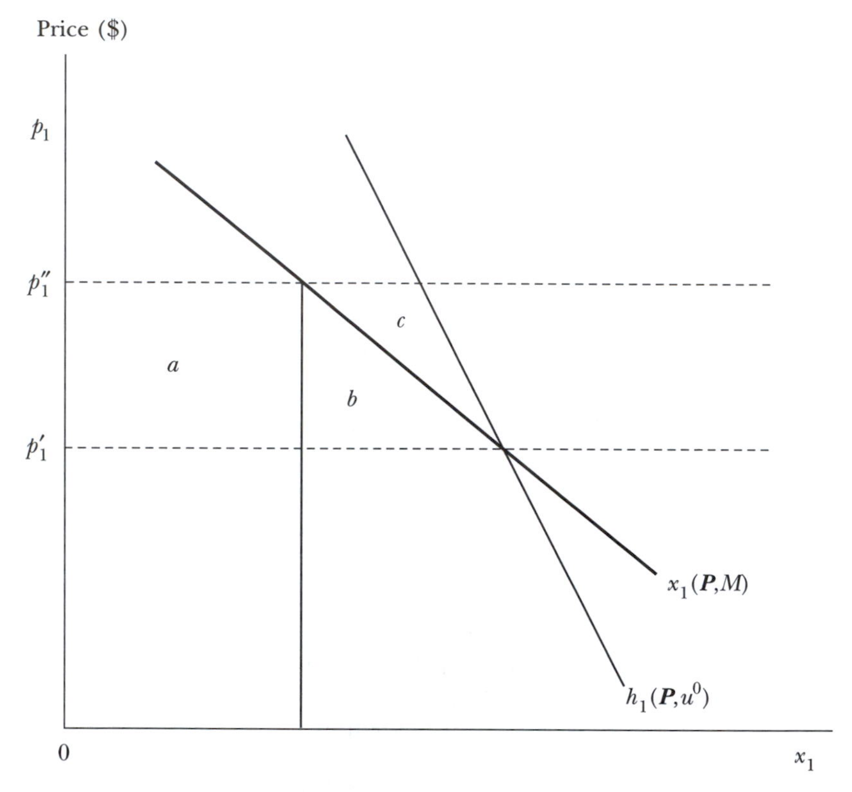

Figure 3-6. The Willig approximation and the error in estimating dead-weight losses

not be calculated. In a comment on Bockstael and McConnell 1980, Hanemann (1980) showed that if the parameters of the Marshallian demand function were known, it was unnecessary to compute the Willig approximation error because the *CV* could be calculated directly. In this idea, Hanemann (1980) apparently anticipated the analysis of Hausman (1981), discussed below.

For some questions, the variable of interest to policymakers is not *CV* but some fraction of *CV*—for example, the dead-weight loss associated with a tax on a commodity. Suppose an excise tax raises the price of a good from p_1' to p_1'', as shown in Figure 3-6. The consumer's loss as measured by *CV* is the area $a + b + c$, but only $b + c$ is an efficiency loss because a is a revenue transfer to the government. If the ordinary demand curve is used to approximate the consumer's loss, the area c is the error. If the Willig conditions are satisfied, c is an acceptably small percentage of S and *CV*, but it can be an unacceptably large percentage of the true dead-weight loss.

The second reason for being cautious about using the Willig approximation is that better methods of welfare measurement now exist. If the demand functions used to calculate S reflect utility-maximizing behavior on the part of individuals, they should satisfy the integrability conditions. And if this is the case, as I explain in the next section, it is possible to calculate CV and EV directly without approximation. On the other hand, if the demand functions do not satisfy the integrability conditions, then it is inappropriate to use the Willig approximations because their derivation was also based on the assumption of utility-maximizing behavior.

Exact Welfare Measurement. It is a basic proposition in the theory of demand that if the ordinary demand functions reflect utility-maximizing behavior, they contain all of the information on preferences that is necessary to reconstruct an ordinal index of utility. Recognition of this dual relationship between preferences and observable demand has resulted in a search for practical procedures for recovering the utility function and CV and EV welfare measures directly from estimated demand functions. One path that has been followed is reflected in the work of Reaume (1973), McKenzie and Pearce (1976, 1982), and McKenzie (1976, 1983). This method involves Taylor's series expansions of the indirect utility function based on terms that are derivatives of the ordinary demand functions. This approach has been criticized as being cumbersome (e.g., in Vartia 1983), potentially inexact because successive terms of a convergent Taylor's series are not necessarily monotonically decreasing (e.g., in Willig 1979, 473), and unnecessary, as I explain below.

More recently, several authors have presented operational procedures for recovering arbitrarily close approximations to CV and EV. One example is Vartia's (1983) algorithm for approximating welfare measures to an arbitrarily small error. More recently, Irvine and Sims (1998) developed a computationally simple approximation to CV and EV that is much more accurate than the ordinary consumer's surplus measure. It is based on the concept of the Slutsky-compensated demand curve, in which the individual is compensated not to the original level of utility (in the case of CV) but enough to enable the purchase of the original bundle of goods. Because this bundle is observable, it is straightforward to calculate the required compensation for any price change and to use the parameters of the ordinary demand function to calculate a welfare measure. Irvine and Sims showed that the error associated with this measure is typically an order of magnitude smaller than the error associated with S.

Hausman (1981) presented a procedure for exact welfare measurement based on the recovery of the parameters of the utility function from data on consumers' demands. His procedure, which was developed for the case of only one price change, involves four steps. The first step involves combining the ordinary demand function and Roy's Identity to obtain a partial differential equation:

$$x_1(\boldsymbol{P}, M) = -\frac{[\partial v(\boldsymbol{P}, M)/\partial p_1]}{[\partial v(\boldsymbol{P}, M)/\partial M]} \tag{3-40}$$

If the utility function is separable so that the demand function contains only its own price argument, and if the demand function is linear, this becomes

$$a - (b \cdot p_1) + (c \cdot M) = -\frac{[\partial v(\boldsymbol{P}, M)/\partial p_1]}{[\partial v(\boldsymbol{P}, M)/\partial M]} \tag{3-41}$$

where the parameters a, b, and c are estimated econometrically and where p_1 and M are deflated by an appropriate index of the other prices. Changes in p_1 and M that involve moving along an indifference curve must satisfy

$$\left[\frac{\partial v(\cdot)}{\partial p_1(t)} \cdot \frac{dp_1(t)}{dt}\right] + \left[\frac{\partial v(\cdot)}{\partial M(t)} \cdot \frac{dM(t)}{dt}\right] = 0 \tag{3-42}$$

where t defines a path of price changes. Rearranging this expression, substituting into equation 3-41, and using the implicit function theorem gives

$$\frac{dM(p_1)}{dp_1} = a - (b \cdot p_1) + (c \cdot M) \tag{3-43}$$

the solution of which is

$$M(p_1) = [k \cdot e^{c \cdot p_1}] - \frac{1}{c}\left[a - (b \cdot p_1) - \frac{b}{c}\right] \tag{3-44}$$

where k is the constant of integration, which depends on the initial level of utility. If units are arbitrarily chosen so that k is the initial utility level, the quasi-indirect utility function and quasi-expenditure function follow directly:

$$u = k^0 \cdot e^{-c \cdot p_1}\left\{\left(M + \frac{1}{c}\right)\left[a - (b \cdot p_1) - \frac{b}{c}\right]\right\} \tag{3-45}$$

and

$$e = [k^0 \cdot e^{c \cdot p_1}] - \frac{1}{c}\left[a - (b \cdot p_1) - \frac{b}{c}\right] \tag{3-46}$$

These expressions are termed "quasi-" functions because they do not contain information about the effects of the prices of other goods on utility or expenditure. Hausman's method depends on the ability to solve the differential equation that is obtained from Roy's Identity. He showed a method of solution for the case when only one price changes and discussed in general terms the solution in the case of multiple price changes.

Because researchers seldom have information on all demands, empirical studies typically use either partial or incomplete demand systems. A partial demand system is a system of the n goods of interest, artificially augmented by a composite numeraire commodity representing the remaining goods. An incomplete demand system includes only the n demands of interest. Separability is assumed between this group of goods and all others, and only the group's budget allocation is included in the system. LaFrance and Hanemann (1989) developed a means for discovering the necessary restrictions implied by weak integrability to obtain exact welfare measures from an incomplete demand system. LaFrance (1993) illustrated the bias that can occur with the usual practice of treating the group's budget allocation as exogenous. In a related literature, Hanemann and Morey (1992) investigated the relationship between true CV and EV measures and those produced by partial demand systems where only the n demands of interest are estimated. These problems arise most naturally when analyzing the effects of price changes for marketed goods, but they can also arise in some nonmarket settings when demands for goods related to the environmental good help reveal its value.

Conclusions

Selection of a welfare measure has long involved questions both of appropriateness and practicality. The Marshallian surplus measure was frequently chosen on the grounds of practicality, even though it was recognized that the measure did not answer any specific well-formed welfare question. Willig's development of the bounds for the errors of approximation in using S gave encouragement to this practice. However, quickly on its heels have come new approaches to exact welfare measurement that offer the opportunity to calculate the more appropriate CV and EV measures directly.

One question related to practicality remains, however. Do we know enough about the functional form of the utility function to implement exact measurement methods? Assuming a functional form for the system of demand functions for purposes of estimation is equivalent to assuming the functional form of the underlying utility function. One approach is to assume a specific functional form for the utility function or indirect utility function and to derive the demand functions for estimation. If this is the approach taken, then plugging the estimated parameters back into the utility function to calculate welfare changes is straightforward, provided that the parameter estimates of the demand function satisfy the integrability conditions. Because researchers have been reluctant to specify the functional form of the utility function, one alternative has been to specify so-called *flexible forms* for the indirect or direct utility function (e.g., Deaton and Muellbauer 1980). Again, if the integrability conditions are satisfied, deriving "exact" welfare measures from the "approximate" flexible functional form of the utility function is straightforward. The alternative is to seek guidance from the data by selecting the functional form for the demand functions on the basis of goodness of fit and consistency with the restrictions imposed by theory.

Welfare Measures for Changes in Factor Prices

If the quantity of the factor that the individual supplies to the market is unaffected by changes in its price, the price change affects only income. In this case, the income change is an exact measure of the welfare effect of the price change. However, if the change in factor price affects the quantity supplied—for example, by altering the choice between labor and leisure—then the welfare measure must reflect both changes in income and changes in the utility derived from the quantity of the factor withheld from the market. *CV* and *EV* measures of welfare change can be defined in a manner consistent with the preceding analysis of product price changes. The relationship between these measures and the area above the individual's factor supply curve is similar to the relationships among *CV*, *EV*, and *S* for product price changes.

Consider the supply of labor where the individual allocates time between income-producing work and utility-yielding leisure. The individual's preferences are described by $u(X, f)$ where f is time spent in leisure (fun). The individual maximizes utility subject to the budget and time constraints:

$$M^* + (p_w \cdot w) - (\boldsymbol{P} \cdot \boldsymbol{X}) = 0 \tag{3-47}$$

$$t^* - w - f = 0 \tag{3-48}$$

or

$$M^* + p_w(t^* - f) - (\boldsymbol{P} \cdot \boldsymbol{X}) = 0 \tag{3-49}$$

where M^* is exogenous income, p_w and w are the wage rate and hours worked, t^* is the total endowment of time, and f is hours of leisure. The

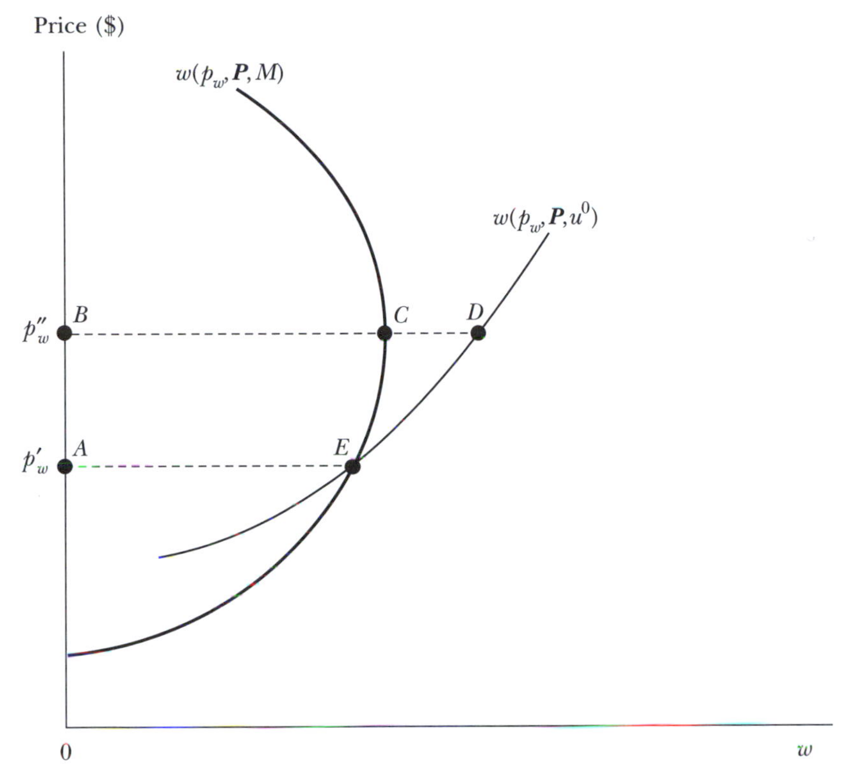

Figure 3-7. The compensating variation for a factor price increase

individual's labor supply function $w = w(p_w, \boldsymbol{P}, M^*)$ is part of the solution to this maximization problem. As is well known, the labor supply curve can be backward bending, at least over some range, if the positive income effect on the demand for leisure outweighs the negative substitution effect of more expensive leisure. See Figure 3-7.

Following the usual procedure, the indirect utility function, $v(\boldsymbol{P}, p_w, M^*)$, and the pseudo-expenditure function, $e^*(\boldsymbol{P}, p_w, u^0)$, can be derived (Just et al. 1982). The pseudo-expenditure function gives the amount of exogenous income (not expenditure) required to sustain any utility level as a function of prices. The compensated labor supply curve shows the quantities supplied at various prices, assuming that exogenous income has been adjusted to maintain the individual at that person's initial utility level, say u^0. It is derived by differentiating the pseudo-expenditure function with respect to p_w. Because the compensated supply curve represents only the substitution effect between work and leisure, it is always upward sloping and more elastic than the ordinary supply curve.

If p_w increases from, say, p_w' to p_w'', the *CV* measure of welfare gain is the integral of the derivative of the pseudo-expenditure function with respect to p_w. Geometrically this is the area to the left of the compensated supply curve between the two price levels, that is, *ABDE* in Figure 3-7. This is greater than the area between the prices bounded by the ordinary supply curve. There is also another compensated supply curve defined for the new utility level after the price change. It goes through point *C*. The *EV* measure of welfare change is the area to the left of this curve (not shown in Figure 3-7) between the price levels. Unless the income elasticity of demand for leisure is zero, *CV* is greater than *EV* for wage increases, while *EV* is greater than *CV* for wage decreases. Given an estimate of the ordinary labor supply function, exact calculation of *EV* and *CV* measures is possible (Hausman 1981).

Welfare Measures for Quantity Changes

Many environmental policy proposals involve changes in either the quantities or the qualities of nonmarket environmental goods and services rather than changes in the price of a marketable good. From the individual's point of view, the most important characteristic of some environmental goods is that they are available only in fixed, unalterable quantities. These quantities act as constraints on each individual's choice of a consumption bundle. The analysis of this class of problems is often referred to as the theory of choice and welfare under quantity constraints (Johansson 1987) or under imposed quantities (Lankford 1988). The imposition of quantity constraints raises some new issues in the theory of choice and welfare measurement. The analysis of these problems has evolved out of the theory of rationing as initially developed by Tobin and Houthakker (1950–1951) and Neary and Roberts (1980).

In this section, I briefly describe the model of individual preferences and choice under imposed quantity constraints. I then derive measures of welfare change for changes in the quantities of imposed goods. These measures are essentially similar to the compensating and equivalent surplus measures for price changes discussed earlier. I also take up the question of measurement, describing what is necessary to obtain exact measures by solving for the expenditure function or indirect utility function, and the potential errors involved in approximating these exact measures by areas under uncompensated inverse demand (bid) curves. Finally, I discuss the relationship between WTP and WTA and the welfare value of changes in q when q is a bad.

The Basic Model

Consider an individual whose utility function has the following form:

$$u(\boldsymbol{X}, \boldsymbol{Q}) \tag{3-50}$$

where $\boldsymbol{X}$ is the vector of private goods quantities ($\boldsymbol{X} = x_1, \ldots, x_i, \ldots, x_n$) and $\boldsymbol{Q}$ is a vector of environmental and resource service flows ($\boldsymbol{Q} = q_1, \ldots, q_j, \ldots, q_m$) that is given to the individual. It is possible that for at least some of the elements in $\boldsymbol{Q}$ there is a positive price charged. Let $\boldsymbol{R} \geq 0$ be the vector of these prices, and $\boldsymbol{P}$ be the vector of prices for $\boldsymbol{X}$. The individual maximizes utility subject to a budget constraint

$$\boldsymbol{P} \cdot \boldsymbol{X} + \boldsymbol{R} \cdot \boldsymbol{Q} = M \tag{3-51}$$

where M is money income. This yields a set of conditional demand functions for the marketed goods:

$$x_i = x_i(\boldsymbol{P}, M - \boldsymbol{R} \cdot \boldsymbol{Q}, \boldsymbol{Q}) \tag{3-52}$$

In general, $\boldsymbol{Q}$ will be an argument in these conditional demand functions, along with the prices of both market and nonmarket goods. The term *conditional* refers to the fact that these functions are conditioned upon the imposed $\boldsymbol{Q}$.

Inserting the conditional demand functions into the utility function gives the conditional indirect utility function

$$v = v(\boldsymbol{P}, M - \boldsymbol{R} \cdot \boldsymbol{Q}, \boldsymbol{Q}) \tag{3-53}$$

Inverting the conditional indirect utility function for the term $M - \boldsymbol{R} \cdot \boldsymbol{Q}$ yields a conditional expenditure function that gives the minimum expenditure on market goods required to produce utility level u, given $\boldsymbol{P}$ and $\boldsymbol{Q}$. This function is

$$e^* = M - \boldsymbol{R} \cdot \boldsymbol{Q} = e^*(\boldsymbol{P}, \boldsymbol{Q}, u) \tag{3-54}$$

The dual to the utility maximization problem can be stated as follows: minimize expenditure subject to the constraint that utility equal or exceed some stated level, say u^0. The solution to this problem gives the restricted expenditure function

$$e = e(\boldsymbol{P}, \boldsymbol{R}, \boldsymbol{Q}, u^0) \tag{3-55}$$

which is the total expenditure on all goods, including $\boldsymbol{Q}$, necessary to achieve u^0 given $\boldsymbol{P}$, $\boldsymbol{R}$, and $\boldsymbol{Q}$. The conditional and restricted expenditure functions are related by the following expression:

$$e = e^* + \boldsymbol{R} \cdot \boldsymbol{Q} \tag{3-56}$$

Welfare measures can be defined in terms of either the conditional or restricted expenditure function. These two forms of expenditure function coincide in the special case where $\boldsymbol{R} = 0$.

Welfare Measures

Lankford (1988) derived welfare measures for the general case with simultaneous changes in $\boldsymbol{P}$, $\boldsymbol{Q}$, $\boldsymbol{R}$, and M. Here I consider only changes in $\boldsymbol{Q}$. Changes in $\boldsymbol{R}$ have the same effect as lump-sum changes in money income equal to $-\Delta\boldsymbol{R} \cdot \boldsymbol{Q}$, so they pose no special problem for welfare measurement. For simplicity, in what follows I will assume that $\boldsymbol{Q}$ consists of only one element, q, with a price of r. I will first present welfare measures for the general case where $r \geq 0$. I will also show how these measures can be simplified when $r = 0$. So that graphic presentations of some of the key points can be made, I will assume that $\boldsymbol{X}$ is the numeraire represented as x with a price of 1. Finally, I will assume that at the given prices and income, the individual would choose more of q if given the option. The case where more q is a bad is taken up below.

To begin with, the marginal value of a small increase in q is the reduction in income that is just sufficient to maintain utility at its original level. If w_q is the marginal value or marginal willingness to pay for a change in q, it is given by the derivative of the restricted expenditure function with respect to q or

$$w_q = -\frac{\partial e}{\partial q} \tag{3-57}$$

The right-hand side of this expression is also equal (in absolute value) to the slope of the indifference curve through the point at which the welfare change is being evaluated. Alternatively, from equation 3-56

$$w_q = -\frac{\partial e^*}{\partial q} - r \tag{3-58}$$

The term $-r$ reflects the fact that the additional q imposes an obligation on the individual to increase the lump-sum payment given by $r \cdot q$ whenever r is positive. Lankford (1988) calls this the *income value* of the change in q. An increase in q has a negative income value.

The quantity-constrained nature of the problem makes it impossible for an individual to adjust the quantity of q to satisfy the conventional optimizing conditions of equality of marginal rates of substitution and price ratios. Therefore, compensating surplus (CS) and equivalent surplus (ES) are the relevant measures of welfare change. The choice between them depends on the same considerations applying to the choice between *CV* and *EV* measures for price changes. There are several ways to present *CS* and *ES* for changes in quantity-constrained goods.

The first way is based on the conditional indirect utility function. The *CS* and *ES* measures are defined implicitly as the solutions to the following expressions:
CS is the solution to

$$v(\boldsymbol{P}, M - r \cdot q^0, q^0) = v(\boldsymbol{P}, M - r \cdot q^1 - CS, q^1) \tag{3-59}$$

and *ES* is the solution to

$$v(\boldsymbol{P}, M - r \cdot q^0 + ES, q^0) = v(\boldsymbol{P}, M - r \cdot q^1, q^1) \tag{3-60}$$

These two measures can also be defined in terms of the expenditure function. Consider first the *CS*. Using the restricted expenditure function,

$$CS = e(\boldsymbol{P}, r, q^0, u^0) - e(\boldsymbol{P}, r, q^1, u^0) = M - e(\boldsymbol{P}, r, q^1, u^0) \tag{3-61}$$

This case is shown in Figure 3-8. The initial position is at point *A*, where the individual consumes q^0 and x^0 and achieves u^0 in utility. The increase in q enables the individual to reach u^1 at point *B*. If income is reduced by *CS*, the individual is pushed back to u^0 at point *C*.

The conditional expenditure function can be used to show that this welfare measure consists of two components. Recalling that the conditional expenditure function gives the expenditure on privately chosen market goods necessary to achieve a specified utility level given the level of q, we have

$$e^*(\boldsymbol{P}, q^0, u^0) = e(\boldsymbol{P}, r, q^0, u^0) - r \cdot q^0 \tag{3-62}$$

Thus

$$\begin{aligned} CS &= e^*(\boldsymbol{P}, q^0, u^0) + r \cdot q^0 - e(\boldsymbol{P}, q^1, u^0) - r \cdot q^1 \\ &= e^*(\boldsymbol{P}, q^0, u^0) - e(\boldsymbol{P}, q^1, u^0) - r(q^1 - q^0) \end{aligned} \tag{3-63}$$

The last term is the *income value* term, reflecting the extra expenditure on q given r. As shown in Figure 3-8, the reduction in the expenditures on x that is given by the first two terms in the second line of equation 3-63 overstates the welfare gain from the increase in q when r is positive. The true welfare gain is found by deducting from Δe^* an amount equal to the unavoidable increase in expenditure on q, $r(q^1 - q^0)$.

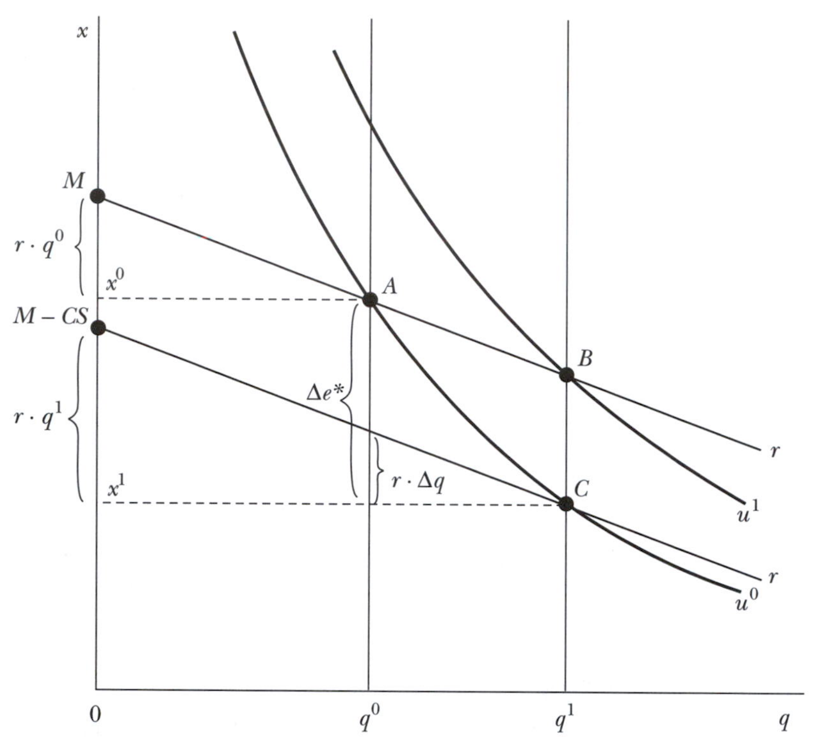

Figure 3-8. The compensating surplus measure for an increase in *q*

The *ES* measure given by the restricted expenditure function is

$$ES = e(\mathbf{P}, r, q^0, u^1) - e(\mathbf{P}, r, q^0, u^0) = e(\mathbf{P}, r, q^0, u^1) - M \qquad (3\text{-}64)$$

This is shown in Figure 3-9. As before, the increase in q enables the individual to reach point B with utility equal to u^1. Alternatively, if income were increased by *ES* while holding q constant, the individual could achieve u^1 at point C.

The second way to present the *ES* measure is based on the conditional expenditure function

$$ES = e^*(\mathbf{P}, q^0, u^1) + r \cdot q^0 - e^*(\mathbf{P}, q^0, u^0) - r \cdot q^0 \qquad (3\text{-}65)$$

Because this welfare measure is defined in terms of income changes holding q constant at q^0, there is no need to take account of the income value of a change in q.

The third way to derive the *ES* and *CS* measures is to make use of equation 3-57. The value of a nonmarginal change in q is the integral of equation 3-57 taken over the relevant range, or

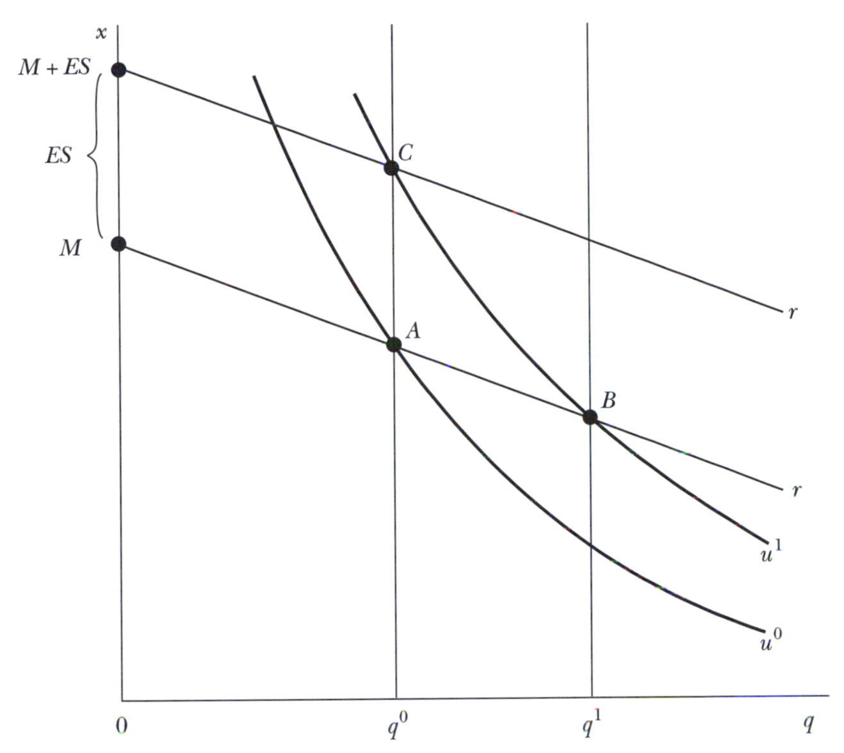

Figure 3-9. The equivalent surplus measure for an increase in *q*

$$W_q = -\int_{q^0}^{q^1} \frac{\partial e\left(\boldsymbol{P}, r, q, u^t\right)}{\partial q} \cdot dq \tag{3-66}$$

This is either a *CS* or an *ES* measure, depending on whether $t = 0$ or 1.

Now assume that $r = 0$. There is no need to take into account the income value term in any of the welfare measures. The budget lines are horizontal, as shown in Figure 3-10. The full and conditional expenditure functions are the same. And the *CS* and *ES* measures are as shown in the diagram.

Another way of looking at the welfare measurement problem is to ask what income levels and prices for q would be necessary to support each quantity-constrained bundle as a freely chosen utility-maximizing bundle. For example, consider the initial position shown in Figure 3-11, with q^0 and the individual at point A realizing utility level u^0. A "virtual" budget line through point A and tangent to u^0 has a slope equal to $-b^0$ and an intercept equal to μ^0. We can interpret b^0 as the virtual price of q and μ^0 as the virtual income associated with point A. Also, b and μ are functions of q, u, and the prices of marketed goods. The function $b(\cdot)$ is the compensated inverse demand function for q. It might also be called either the marginal

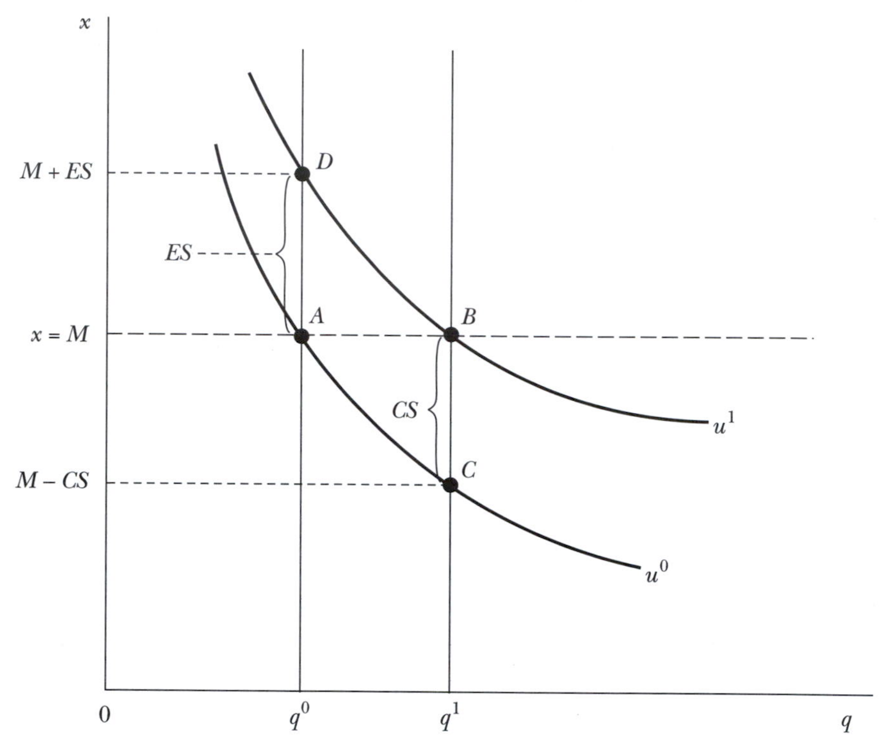

Figure 3-10. Compensating and equivalent surpluses for a change in q when $r = 0$

WTP function or the virtual price function. Other "virtual" budget lines could be drawn tangent to indifference curves at points such as *B*, *C*, and *D* to show the dependence of b on q and u. Also,

$$b = -\frac{\partial e}{\partial q} \tag{3-67}$$

The compensated inverse demand function can also be used to define welfare measures. Because the virtual price is equal to the slope of the indifference curve at that point, it is also equal to $\partial e/\partial q$ evaluated at that point. Thus

$$W_q = \int_{q^0}^{q^1} b\left(\boldsymbol{P}, q, u^t\right) dq \tag{3-68}$$

This is either a *CS* or an *ES* measure, depending on whether $t = 0$ or 1.

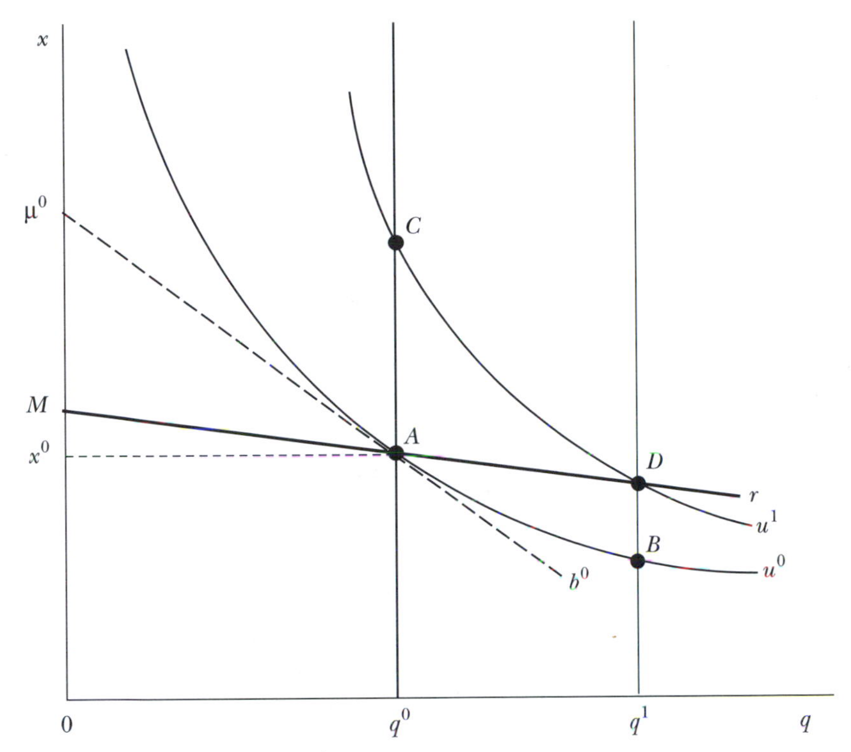

Figure 3-11. Virtual income and the marginal WTP function

Exact and Approximate Welfare Measures

The question to be considered now is whether there is a relationship describing the individual's behavior with respect to choices of q that is, at least in principle, observable and that can be used to solve for the indirect utility function or expenditure function in a manner similar to the Hausman (1981) method for exact welfare measurement for price changes. Lankford (1988) described techniques for recovering the expenditure function or indirect utility function when the individual faces a quantity constraint. However, these techniques require information that is not often available, especially in the case of public goods and nonmarketed environmental and resource service flows.

What these techniques have in common is the requirement for information that reflects the choices of an individual who is maximizing utility subject only to price and income constraints. Simply put, in the absence of quantity constraints and when price ratios are known, marginal rates of substitution can be inferred from utility-maximizing choices. However, as is clear from Figures 3-8 through 3-11, the presence of quantity constraints means that marginal rates of substitution are not in general equal to price ratios; so marginal rates of substitution are not directly revealed.

The Lankford techniques require some knowledge of what the individual would choose to do if he or she could freely choose q at exogenous prices. The required data might be available, for example, in the case of wartime rationing, in which the quantity-constrained goods were previously available in unconstrained markets. But these data are not likely to be available for environmental and resource services. However, Ebert (1998) showed how, at least in principle, adding information on the marginal WTP functions for nonmarket goods to estimates of market demands makes it possible to recover the underlying preferences.

A related question is whether there is an observable counterpart to the Marshallian measure for price changes and, if so, what its relationship is to the exact *ES* and *CS* measures for quantity changes. Randall and Stoll (1980) analyzed the difference between *ES* and *CS* measures and their relationship to a Marshallian measure. However, their Marshallian surplus is somewhat unusual, and it is not directly observable. To explain this, we need to define two other functions. The first is a bid function conditional on M and r. Assume that the individual receives q and pays r per unit. This bid function gives the maximum willingness to pay for one additional unit of q as a function of q, while holding constant the other prices and the income available to allocate between additional q and $\mathbf{X}$, $(M - r \cdot q)$. This bid function can be written as

$$b^* = b^*(\mathbf{P}, q, M - r \cdot q) \tag{3-69}$$

This equation gives the slope of the indifference curve passing through the point defined by q and $M - r \cdot q = x$. At point A in Figure 3-11, $b^* = b^0$, but at other levels of q, $b^* \neq b^0$. For example, at q^1 in Figure 3-11, $b^*(\cdot)$ would give the slope of u^1 at point D, whereas $b(\cdot)$ as given by equation 3-67 gives the slope of u^0 at that level of q; and $b^* > b$. The $b^*(\cdot)$ function cannot be inverted to get the optimal choice of q as a function of b except in the special case where $b^* = r$. For example, at q^1 this function would yield the slope of u^1 at point D, but this point would not be freely chosen given M and r and the preferences described by these indifference curves.

The second function to be defined also determines a bid, say b^{**}, but makes this bid depend on q and virtual income, $\mu = M + (b - r)q$. This income-compensated bid function is given by

$$b^{**} = b^{**}[\mathbf{P}, q, M + (b^{**} - r)q] \tag{3-70}$$

Given q^0, Figure 3-11 shows that $\mu = \mu^0$ and $b^{**} = b^0$ at point A. If q increases to q^1, the compensated bid function gives the new b^{**} and virtual income

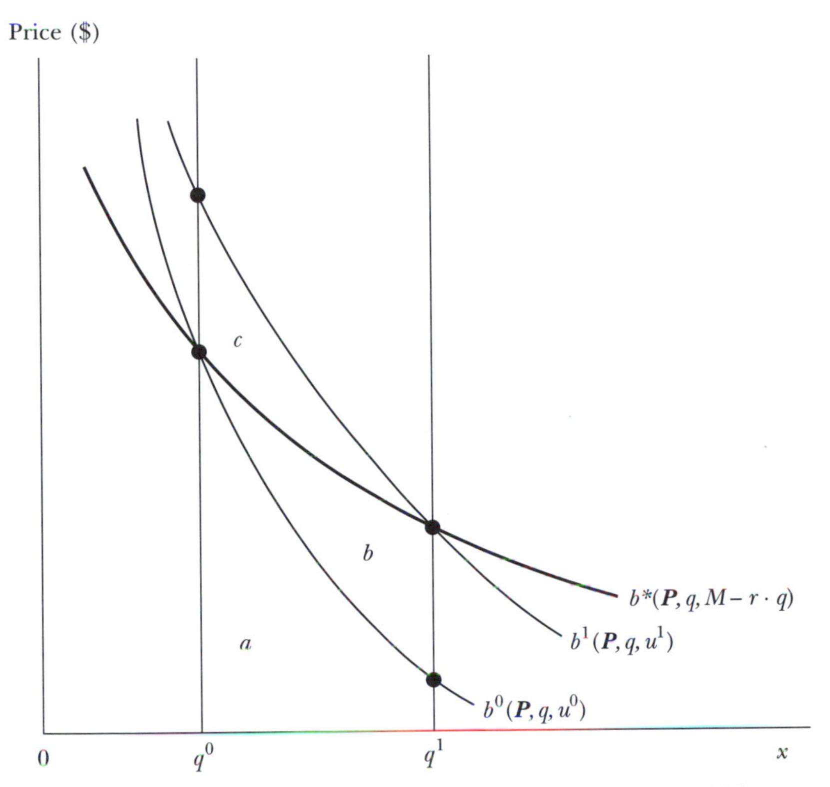

Figure 3-12. The compensated inverse demand curves and the bid curve for *q*

necessary to support point *D* as an optimal choice. In other words, at point *D*, $b^{**} = b^1$, the slope of u^1 at point *D*. In principle, this function could be inverted to find the optimal choice of *q* given b^{**}. However, this would not be a true Marshallian demand function because it assumes compensation for the income value of the Δq. Also, it cannot be estimated from observable data because only *M*, *q*, and *r* are observed, not $M + (b^{**} - r)q$.

It is interesting to note that although these two forms of bid function are defined differently and have different forms, they will compute the same bid for given *q*, *M*, and preferences. They both compute the slope of the indifference curve passing through the intersection of the $M - r \cdot q$ budget line and the given *q*. Yet neither function can be given a strictly Marshallian interpretation.

The relationship between the Hicks-compensated inverse demand function (equation 3-67) and the bid function $b^*(\cdot)$ is shown in Figure 3-12. At q^0, $b^0(\cdot)$ and $b^*(\cdot)$ coincide. But as *q* increases, $b^0(\cdot)$ diverges from $b^*(\cdot)$. This is because the first curve holds utility constant while the second curve holds income (*M*) constant.

As Lankford (1988, 56–61) shows, if the income-compensated bid function (equation 3-70) is known, it is possible to solve for the conditional expenditure function and through that to obtain the *CS* or the *ES* welfare measures exactly. *CS* is the area under $b^0(\cdot)$ between q^0 and q^1. *ES* is the area under $b^1(\cdot)$ over the same interval.

Randall and Stoll (1980) employed a line of reasoning similar to Willig's (1976) to analyze the errors involved in using the area under the $b^*(\cdot)$ function as an approximation to either *CS* or *ES* and the likely magnitude of the difference between *CS* and *ES*. As I have done in this section, they assumed that there is only one quantity-constrained good. Referring to Figure 3-12, for a move from q^0 to q^1 the three alternative welfare measures are given by the following geometric areas:

$$CS = a$$
$$S = a + b$$
$$ES = a + b + c \tag{3-71}$$

Randall and Stoll show that when

$$\frac{S}{M} \cdot \frac{E_q}{2} \leq 0.05 \tag{3-72}$$

the error bounds are

$$\frac{S}{M} \cdot \frac{\underline{E}_q}{2} \leq \frac{ES - S}{S} \leq \frac{S}{M} \cdot \frac{\bar{E}_q}{2} \tag{3-73}$$

and

$$\frac{S}{M} \cdot \frac{\underline{E}_q}{2} \leq \frac{S - CS}{S} \leq \frac{S}{M} \cdot \frac{\bar{E}_q}{2} \tag{3-74}$$

where *S* is the area under $b^*(\cdot)$, the lower and upper bars denote the smallest and largest values of E_q when E_q is not constant over the relevant range, and E_q is the price flexibility of income, defined as

$$E_q \equiv \frac{\partial b^* \left(\boldsymbol{P}, q, M\right)}{\partial M} \cdot \frac{M}{b^*} \tag{3-75}$$

General expressions for the error bounds are also presented. This result implies that using the area under $b^*(\cdot)$ to approximate *CS* or *ES* is an acceptable procedure. However, recall that $b^*(\cdot)$ is not observable; and if it is known somehow, it should be possible to calculate *CS* or *ES* directly, making the approximation unnecessary.

The result also seems to imply that *CS* (or willingness to pay for a change) and *ES* (or willingness to accept compensation to forgo a change) will be approximately equal in many circumstances. Hanemann (1991) showed that this is not the case when there is only one quantity-constrained good.

Willingness To Pay versus Willingness To Accept Compensation

The results of the preceding section imply that *CS* or willingness to pay for a change (WTP) and *ES* or willingness to accept compensation to forgo a change (WTA) will be approximately equal in most circumstances. However, there is a substantial body of evidence from both stated preference studies and laboratory experiments in valuation that differences between WTP and WTA for the same good can be quite large (Horowitz and McConnell 2002). These large differences are troubling. Efforts at explaining these differences have generally taken one of two paths.

On the first path, theoretical arguments for small differences between true WTP and true WTA are accepted as correct, and explanations for the disparity between revealed WTP and revealed WTA are sought in the consequences for behavior of lack of information and uncertainty about preferences. For example, Hoehn and Randall (1987) modeled the strategy of a rational individual who was risk-averse and uncertain about the true value of the postulated environmental change. They showed that in the context of stated preference studies, on average the individual's stated WTA would be greater than her true WTA, and stated WTP would be less than true WTP. Thus the difference between stated WTA and WTP would be greater than the true difference, on average. Kolstad and Guzman (1999) and Zhao and Kling (2001) also modeled rational behavior of uncertain individuals in the context of other preference-revealing mechanisms and found similar results.

The second path involves examining the theory of preferences and value more closely to see whether theory predicts the large disparities between true WTA and WTP. One example of this is in the work of Hanemann (1991, 1999). He showed that the price flexibility of income can be expressed as the ratio of two other terms:

$$E_q = \frac{E_M}{\sigma_q} \tag{3-76}$$

where σ_q is the aggregate Allen–Uzawa elasticity of substitution between q and the composite commodity $\mathbf{X}$, and E_M is the income elasticity of

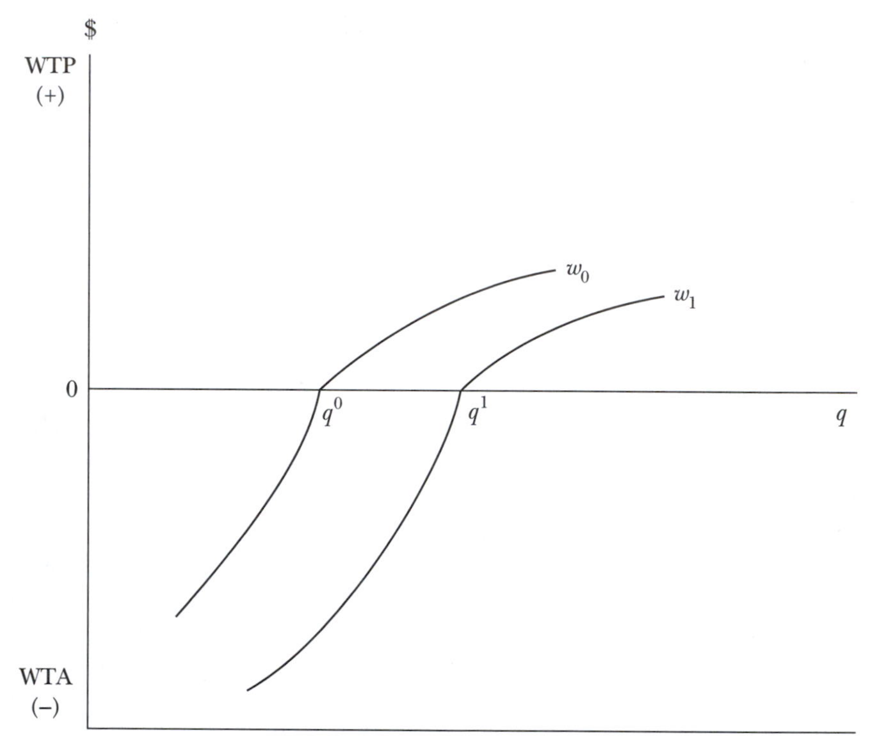

Figure 3-13. The value function and the endowment effect

demand for q. If the elasticity of substitution (a measure of the curvature of the indifference curve between q and private goods) is low, σ_q can be close to zero. This can lead to a high value for E_q and a large difference between *CS* and *ES*. However, Hanemann's analysis does not explain the persistent differences between the two measures in experiments with simulated markets involving commonplace goods such as lottery tickets, coffee mugs, and pens (see Knetsch and Sinden 1984; Kahneman et al. 1990).

Thaler (1980) proposed that the reconciliation of theory with observation can be brought about by postulating an "endowment effect" on individuals' valuation functions and a kink in this function at the status quo point. He suggests that this is a reasonable extension and generalization of the prospect theory of Kahneman and Tversky (1979) to choices not involving uncertainty. The idea of the endowment effect and the differential valuation of gains and losses can be shown with the aid of Figure 3-13. The horizontal axis shows the quantity of an environmental good q. The vertical axis shows the compensating welfare measure for changes in q. This measure is positive (WTP) for increases and negative (WTA) for decreases from some status quo point. Suppose that the status quo is q^0. The associated valuation function w_0 shows the monetary payment (compensation) that holds utility constant for a given increase (decrease) in q from q^0. This function is kinked at the status quo point of q^0, showing that the marginal valuation of increases in q is substantially lower than the marginal valuation of losses from q^0. A change in the endowment of q from q^0 to q^1 shifts the valuation function. As Figure 3-13 shows, the willingness to pay for an increase from q^0 to q^1 is substantially less than the required compensation for the decrease from q^1 to q^0.

In conclusion, although the observed large differences between WTP and WTA can be explained by replacing the standard utility model with one that incorporates an endowment effect, it is not clear that this is necessary. These differences can also be explained by the absence of close substitutes in the case of unique and perhaps irreplaceable resources and as the rational response to uncertainty and the high cost of information about preferences.

When q Is a Bad

There are two ways in which more q could be a bad, rather than a good, for an individual. The first way is when r is greater than zero and the individual would prefer to have less than the quantity being imposed given that price. This signifies that $r > b > 0$. The welfare measures *ES* and *CS* are still defined in the same way, but now they are negative for increases in q and positive for decreases in q.

The second way in which q can be a bad is the more fundamental one; it is when the marginal utility of q is negative. Even at a zero price, the individual would prefer to receive a smaller quantity. This means that b is negative. In both cases, the welfare measures *ES* and *CS* are defined in the same way, and again they are negative for increases in q and positive for decreases in q. Also, all the discussion of exact welfare measurement techniques and approximations carries over with appropriate changes to the case of q as a bad.

Aggregation and Social Welfare

Assume now that we have obtained measures of the welfare changes, either plus or minus, for all individuals. How can we use that information to make choices about public policy alternatives? To put the question in its most profound sense, what is the appropriate relationship between the welfare of individuals and the social welfare? What follows is a brief review of alternative social welfare criteria. Because the main concern of this book is with measurement, the question of social welfare criteria—that is, how to use the measures—is off the main track. For a more extensive discussion of the problem, see Mishan 1960, especially Section III, and Boadway and Bruce 1984.

In the literature on welfare economics there are basically four ways to approach this question. The first approach to the question is the so-called *Pareto criterion.* Only policy changes that make at least one person better off (that is, an individual experiences a positive welfare change) and make no individual worse off (that is, no individual experiences a negative welfare change) pass this criterion. This criterion deliberately rules out any attempts to add up or otherwise make commensurable the welfare measures of different individuals. Because virtually all actual public policy proposals impose net costs on at least some individuals, most policy actions by the state could not be accepted under this criterion. This would be particularly true in the environmental area, where environmental management costs are often channeled through the production sector while benefits accrue to households in the form of increased levels of environmental services. It is unlikely that this would result in a pattern of incidence of benefits and costs in which no one would lose. The restrictive features of the Pareto criterion have stimulated an ongoing search for a welfare criterion that would justify the state doing certain things that at least some people feel it should be able to do.

The second approach to the question was proposed in slightly different forms by Kaldor (1939) and Hicks (1939). These are the two different forms of a potential compensation test discussed earlier, in the section on Welfare Measures for Changes in Prices. Let us review these tests in the present context of aggregation and social welfare.

As noted earlier, the Kaldor version of the test asks whether those who gain by the policy can fully compensate for the welfare losses of those who lose by the policy. The Kaldor version of the test would be satisfied if the sum of all individual *CV* and *CS* measures of welfare changes were greater than zero. The criterion is essentially one of potential Pareto improvement because if the compensation were actually paid, no one would lose from the policy.

The Hicks version of the potential compensation test asks whether those who lose from the policy could compensate the gainers for a decision not to proceed with the policy. If the answer is yes, the policy should be rejected according to the Hicks criterion. If the policy were rejected and compensation were actually paid, those who would have gained from the policy would be just as well off as if the policy had been adopted, and those who would have lost are at least as well off as they would have been with the policy. The Hicks version of the test takes acceptance of the project as its reference point. In effect, it is a decision to forgo the project that creates the gains and losses that are relevant to the Hicksian version of potential Pareto improvement criterion.

In either the Kaldor case or the Hicks case, should the compensation actually be paid? If one thinks the answer should be yes, then the compensation test is transformed into a variation of the Pareto criterion, in which the state serves to enforce the taxes and transfer payments that are necessary to ensure that no one actually experiences a welfare loss, assuming that such taxes and transfers would be costless. If one thinks the answer should be no, this is equivalent to, in effect, assuming that all individual welfare changes are commensurate and can be summed together into an aggregate measure of welfare change. This is the efficiency criterion of the new welfare economics. According to the efficiency criterion, the objective of social policy is to maximize the aggregate value of all the goods and services people receive, including environmental and resource services. One justification for the Hicks–Kaldor potential compensation test is that a large number of efficient projects will spread benefits sufficiently widely so that everyone is a net gainer from the set of projects taken as a whole, even though some might be losers on individual projects. See Polinsky 1972 for an interesting development of this line of reasoning.

Alternatively, one might believe that whether compensation should be paid depends upon who has to pay and who gets the benefits. This requires consideration of the equity (fairness) in the distribution of income as an element in the evaluation of social policy. The third approach to the question of social welfare criteria, proposed by Little (1957), makes explicit the concern for equality. He proposed a twofold test. First, does the policy pass the Kaldor test? And second, does the resulting change improve the distribution of income? The Little criterion legitimizes a concern with the distributional effects of changes in resource allocation, but it does not resolve the question of what constitutes an improvement.

The fourth approach to the question involves an attempt to make specific social judgements regarding equity and to introduce equity considerations systematically into the evaluation of social policy. The most common proposal calls for the establishment of a social welfare function that gives different weights to individual welfare changes according to the relative deservingness of the different individuals (Eckstein 1961; Haveman and Weisbrod 1975). Of course, the main problem with the social welfare weight approach is the determination of the weighting function (Freeman 1971).

Nevertheless, willingness to make explicit value judgements about equity makes it possible to consider a wider range of policy choices. For example, if one opts for the Pareto criterion or the potential compensation version of the Hicks–Kaldor test, one rules out the possibility of accepting a project that has a sum of individual welfare changes that is less than zero but that would substantially improve the distribution of income. An example of such a policy would be one that imposes a welfare loss of $1,000 on a millionaire while bringing benefits of $99 to each of 10 impoverished orphans. A welfare-weighting function could approve negative sum policies like this, provided that the weights given to the beneficiaries were sufficiently greater than the welfare weights of the losers. In addition, neither of these criteria would reject a project that imposes costs on no one but distributes benefits only to the richest in our society. Some might make the value judgement that this, in itself, is undesirable. A social welfare function that included some measure of inequality of the aggregate distribution as an argument might reject inequality-creating projects like this. It would also be likely to accept negative-sum projects that reduced inequality.

The potential compensation test criterion is perhaps the most controversial feature of standard welfare economics. It has been criticized as being incompatible with the Pareto criterion because it allows for a ranking of projects that are Pareto noncomparable. On the other hand, its application has been rationalized on the grounds that if a large enough number of potential Pareto improvment projects are undertaken, benefits and costs will be spread sufficiently widely that everyone will be a net gainer from the set of projects as a whole, even though some might be losers on some individual projects. Thus benefit–cost analysis as a basis for policy choice rests on a shaky foundation. However, this has not deterred governments from using it for some kinds of policy choices and economists from advocating greater use of it in a wider range of environmental and resource policy questions. Whether this foundation can take the strain associated with its use in emerging environmental policy issues is an important question.

Summary

In this chapter, I have explained the derivation of the compensating and equivalent measures of individual welfare change for changes in prices and imposed quantities. The compensating and equivalent measures answer different kinds of policy-relevant questions because they make different implicit assumptions about the relevant status quo. It is interesting to examine some hypothetical examples.

Suppose that the question is whether to locate a landfill in a particular neighborhood. The neighbors are likely to oppose this proposal. Suppose that it is accepted that the neighbors have a right to an undisturbed neighborhood. Then the relevant measure of the harm for locating the landfill in their neighborhood would be the sum of their compensating measures of loss (*CV* and *CS*). The appropriate measure of the gain to those who would use the landfill would be their willingness to pay to locate it in this neighborhood (also a compensating measure). Alternatively, if it is argued that the larger society has a right to locate the landfill anywhere, then what

Table 3-1. Implied Property Rights and Associated Welfare Measures

Implicit "rights"	*Policy question*	*Gainers*	*Losers*
To the present polluter	Require cleanup?	Neighbors, compensating measure (WTP)	Polluter, compensating measure (WTA)
To the potential polluter	Allow pollution?	Polluter, equivalent measure (WTA)	Neighbors, equivalent measure (WTP)
To the neighbors	Require cleanup?	Neighbors, equivalent measure (WTA)	Polluter, equivalent measure (WTP)
To the neighbors	Allow pollution?	Polluter, compensating measure (WTP)	Neighbors, compensating measure (WTA)

is relevant is the neighbors' willingness to pay to keep it out of their neighborhood. This is an equivalent measure of the potential loss (*EV* and *ES*). For the users of the landfill, the value of locating the landfill in this neighborhood is what its users would require to compensate them for locating it in a less desirable place, an equivalent measure of benefit.

Suppose, instead, that the offending facility is a polluting factory that has been in the neighborhood for a long time. If the neighbors are deemed to have a right to a clean neighborhood, then the appropriate reference point for welfare measurement is their utility levels after the factory has stopped its pollution. This implies an equivalent measure of welfare change (*EV* and *ES*). Specifically, this is the compensation that the neighbors would require to forgo having the pollution stopped and the factory owners' willingness to pay to continue to pollute. Alternatively, if the factory has a right to pollute, compensating measures of the gain from stopping the pollution are appropriate (*CV* and *CS*).

In each case, the appropriate welfare measure can be found by examining the nature of the social transaction that is implied by the policy decision at hand and by the implicit rights to the services of the environment presumed to be held by the various parties to the transaction. The results for the examples discussed here can be summarized in Table 3-1.

References

Bergson, Abram. 1966. *Essays in Normative Economics.* Cambridge, MA: Harvard University Press.

Boadway, Robin, and Neil Bruce. 1984. *Welfare Economics.* Oxford, England: Basil Blackwell.

Bockstael, Nancy E., and Kenneth E. McConnell. 1980. Calculating Equivalent and Compensating Variation for Natural Resource Facilities. *Land Economics* 56(1): 56–63.

Currie, John M., John A. Murphy, and Andrew Schmitz. 1971. The Concept of Economic Surplus. *Economic Journal* 81(324): 741–799.

Deaton, Angus, and John Muellbauer. 1980. *Economics and Consumer Behavior.* Cambridge, U.K.: Cambridge University Press.

Ebert, Udo. 1998. Evaluation of Nonmarket Goods: Recovering Unconditional Preferences. *American Journal of Agricultural Economics* 80(2): 241–254.

Eckstein, Otto. 1961. A Survey of the Theory of Public Expenditures Criteria. In National Bureau of Economic Research, *Public Finances: Needs, Sources, and Utilization.* Princeton, NJ: Princeton University Press.

Freeman, A. Myrick, III. 1971. Project Design and Evaluation with Multiple Objectives. In *Public Expenditures and Policy Analysis,* edited by R. Haveman and J. Margolis. Chicago, IL: Markham.

Hanemann, W. Michael. 1980. Measuring the Worth of Natural Resource Facilities: Comment. *Land Economics* 56(4): 482–490.

———. Willingness To Pay and Willingness To Accept: How Much Can They Differ? *American Economic Review* 81(3): 635–647.

———. The Economic Theory of WTP and WTA. In *Valuing Environmental Preferences: Theory and Practice of the Contingent Valuation Method in the US, EU, and Developing Countries,* edited by Ian J. Batemen and Kenneth G. Willis. Oxford, U.K.: Oxford University Press.

Hanemann, W. Michael, and Edward Morey. 1992. Separability, Partial Demand Systems, and Consumer's Surplus Measures. *Journal of Environmental Economics and Management* 22(3): 241–258.

Hausman, Jerry A. 1981. Exact Consumer's Surplus and Dead Weight Loss. *American Economic Review* 71(4): 662–676.

Haveman, Robert H., and Burton A. Weisbrod. 1975. The Concept of Benefits in Cost–Benefit Analysis: With Emphasis on Water Pollution Activities. In *Cost–Benefit Analysis and Water Pollution Policy,* edited by Henry M. Peskin and Eugene P. Seskin. Washington, DC: The Urban Institute.

Hicks, John R. 1939. The Foundations of Welfare Economics. *Economic Journal* 49(196): 696–712.

———. 1943. The Four Consumer Surpluses. *Review of Economic Studies* 11(1): 31–41.

Hoehn, John P., and Alan Randall. 1987. A Satisfactory Benefit Cost Indicator from Contingent Valuation. *Journal of Environmental Economics and Management* 14(3): 226–247.

Horowitz, John K., and Kenneth E. McConnell. 2002. A Review of WTA/WTP Studies. *Journal of Environmental Economics and Management* 44(3): 426–447.

Hurwicz, Leonid, and Hirofumi Uzawa. 1971. On the Integrability of Demand Functions. In *Preferences, Utility and Demand,* edited by John S. Chipman, Leonid Hurwicz, Marcel K. Richter, and Hugo F. Sonnenschein. New York: Harcourt Brace Jovanovich.

Irvine, Ian J., and William A. Sims. 1998. Measuring Consumer Surplus with Unknown Hicksian Demands. *American Economic Review* 88(1): 314–322.

Johansson, Per-Olov. 1987. *The Economic Theory and Measurement of Environmental Benefits.* Cambridge, U.K.: Cambridge University Press.

Just, Richard E., Darrell L. Hueth, and Andrew Schmitz. 1982. *Applied Welfare Economics and Public Policy.* Englewood Cliffs, NJ: Prentice-Hall.

Kahneman, Daniel, Jack L. Knetsch, and Richard H. Thaler. 1990. Experimental Tests of the Endowment Effect and the Coase Theorem. *Journal of Political Economy* 98(6): 1325–1347.

Kahneman, Daniel, and Amos Tversky. 1979. Prospect Theory: An Analysis of Decisions under Risk. *Econometrics* 47(1): 263–291.

Kaldor, Nicholas. 1939. Welfare Propositions of Economics and Interpersonal Comparisons of Utility. *Economic Journal* 49(196): 549–552.

Knetsch, Jack L., and John A. Sinden. 1984. Willingness To Pay and Compensation Demanded: Experimental Evidence of an Unexpected Disparity in Measures of Values. *Quarterly Journal of Economics* 99(3): 507–521.

Kolstad, Charles D., and Rolando M. Guzman. 1999. Information and the Divergence Between Willingness To Accept and Willingness To Pay. *Journal of Environmental Economics and Management* 38(1): 66–80.

Krutilla, John V. 1967. Conservation Reconsidered. *American Economic Review* 57(4): 777–786.

LaFrance, Jeffery T. 1993. Weak Separability in Applied Welfare Analysis. *American Journal of Agricultural Economics* 75(3): 770–775.

LaFrance, Jeffery T., and W. Michael Hanemann. 1989. The Dual Structure of Incomplete Demand Systems. *American Journal of Agricultural Economics* 71(2): 262–274.

Lankford, R. Hamilton. 1988. Measuring Welfare Changes in Settings with Imposed Quantities. *Journal of Environmental Economics and Management* 15(1): 45–63.

Little, I.M.D. 1957. *A Critique of Welfare Economics.* 2nd ed. Oxford, U.K.: Clarendon Press.

Marshall, Alfred. 1920. *Principles of Economics: An Introductory Volume.* 8th ed. London: Macmillan.

McKenzie, George W. 1976. Measuring Gains and Losses. *Journal of Political Economy* 84(3): 641–646.

———. 1983. *Measuring Economic Welfare: New Methods.* Cambridge, U.K.: Cambridge University Press.

McKenzie, George W., and I.F. Pearce. 1976. Exact Measures of Welfare and the Cost of Living. *Review of Economic* Studies. 43(4): 465–468.

———. 1982. Welfare Measurement: A Synthesis. *American Economic Review* 72(4): 669–682.

Mishan, Ezra J. 1960. A Survey of Welfare Economics, 1939–1959. *Economic Journal* 70(278): 197–256.

———. 1976. The Use of Compensating and Equivalent Variations in Cost–Benefit Analysis. *Economica* 43(170): 185–197.

Morey, Edward R. 1984. Confuser Surplus. *American Economic Review* 74(1): 163–173.

Neary, J.P, and K.W.S. Roberts. 1980. The Theory of Household Behavior Under Rationing. *European Economic Review* 13(2): 25–42.

Polinsky, A. Mitchell. 1972. Probabilistic Compensation Criteria. *Quarterly Journal of Economics* 86(3): 407–425.

Randall, Alan, and John R. Stoll. 1980. Consumer's Surplus in Commodity Space. *American Economic Review* 70(3): 449–455.

Reaume, David M. 1973. Cost–Benefit Techniques and Consumer Surplus: A Clarificatory Analysis. *Public Finance* 28(2): 196–211.

Samuelson, Paul A. 1942. Constancy of the Marginal Utility of Income. In *Studies in Mathematical Economics and Econometrics, in Memory of Henry Schultz,* edited by Oscar Lange, Francis McIntyre, and Theodore O. Yntema. Chicago, IL: University of Chicago Press; reprinted in *The Collected Scientific Papers of Paul A. Samuelson.* vol. 1, edited by Joseph E. Stiglitz. Cambridge, MA: MIT Press, 1966.

Silberberg, Eugene. 1972. Duality and the Many Consumer's Surpluses. *American Economic Review* 62(5): 942–952.

———. 1978. *The Structure of Economics: A Mathematical Analysis.* New York: McGraw-Hill.

Thaler, Richard. 1980. Toward a Positive Theory of Consumer Choice. *Journal of Economic Behavior and Organization* 1(1): 39–60.

Tobin, James, and Hendrik S. Houthakker. 1950–1951. The Effects of Rationing on Demand Elasticities. *Review of Economic Studies* 18(3): 140–153.

Varian, Hal R. 1984. *Microeconomic Analysis.* 2nd ed. New York: Norton.

Vartia, Yrjo O. 1983. Efficient Methods of Measuring Welfare Change and Compensated Income in Terms of Ordinary Demand Functions. *Econometrica* 51(1): 79–98.

Willig, Robert D. 1976. Consumer's Surplus without Apology. *American Economic Review* 66(4): 589–597.

———. 1979. Consumer's Surplus without Apology: Reply. *American Economic Review* 69(3): 471–474.

Zhao, Jinhua, and Catherine L. Kling. 2001. A New Explanation for the WTP/WTA Disparity. *Economics Letters* 73(3): 293–300.

CHAPTER

4

Revealed Preference Models of Valuation: Basic Theory

For market goods, welfare effects due to changes in prices have been defined in terms of the area under the appropriate Hicks-compensated demand curve. For nonmarket goods, welfare effects due to changes in quantities have been defined in terms of the area under the marginal willingness-to-pay (WTP) curve for the good or service. The marginal WTP curves exist for public goods and nonmarket goods such as the services of the environment, but they cannot be estimated from direct observations of transactions in these goods. Given the absence of markets for public goods and environmental goods, how can information on demand and benefits be obtained?

As described in Chapter 2, there are basically two approaches to obtaining demand and value information for changes in the quantities of nonmarket goods. They are the revealed preference methods, which involve the estimation of value from observations of behavior in the markets for related goods, and the stated preference methods, which derive values from responses to hypothetical questions. In this chapter, we shall explore some of the possible relationships between demands for private goods and demands for environmental services in an effort to determine under what circumstances the demands for environmental services can be inferred from information on market transactions for the related private good. We will use q to denote some parameter of environmental or resource quality. The task is to estimate in monetary terms the changes in individuals' welfares associated with changes in q. The basic thesis of this chapter is that the degree to which inferences about the benefits of increases in q can be drawn from market observations, and the appropriate techniques to be used in drawing these inferences, both depend on the way in which q

enters individual utility functions. Broadly speaking, q can affect an individual's utility three ways:

1. q can produce utility indirectly as a factor input in the production of a marketed good that yields utility;
2. q can be an input in the household production of utility-yielding commodities; or
3. q can produce utility directly by being an argument in an individual's utility function.

In the third case, q can interact with one or more market goods in the individual's preference structure in many ways. For example, there may be a substitution or complementary relationship between q and some private good. If the nature of the household production process or the forms of interaction between q and private goods can be specified, it may be possible to infer the value of q to the individual from observations of choices of the related market goods.

The next section of this chapter explores the case where q is a factor of production for a market good. The remainder of the chapter is devoted to examining models of the ways in which q can affect utility more directly. The exploitation of possible relationships between environmental goods and private goods leads to several empirical techniques for estimating environmental and resource values. These techniques have the following characteristics: (1) they are consistent with the basic theory of demand and consumer preferences; (2) they provide a means for estimating the indirect utility function, the expenditure function, or the compensated demand function for the environmental service; and (3) they are practical in the sense of imposing realistic data and computational requirements.

Environmental Quality as a Factor Input

When q is a factor of production, changes in q lead to changes in production costs, which in turn affect the price and quantity of output or the returns to other factor inputs, or both. The benefits of changes in q can be inferred from these changes in observable market data. There are several examples where q can be interpreted as a factor input. The quality of river water diverted for irrigation affects the agricultural productivity of irrigated land. The quality of intake water may influence the costs of treating domestic water supplies and the costs of production in industrial operations that use water for processing purposes. Agricultural productivity is impaired by some forms of air pollution. To the extent that air pollution causes material damages, it can affect the costs of production for a wide variety of goods and services.

Assume that good x is produced with a production function

$$x = x(k, w, \ldots, q) \tag{4-1}$$

where k and w are capital and labor (work), respectively, and where the marginal product of q is positive. With given factor prices, and assuming cost-minimizing behavior, there is a cost function

$$C = C(p_w, p_k, x, q) \tag{4-2}$$

Because q affects the production and supply of a marketed good, the benefits of changes in q can be defined and measured in terms of changes in market variables related to the x industry. A change in q will cause shifts in both cost curves and factor demand curves. The consequences of these shifts depend on conditions in factor and product markets. Changes in q can produce benefits through two channels. The first is through changes in the price of x to consumers. The second is through changes in the incomes and profits received by owners of factor inputs used in x production.

To illustrate the first channel, assume that x is produced in a competitive industry under conditions of constant cost, that is, factor supplies to this industry are infinitely elastic. Assume that the change in q affects the cost curves of a significant proportion of producers in the market. As a result, the supply curve shifts downward, causing a fall in the price and an increase in total quantity. The benefit of the price reduction accrues to consumers and can be measured by the methods described in Chapter 3.

To illustrate the second channel—changes in the incomes received by factors of production—consider only one producer who is a price taker in all markets. If the change in q affects only this producer, output price will not be changed. Because the change in q affects the marginal costs of production, the firm's marginal cost and supply curves are shifted down. In this case the benefit is equal to the increase in quasi-rents to the firm. This benefit will accrue to the owner of a fixed factor—land, for example—or to the residual income claimant as profit. In either case, benefits can be measured by changes in profits and fixed factor incomes. However, if the producers affected by changes in q face less than perfectly elastic factor supply curves, at least some of the benefits will be passed on to factors through changes in factor prices and incomes. The factors' shares of benefits can be approximated by the areas to the left of factor supply curves.

The effects of these two channels are combined in Figure 4-1. When the supply curve of the industry is shifted down to S'', price decreases to p''. The benefit to consumers of x is approximated by the change in consumer's surplus, the area $p'BCp''$. Part of this benefit, $p'BFp''$, is at the expense of a reduction in producer and factor surpluses, so the net gain

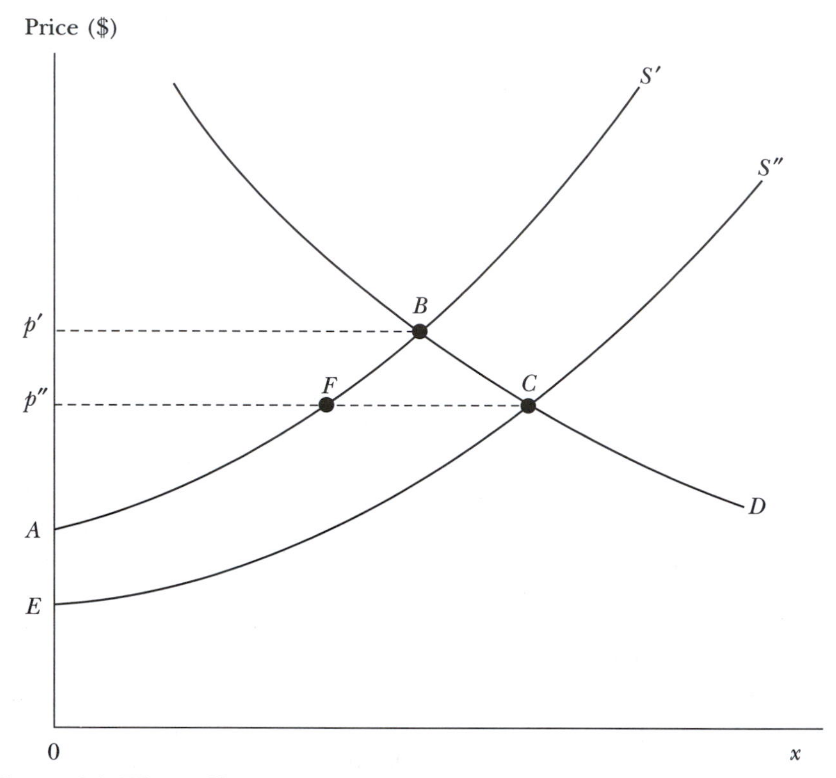

Figure 4-1. The welfare measure when q affects the production of x

from the lower price is *BCF.* The lower supply curve results in factor surpluses and quasi-rents equal to $p''CE$. The net increase to producers and factors is *AFCE*, so total benefits are equal to *ABCE.*

Implementation of these measures requires knowledge of the effects of changes in q on the cost of production, the supply conditions for output, the demand curve for good x, and factor supplies. Two special cases make the estimation of benefits relatively straightforward.

The first is the case where q is a perfect substitute for other inputs in the production of a good. An increase in q leads to a reduction in factor input costs. If the substitution relationship is known, the decrease in per-unit production costs is readily calculated. For example, if water-quality improvement results in a decrease in chlorination requirements for drinking water supplies, the decrease in chlorination costs per unit of output can be readily calculated. Where the change in total cost does not affect marginal cost and output, the cost saving is a true measure of the benefit of the change in q. If the change in q affects marginal cost, the benefits should include the effect of the lower cost on output and price. However, if the percentage reduction in marginal costs is small, or the marginal cost curve is inelastic, or both, the corresponding increase in output would be relatively small. Thus, the decrease in total cost could still be used to provide a rough approximation of true benefits. This approach, sometimes referred to as the *damage function approach*, has been the basis of a number of estimates of the materials, household-cleaning, and agricultural crop-loss benefits of air pollution control, and of the benefits to municipalities, industries, and households of reduced contamination of intake water supplies.

The second case that makes the estimation of benefits relatively straightforward is where knowledge of cost, demand, and market structure suggests that the benefits of a change in q will accrue to producers. Then benefits may be estimated from observed or predicted changes in the net income of certain factor inputs. If the production unit in question is small relative to the market for the final product and for variable factors, it can be assumed that product and variable factor prices will remain fixed after the change in q. The increased productivity then accrues to the fixed factors of production in profit or quasi-rent.

More generally, however, estimates of the value of q require knowledge of the cost and demand functions. In some studies it has been possible to use econometric methods to estimate a cost function that includes an environmental quality variable (e.g., Mjelde and coauthors 1984). Other studies have used various simulation approaches to model the behavior of producers and their responses to changes in an environmental variable. Models and techniques for valuing the effects of q on production are discussed in more detail in Chapter 9.

An Individual's Demand for Environmental Quality

To analyze those cases where q affects individuals directly I first review the basic model of individual preference and demand, with environmental quality included as an argument in the utility function. I then consider the implications of different forms of utility functions for estimation of the demand for q.

Consider a single individual who has a utility function

$$u = u(\boldsymbol{X}, q) \tag{4-3}$$

where $\boldsymbol{X}$ is a vector of private goods quantities ($\boldsymbol{X} = x_1, \ldots, x_i, \ldots, x_n$). By entering environmental quality as an argument in the utility function, we are assuming that the individual perceives the effects of changes in environmental quality. For example, if high ozone levels cause respiratory irritation, the individual must be aware of the irritation, so that he or she feels better when it is reduced. He or she need not know the cause of the irritation or the actual levels of air pollution. If the individual is not aware of the effects of changes in q, the indirect methods of benefit estimation cannot be applied. For example, individuals may not perceive the effects of long-term exposure to air pollutants on their probability of chronic illness or death. If that is the case, changes in q will not affect their behavior and observations of market behavior will yield no information about the value of reducing risks to health.

Assume that the individual maximizes utility subject to a budget constraint

$$\sum_i p_i \cdot x_i = M \tag{4-4}$$

where M is money income. The individual takes q as given and does not have to pay a price for this "imposed" quantity. The solution to this problem yields a set of ordinary demand functions

$$x_i = x_i(\boldsymbol{P}, M, q) \tag{4-5}$$

where $\boldsymbol{P}$ is the vector of private goods prices ($\boldsymbol{P} = p_1, \ldots, p_i, \ldots, p_n$). Note that in general q could be an argument in all private goods demand functions.

The dual to the utility-maximization problem can be stated as follows: minimize expenditure ($\Sigma^i p_i \cdot x_i$) subject to the constraint that utility must equal or exceed some stated level, say u^0. The solution to this problem gives the expenditure function

$$e(\boldsymbol{P}, q, u^0) = M \tag{4-6}$$

The expenditure function has a number of useful properties for applied welfare analysis. First, as was shown in Chapter 3, the derivative of the expenditure function with respect to any price gives the Hicks-compensated demand function for that good, that is,

$$\frac{\partial e}{\partial p_i} = h_i\left(\boldsymbol{P}, q, u^0\right) \tag{4-7}$$

Similarly, the derivative of equation 4-6 with respect to q (with the appropriate change of sign) gives the Hicks-compensated inverse demand function or marginal willingness to pay for changes in q. Let w_q be the marginal willingness to pay or marginal demand price for q. Then

$$w_q = -\frac{\partial e\left(\boldsymbol{P}, q, u^0\right)}{\partial q} \tag{4-8}$$

An alternative expression for the marginal WTP can be obtained by setting the total differential of the indirect utility function equal to zero and solving for the compensating change in income associated with the change in q. Specifically,

$$u = v(M, \boldsymbol{P}, q) \tag{4-9}$$

$$du = \left(\frac{\partial v}{\partial M} dM \right) + \left(\frac{\partial v}{\partial q} dq \right) = 0 \tag{4-10}$$

and

$$\frac{dM}{dq} = -\frac{(\partial v / \partial q)}{(\partial v / \partial M)} \tag{4-11}$$

where $d\boldsymbol{P}$ is zero by assumption.

If the value of the derivative of the right-hand side of equation 4-8 can be inferred from observed data, then we have a point estimate of the marginal willingness to pay for q. If this derivative can be estimated as a function of q, then we have the marginal WTP function for q. Let W_q represent the benefit to the individual of a nonmarginal increase in the supply of q. W_q is the integral of this function, or

$$\begin{aligned} W_q &= -\int_{q'}^{q''} \frac{\partial e(\boldsymbol{P}, q, u)}{\partial q} dq \\ &= e(\boldsymbol{P}, q', u) - e(\boldsymbol{P}, q'', u) \end{aligned} \tag{4-12}$$

This is either a compensating surplus (CS) or equivalent surplus (ES) measure of welfare change, depending on the level of utility at which equation 4-12 is evaluated. The question to be discussed in the next section is whether there are any circumstances in which information about equations 4-8 or 4-12 can be derived from observations of market prices and quantities for private goods.

The Structure of Preferences and Measures of Value

The main purpose of this section is to describe the available techniques for revealing these welfare measures or approximations of them, using observable data on related behavior and individual choice. My strategy will be to explore credible a priori assumptions that support restrictions on the form of the utility function and/or demand functions for market goods or household-produced goods that, in turn, aid in revealing the individual's preferences for environmental quality. Different types of restrictions have different implications for the measurability of the demand for environmental quality. Developing a careful taxonomy of the methods for teasing out the welfare effects of interest helps reveal the broader and more general basis for the welfare economics of environmental valuation.

Several ways may be used to give additional structure to the general model of preferences and choice discussed above. Each of these alternatives involves some assumption about the structure of preferences and/or the constraints on individual choice. Each of the assumptions implies some connection between observable demands for market goods and the values of environmental services and public goods. Each assumption provides a basis for inferring the marginal willingness to pay for q from observations of the relationships between q and the demands for market goods. I will also examine what is required to obtain values of nonmarginal changes in q because these are required for most real-world policy questions. For each of the alternatives, I describe the specific restrictions and discuss how inferences about value can be drawn from observations of individual choices.

The relationships between q and other goods that have been found to be of use involve, broadly speaking, either substitution or complementarity relationships between q and other goods. Exactly how these relationships work out methodologically, however, depends on several other considerations. I first consider cases in which the environmental good is a substitute for a marketed good that enters the utility function. A fundamentally equivalent construct is one in which the environmental good is an input into a household production function and has marketed-good substitutes in the production process. The latter is perhaps the more general and also the more useful way of conceptualizing the problem.

In the second category are those models in which the environmental good is in some way complementary to another good. The complementarity is often most usefully conceived such that the environmental good is a quality characteristic of the related good. There are two often-used derivatives of this construct. In one, the related good is itself a nonmarket good, produced by the household using a household production process. The second is one in which the related good is marketed, but units of the good are heterogeneous and quality-differentiated. Because the good is marketed, the prices of units with higher levels of quality embodied in them are bid up. For the above constructs, discrete choice versions have been developed.

Several of the alternatives involve making some assumption about the separability of the utility function. Thus it will be useful to first review the concept of separability and the implications of various forms of separability for observable demands for market goods. For a review of concepts of separability, see Goldman and Uzawa 1964; Katzner 1970; or Deaton and Muellbauer 1980.

Separability refers to the possible effect of partitioning the goods entering into the utility function into subsets, and the relationships among these subsets. Suppose that there are three types of market goods, so that we can write

$$u = u(\boldsymbol{X}, \boldsymbol{Y}, \boldsymbol{Z}) \tag{4-13}$$

where $\boldsymbol{X} = x_1, \ldots, x_i, \ldots, x_n$; $\boldsymbol{Y} = y_1, \ldots, y_j, \ldots, y_m$; and $\boldsymbol{Z} = z_1, \ldots, z_k, \ldots, z_p$. A utility function is *weakly separable* if the marginal rate of substitution (MRS) between any pair of goods within the same subset is independent of the quantities of goods in any other subset. The following utility function is weakly separable:

$$u = v[u^1(\boldsymbol{X}), u^2(\boldsymbol{Y}), u^3(\boldsymbol{Z})] \tag{4-14}$$

This means that the demand functions for goods in one subset can be written as depending only on the prices of goods in that subset and the expenditure share of that subset. Thus, if separability can be assumed, it simplifies the task of econometric estimation of demand functions.

A utility function is *strongly separable* if the MRS between two goods in different subsets is independent of the quantity of any good in any other subset. Specifically, with strong separability the $\text{MRS}_{x_i y_j}$ is independent of z_k. Any additive utility function is strongly separable, and any strongly separable utility function is additive in some monotone transform (Deaton and Muellbauer 1980). The following utility function is strongly separable:

$$u = v[u^1(\boldsymbol{X}) + u^2(\boldsymbol{Y}) + u^3(\boldsymbol{Z})] \tag{4-15}$$

If the utility function is not separable, there are no restrictions on the terms that are arguments in marginal rates of substitution or in the demand functions for individual goods. The MRS between any pair of goods depends on the quantities of all goods. The demand function for any good depends upon the prices of all goods as well as income.

In the rest of this section, I examine three major forms of relationships between q and market goods. These are (1) some form of substitution relationship between q and one or more market goods, (2) some form of complementary relationship, and (3) the case of differentiated goods where the amount of q embodied in or attached to a market good is one of its differentiating characteristics. Because some of these relationships can be implicitly or explicitly modeled as involving some form of household production, I begin with a brief overview of the household production framework for modeling individual preferences and choice.

The Household Production Framework

The household production function model provides a framework for examining interactions between demands for market goods and the availability of a public good such as environmental quality. This framework is based on the assumption that there is a set of technical relationships among goods used by households in the implicit production of utility-yielding final services. In the household production function literature, the utility-yielding final services are often termed *commodities*, while market goods are simply *goods*. The terminology used in this chapter is more descriptive of the relationship between goods being bought in the market and the final service flows that yield utility to individuals.

Examining the household production technology is one approach to gaining knowledge of the relationship between demands for market goods and the value of environmental quality change. In the household production framework, utility is a function of the level of final service flows:

$$u = u(\mathbf{Z}) = u(z_1, \ldots, z_j, \ldots, z_m) \tag{4-16}$$

$\mathbf{Z}$ is produced according to a technology common to all households and assumed to be known.

$$z_j = z_j(\mathbf{X}, \mathbf{Q}), \text{ for } j = 1, \ldots, m \tag{4-17}$$

where $\mathbf{X}$ represents a vector of market goods available at prices $\mathbf{P}_x$, and $\mathbf{Q}$ represents a vector of environmental quality attributes. The cost of household production depends on the technology, input prices, and exogenously determined $\mathbf{Q}$, that is, $C(\mathbf{Z}) = C(\mathbf{Z}, \mathbf{P}_x, \mathbf{Q})$.

Formally, the individual choice problem is to maximize equation 4-16 subject to the constraints provided by equation 4-17, the given level of $\mathbf{Q}$, and the budget constraint $\Sigma^i \mathbf{P}_x \cdot \mathbf{X} \leq M$. This problem can be solved by a two-step procedure in which the first step is to combine market goods and $\mathbf{Q}$ to minimize the costs of producing the z_j. This step determines the marginal costs of final services C_z. These are essentially implicit prices, but unlike market prices, they are not parametric to the individual unless marginal costs are constant. The second step is to maximize equation 4-16 subject to the budget constraint $C(\mathbf{Z}) \leq M$. The observable manifestation of the solution to this problem is a set of derived market goods demands:

$$\mathbf{X} = \mathbf{X}(\mathbf{P}_x, \mathbf{Q}, M) \tag{4-18}$$

These demand functions will reflect both the role of $\boldsymbol{Q}$ in the household production technology (equation 4-17) and the preferences over the final service flows. As we will see, some of the existing models for interpreting demands for market goods (for example, the averting behavior and defensive expenditures model and the weak complementarity model) can be interpreted as simpler versions of a household production function model. In fact, as Smith argued (1991), the household production framework may be a useful way of thinking about whether a specific market good is likely to be a substitute or a complement for environmental quality. He uses the household production framework as an organizing principle to examine a number of models for nonmarket valuation and to interpret several empirical applications of these models. As we consider various forms of complementarity and substitutability relationships, I will show how they can also be interpreted as applications of the household production function model.

Substitutes

Mäler (1974, 116–118) showed that the marginal willingness to pay for q can be expressed in terms of the price of any private good and the marginal utilities of that good and q. The expression is

$$w_q = -\frac{\partial e\left(\boldsymbol{P}, q, u^0\right)}{\partial q} = -p_i \frac{(\partial u / \partial q)}{(\partial u / \partial x_i)} = p_i \cdot \text{MRS}_{qx_i} \tag{4-19}$$

This would be a useful practical result if it were possible to derive simple expressions for the marginal rates of substitution. Of course, if q could be purchased at a given price the marginal rate of substitution between q and x_1 could be inferred from the price ratio. However, when the level of q is determined exogenously, the marginal rate of substitution can only be determined through knowledge of the utility function or household production function.

Consider the case where some assumption about the separability of the utility function is used to isolate q and a good z from the other determinants of utility (see Mäler 1974, 178–183). Suppose the utility function is weakly separable and is of the following form:

$$u = v\left\{u^1(\boldsymbol{X}),\ u^2(\boldsymbol{Y}),\ \left[c \cdot z^{-\alpha} + (1-c)q^{-\alpha}\right]^{-1/\alpha}\right\} \tag{4-20}$$

Given the separability assumption, the marginal rate of substitution between z and q is independent of the quantities of $\boldsymbol{X}$ and $\boldsymbol{Y}$. In this case, the marginal rate of substitution is

$$\text{MRS}_{zq} = \frac{c}{1-c}\left(\frac{q}{z}\right)^{\alpha-1} = \frac{c}{1-c}\left(\frac{q}{z}\right)^{1/\sigma} \tag{4-21}$$

where σ is the elasticity of substitution, which is constant. From equation 4-19 we have

$$w_q = -p_z \cdot \left(\frac{c}{1-c}\right)\left(\frac{q}{z}\right)^{1/\sigma} \tag{4-22}$$

Because in general the marginal rate of substitution depends on the ratio q/z, we need to know both the elasticity of substitution, σ, and c to compute w_q. This requires knowledge of the utility function.

Perfect Substitutes. There is one special case where the expression for w_q reduces to a usable term. If z and q are perfect substitutes in consumption, the elasticity of substitution between them is infinite, and the expression for the demand price for q reduces to $p_z \cdot s$, where s is the substitution ratio between z and q [$s = c/(1 - c)$]. If perfect substitutability can be assumed, s (or c) should be computable from known or observable technical consumption data or from the household production function.

The perfect substitutability assumption lies behind the simplest application of the defensive expenditure technique for estimating the benefits of pollution control. Defensive expenditures are made either to prevent or to counteract the adverse effect of pollution. They are also referred to as averting expenditures (e.g., Courant and Porter 1981). In effect, a defensive expenditure is spending on a good that is a substitute for higher q. An increase in q is assumed to lead to a decrease in spending on the substitute.

As equation 4-19 shows, the marginal change in the spending on z is the correct measure of the marginal willingness to pay for the change in q. For nonmarginal changes in q, the benefit is $p_z \cdot \Delta q$. However, this will not necessarily be the same as the observed change in spending on q. The intuition behind this statement is straightforward. The benefit of a nonmarginal change in q is the reduction in the spending on z that is required to keep the individual on the original indifference curve. However, in general the individual will not actually reduce spending on z by this amount. There is an income effect on z as well as a substitution effect. The increase in q means that the same level of utility can be maintained with a smaller expenditure on z. As a consequence, the individual will reallocate expenditure among all goods, including z, so as to maximize the increase in total utility. This will result in increases in the expenditures on all goods, with positive income elasticities of demand. Hence, the observed decrease in spending on z will be less than that necessary to hold utility constant. The reduction in defensive spending will be an underestimate of the benefits of higher q. In fact, as we will see below, actual spending on z could increase.

A General Model of Substitution. Because perfect substitutes represent a special case, it will be useful to explore a more general model in which q and a market good are less than perfect substitutes and in which the substitution relationship arises because q and the market good contribute to utility through the same mechanism. Following the analysis of Courant and Porter (1981), this case is modeled in the household production framework. For other analyses dealing with this type of model, see Shibata and Winrich 1983; Harford 1984; and Harrington and Portney 1987. Suppose that clean air and soap are substitutes in the production of cleanliness. Let

$$u = u(z, \mathbf{X}) \tag{4-23}$$

where z is cleanliness and $\mathbf{X}$ is a vector of market goods with prices normalized to 1. Suppose that z is produced by households by combining a market good y and air quality according to the production function

$$z = z(y, q) \tag{4-24}$$

with positive partial derivatives for both arguments. Also, assume that y contributes to utility only through its contribution to cleanliness. This is equivalent to assuming that the utility function is weakly separable, with y and q in one group.

For any given q, the individual chooses $\mathbf{X}$ and y so as to maximize:

$$u = u[\mathbf{X}, z(y, q)] \tag{4-25}$$

subject to the budget constraint

$$M - \mathbf{X} - p_y \cdot y = 0 \tag{4-26}$$

the household production technology, and the exogenous q. The first-order conditions are

$$\begin{aligned} &\frac{\partial u}{\partial \mathbf{X}} - \lambda = 0 \\ &\frac{\partial u}{\partial z} \cdot \frac{\partial z}{\partial y} - \lambda \cdot p_y = 0 \end{aligned} \tag{4-27}$$

Substituting the expressions for the demands for $\mathbf{X}$ and y into the utility function yields the indirect utility function:

$$u = v(\boldsymbol{P}, p_y, M, q) \tag{4-28}$$

By setting the total differential of this expression equal to zero, assuming prices do not change, and rearranging terms, we obtain the following expression for the marginal value of increasing q:

$$w_q = -\frac{dM}{dq} = \frac{(\partial v / \partial q)}{\lambda} = \frac{\left(\frac{\partial u}{\partial z} \cdot \frac{\partial z}{\partial q}\right)}{\lambda} \tag{4-29}$$

where λ is the marginal utility of income and the last term in parentheses is an expansion of $\partial v/\partial q$, making use of the chain rule. Substituting the first-order condition 4-27 gives the marginal willingness to pay for q as a function of the price of the private good and the marginal productivities of y and q:

$$w_q = p_y \cdot \frac{(\partial z / \partial q)}{(\partial z / \partial y)} \tag{4-30}$$

This is the reduction in spending on y holding z constant, that is, moving along a given isoquant in the household production function. This marginal value can be calculated if the household production function is known.

If y and q jointly produce two goods (or help to avoid two bads), this result does not hold. Suppose $u = u[\boldsymbol{X}, z_1(y, q), z_2(y, q)]$. Using the procedure described above, we obtain

$$w_q = p_y \left\{ \frac{[(\partial u / \partial z_1) \cdot (\partial z_1 / \partial q)] + [(\partial u / \partial z_2) \cdot (\partial z_2 / \partial q)]}{[(\partial u / \partial z_1) \cdot (\partial z_1 / \partial y)] + [(\partial u / \partial z_2) \cdot (\partial z_2 / \partial y)]} \right\} \tag{4-31}$$

The unobservable utility terms do not cancel out.

The relationship between the actual change in spending on y and marginal WTP can be obtained by totally differentiating the household production function with respect to q and using the first-order conditions for the optimum choice of y to obtain:

$$\frac{dz}{dq} = \left(\frac{\partial z}{\partial y} \cdot \frac{\partial y^*}{\partial q}\right) + \frac{\partial z}{\partial q} \tag{4-32}$$

where $y^*(M, p_q, q)$ is the derived demand function. Rearranging and multiplying both sides by the second line of equation 4-27 gives

$$p_y \cdot \frac{(\partial z / \partial q)}{(\partial z / \partial y)} = w_q = \left[\frac{(\partial u / \partial z)}{\lambda} \cdot \frac{dz}{dq} \right] - \left(p_y \cdot \frac{\partial y^*}{\partial q} \right) \tag{4-33}$$

If spending on y decreases ($\partial y^*/\partial q < 0$), then the marginal willingness to pay for q is the sum of the observed reduction in spending and the willingness to pay for the induced increase in the final service flow z. The actual savings in spending on y is an underestimate of the marginal value in this case. If the lower implicit cost of household production of z leads to a sufficiently large increase in the consumption of z, then it is possible that spending on y could increase. Thus observed changes in spending on substitute goods are not reliable indicators of marginal willingness to pay for changes in q.

Similar results are obtained for the case where z requires several private goods as well as q for its production. The marginal value of a change in q can be measured by the reduction in spending on any of the market goods for movements along the original z isoquant. Observed changes in spending on market goods are not reliable measures of w_q.

An alternative formulation is to let z represent a bad, with y and q mitigating the adverse effects of z. In other words, u_z, z_q, and $z_y < 0$, where subscripts indicate partial derivatives. Similar conclusions can be reached about WTP and its relationship to changes in mitigating expenditures. This alternative model is typically referred to as an *averting behavior model* or a *mitigating behavior model*.

Bartik (1988) showed how lower and upper bounds on the welfare gains of nonmarginal changes in q can be calculated from knowledge of only the household production function. The lower bound on the compensating surplus measure of benefits of an increase in q is the reduction in the expenditures on y necessary to reach the initial level of z, other things being equal. More formally, the lower bound is given by

$$\text{lower bound} = p_y(y' - y'') \tag{4-34}$$

This can be shown to be a lower bound on CS by defining a restricted expenditure function, $e^*(q, z', u^0)$, where z' is the initial level of z and price terms have been omitted for simplicity. The welfare gain associated with an increase in q from q' to q'' holding z at z' is

$$CS^* = e^*\left(q', z', u^0\right) - e^*\left(q'', z', u^0\right) \tag{4-35}$$

The first term on the right-hand side is $M = \boldsymbol{X}' + p_y \cdot y'$. The increase in q means that less must be spent on y to produce z'. And because $u(\boldsymbol{X}', z') = u^0$, the second term is $\boldsymbol{X}' + p_y \cdot y''$. So

$$CS^* = \mathbf{X}' + (p_y \cdot y') - \mathbf{X}' - (p_y \cdot y'') = p_y (y' - y'') \tag{4-36}$$

This has to be less than the true *CS* because relaxing the constraint holding z at z' would allow the individual to increase utility, unless the marginal utility of z were already zero. In a similar fashion, Bartik showed that an upper bound on the equivalent surplus measure of welfare gain is the decrease in spending on y that is possible while holding z constant at its new equilibrium level. And because in general $ES > CS$, the constrained changes in expenditure bound both of the true measures of welfare change.

We can also consider the case where q contributes to utility directly in addition to its contribution to z. Taking the total differential of the indirect utility function, solving for the compensating change in income, and substituting in the first-order condition for the choice of y gives us

$$w_q = -\frac{dM}{dq} = \frac{(\partial u / \partial q)}{\lambda} + \left[p_y \cdot \frac{(\partial z / \partial q)}{(\partial z / \partial y)} \right] \tag{4-37}$$

Comparing this expression with equation 4-30, we see that the marginal willingness to pay for q now includes an unobservable marginal utility term for the direct effect of q on utility. Similarly, following the same procedures used to derive equation 4-33, we have

$$w_q = \left[\frac{(\partial u / \partial z)}{\lambda} \cdot \frac{dz}{dq} \right] + \frac{(\partial u / \partial q)}{\lambda} - \left(p_y \cdot \frac{\partial y^*}{\partial q} \right) \tag{4-38}$$

Again, if spending on y decreases, it is an underestimate of marginal WTP because it neglects both the utility value of the induced increase in z and the direct utility value of the increase in q.

Perfect Complements

Suppose that the single environmental service q is a perfect complement to a market good, say x_1. By perfect complementarity I mean that x_1 and q must be consumed in fixed proportions, for example, $x_1 / q = a$. This means that to effectively use the services of one unit of q, the individual must purchase $1/a$ units of x_1. In the context of the household production model, this could be interpreted as implying a production function for x_1 of the following form:

$$z_1 = \min(x_1 / a, q) \tag{4-39}$$

If this is the case, there are conditions under which the benefits of changes in q can be estimated from knowledge of the demand function for the complementary market good. As long as p_1 is less than some critical value, say $p_1{}^*$, the quantity demanded of x_1 will be independent of its price and determined solely by the availability of q. If q is increased by one unit, the individual will purchase $1/a$ additional units of x_1 and experience a higher level of utility. Thus, the marginal willingness to pay for q will be positive. However, if p_1 is greater than $p_1{}^*$, the individual will purchase fewer units of x_1 than are required to fully use the available q. Thus, at p_1 greater than $p_1{}^*$, the marginal utility and marginal willingness to pay for q are zero.

Mäler (1974, 180–183) showed that if the demand functions for x_1 and for other market goods are known, it is possible to compute the expenditure function and the demand price for q when p_1 is less than $p_1{}^*$. The exact expression for the demand price for q depends upon the specification of the true demand curves. There is no simple generalization of the technique. Also, as Mäler appears to realize, it is difficult to imagine examples of perfect complementarity between an environmental service and a market good. However, as we will now see, there is a less restrictive form of complementarity that also makes possible the calculation of the marginal willingness to pay for q, and that appears to have real-world applications.

Weak Complementarity

Suppose that q enhances the enjoyment the individual derives from consuming x_1 and that an increase in q increases the quantity demanded of x_1, other things being equal. One reasonable interpretation of these assumptions is that q is an exogenously determined characteristic of x_1. Examples of environmental services that might fit these assumptions include water quality as a characteristic associated with visits to a lake, and the number of fish caught as a characteristic of fishing trips to a stream. The model can also be applied to estimate the value of a network where the purchase of a market good, for example, a cell phone, is required to connect to the network (Hahn et al. 2000).

In this section I show that where the "enjoyment" of q requires the purchase of a market good or where q can be treated as a characteristic of a market good, it is possible to identify a measure of the value of a change in q that is based on the demand for the market good. I describe this measure and the conditions under which it is a valid welfare measure. Then I discuss some issues concerning implementation of the measure.

Let us assume that x_1 and q are Hicksian complements, that is, that the compensated demand function

$$h_1 = h_1(\boldsymbol{P}, q, u^0) \tag{4-40}$$

is characterized by $\partial h_1/\partial q > 0$. This implies that the Marshallian demand for x_1,

$$x_1 = x_1(\boldsymbol{P}, q, M) \tag{4-41}$$

also has $\partial x_1/\partial q > 0$.

Suppose that the complete system of demand equations for this individual has been estimated econometrically and that the system satisfies the Slutsky conditions for integrability. To derive a valid welfare measure from this system of demand functions, one must be able to integrate this system to solve for the underlying utility function and expenditure function. However, Mäler (1974, 183–189) showed that in general it is not possible to solve completely for the utility and expenditure functions with the information given. Mathematically, the result of the integration contains unknown terms that are themselves functions of q and the constants of integration. It is necessary to impose additional conditions on the problem to solve for the unknown terms and determine the constants of integration. The additional conditions involve what Mäler called weak complementarity.

The Conditions. Weak complementarity requires that the marginal utility or marginal demand price of q be zero when the quantity demanded of the complementary private good x_1 is zero. Mathematically, weak complementarity involves two conditions. The first is that x_1 be nonessential, in the sense that there be a price p_1 for x_1 such that the compensated demand

$$h_1\left(p_1^*, q, u^0\right) = 0 \tag{4-42}$$

where the term for all other prices has been omitted for simplicity. This price is often called the *choke price*. The choke price will usually be an increasing function of q. Graphically, the compensated demand curve has a vertical intercept, and there is some level of expenditure on other goods that will sustain u^0 even when x_1 is zero.

The second condition is that for values of p_1 at or above p_1*, the derivative of the expenditure function

$$e = e\left(p_1^*, q, u^0\right) \tag{4-43}$$

be zero, that is,

$$\frac{\partial e}{\partial q} = 0 \tag{4-44}$$

This condition means that at or above the choke price, the marginal utility or marginal willingness to pay for q is zero. Changes in q have no welfare significance unless the price of x_1 is low enough so that its compensated demand would be positive.

These two conditions together establish an initial position for the individual that can be used to determine the constants of integration. For proof and a demonstration with a numerical example, see Mäler (1974, 183–189). Results similar to Mäler's were obtained by Bradford and Hildebrandt (1977). These conditions describe a position for the individual that satisfies the first-order conditions of a utility-maximization problem in which the individual can choose the quantities of all goods, including q, at given prices, including a zero price for q. This is analogous to defining a position in which the virtual prices for imposed quantities (discussed in Chapter 3) are equal to the actual prices.

The Welfare Measure. Given the conditions for weak complementarity, the compensating surplus for a change in q can be measured by the area between the two compensated demand curves for x_1. The intuition for this measure can be illustrated graphically. Assume that the compensated demand curve for x_1 has been found for environmental quality level q'. In Figure 4-2 this demand curve is labeled $h_1(q')$. Assume that the price of x_1 is given at p_1' and does not change throughout the analysis. The compensating surplus associated with the use of x_1 is the area *ABC* under the demand curve. Now assume that quality is improved to q''. The increase in the quality associated with the use of x_1 is assumed to increase the demand for x_1, thus shifting the demand curve outward to $h_1(q'')$. The calculation of the benefit associated with this change is straightforward and can be divided into three steps.

1. Given the initial demand curve $h_1(q')$, postulate a hypothetical increase in price from p_1' to p_1^*, the choke price. To leave the individual no worse off, that person must be compensated by the area *ABC*.
2. Now postulate the improvement in quality and the shift in the demand curve to $h_1(q'')$. Given the weak complementarity assumption, utility is unaffected because the consumption of x_1 is zero. Therefore there is no need for compensation, either positive or negative.
3. Now postulate a return to the old price of p_1'. The individual is made better off by the area *ADE*. To restore the individual to the original welfare position, he or she must be taxed by this amount. The net effect of these changes is a gain to the individual (in the absence of the hypothetical compensating payments) of the area *BCED* (= *ADE* – *ABC*). This is the benefit of the change in q.

The role of the two conditions defining weak complementarity can now be made clear. If there were no choke price, there would be no finite com-

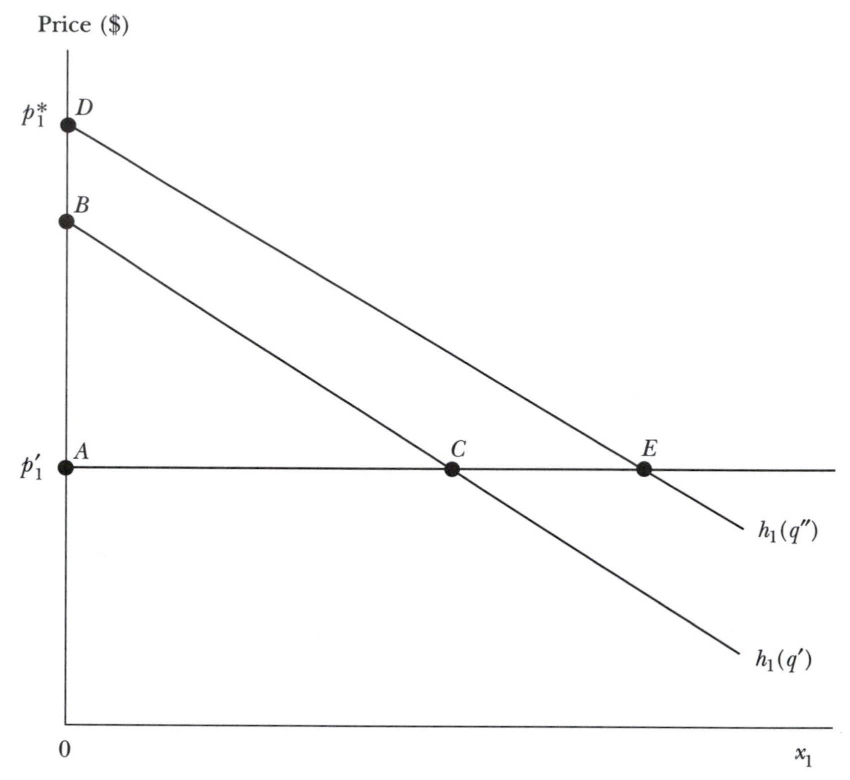

Figure 4-2. The welfare measure for an increase in q when q and x_1 are weak complements

pensating variations (*CV*s) in steps 1 and 3. If the derivative of the expenditure function were not zero at the choke price, there would be a welfare change (presumably positive) in step 2 associated with the increase in q, even if the quantity demanded of x_1 were zero. In this case, the area *BCED* would be an underestimate of the benefits of increasing q. If the consumption of x_1 is interpreted as use of an environmental resource (for example, trips to a recreation site), then whatever welfare change is associated with step 2 must be a nonuse value. Nonuse values are the topic of Chapter 5.

This result can also be established more rigorously. Recall from Chapter 3 that the compensating welfare measure for a change in q is

$$CS_q = W_q = e(p_1, q', u^0) - e(p_1, q'', u^0) \tag{4-45}$$

The area between the two compensated demand curves is $CV_{q''} - CV_{q'}$ where these terms are the *CV*s associated with the ability to purchase x_1 at different levels of q. Specifically,

$$CV_{q'} = e\left(p_1^*, q', u^0\right) - e\left(p_1, q', u^0\right) \tag{4-46}$$

and

$$CV_{q''} = e\left(p_1^*, q'', u^0\right) - e\left(p_1, q'', u^0\right) \tag{4-47}$$

Using these expressions to calculate the area between the compensated demand curves gives

$$W_q = e\left(p_1^*, q'', u^0\right) - e\left(p_1, q'', u^0\right) - e\left(p_1^*, q', u^0\right) + e\left(p_1, q', u^0\right) \tag{4-48}$$

If the second condition for weak complementarity is satisfied, the first and third terms in equation 4-48 sum to zero, and we have

$$W_q = e\left(p_1, q', u^0\right) - e\left(p_1, q'', u^0\right) \tag{4-49}$$

Weak Complementarity with More than One Good. So far, we have assumed that q is complementary to only one market good. It is possible for q to be complementary with several goods at the same time. For example, if q is water quality in a lake, it could be complementary to several market activities such as fishing, boating, and swimming. As Bockstael and Kling (1988) show, the analysis of weak complementarity can be carried over to the multigood case in a straightforward manner.

Suppose there are two market goods that are complementary to q, with Hicksian demand functions as follows:

$$h_1 = h_1(p_1, p_2, q, u^0) \tag{4-50}$$

and

$$h_2 = h_2(p_2, p_1, q, u^0) \tag{4-51}$$

Using the expenditure function, the compensating surplus (CS) for a change in q is

$$CS = e\left(p_1, p_2, q', u^0\right) - e\left(p_2, p_1, q'', u^0\right) \tag{4-52}$$

If the conditions for weak complementarity are satisfied, CS can be estimated from areas between the compensated demand curves for the two market goods in the following manner. First, calculate the area between the compensated demands for x_1, holding p_2 at the observed price level. Then calculate the area between the compensated demands for x_2 evaluated at the choke price (p_1*) for x_1. Mathematically, this involves calculating the following expression:

$$
\begin{aligned}
CS = &\int_{p_1'}^{p_1^*} h_1\left(p_1, p_2', q'', u^0\right) dp_1 - \int_{p_1'}^{p_1^*} h_1\left(p_1, p_2', q', u^0\right) dp_1 \\
&+ \int_{p_2'}^{p_2^*} h_2\left(p_1^*, p_2, q'', u^0\right) dp_2 - \int_{p_2'}^{p_2^*} h_2\left(p_1^*, p_2, q', u^0\right) dp_2
\end{aligned} \tag{4-53}
$$

The first and third integrals give the value to the individual of consuming x_1 and x_2 when q is equal to q'' evaluated over a consistent path of integration. In other words, starting from (p_1', p_2''), p_1 is increased to p_1* and p_2 is increased to p_2*. Similarly, the second and fourth integrals give the value to the individual of consuming x_1 and x_2 when q is equal to q'. The result of this calculation is independent of the order in which the prices are changed.

When each of the integrals in equation 4-53 is expressed as a difference in expenditures, we have:

$$
\begin{aligned}
CS = &\, e\left(p_1^*, p_2', q'', u^0\right) - e\left(p_1', p_2', q'', u^0\right) - e\left(p_1^*, p_2', q', u^0\right) \\
&+ e\left(p_1', p_2', q', u^0\right) + e\left(p_1^*, p_2^*, q'', u^0\right) - e\left(p_1^*, p_2', q'', u^0\right) \\
&- e\left(p_1^*, p_2^*, q', u^0\right) + e\left(p_1^*, p_2', q', u^0\right)
\end{aligned} \tag{4-54}
$$

When like terms are canceled out, and if weak complementarity is satisfied, this reduces to equation 4-52.

Bockstael and Kling (1988) also considered the econometric problems that often arise when trying to include the prices of other goods in the estimation of demand functions—for example, unobservable prices of substitutes or highly correlated substitute prices. They also considered the effects of these problems on calculated welfare changes. They reached the surprising conclusion that if the price of a market good is perfectly correlated with the price of a substitute, and if the substitute price is omitted from the estimating equation, then the misspecified demand function can still yield a correct measure of welfare change.

Welfare Measures with Ordinary Demand Functions. The primary empirical requirement for using weak complementarity is that we be able to obtain an econometric estimate of the demand function for the private good as a function of prices, income, and q. Because the compensated demand curves are not directly observable, we must ask whether Willig's (1976) bounds on consumer's surplus as an approximation to *CS* and *ES* apply in this situation. Unfortunately, because these bounds were derived for differences in areas

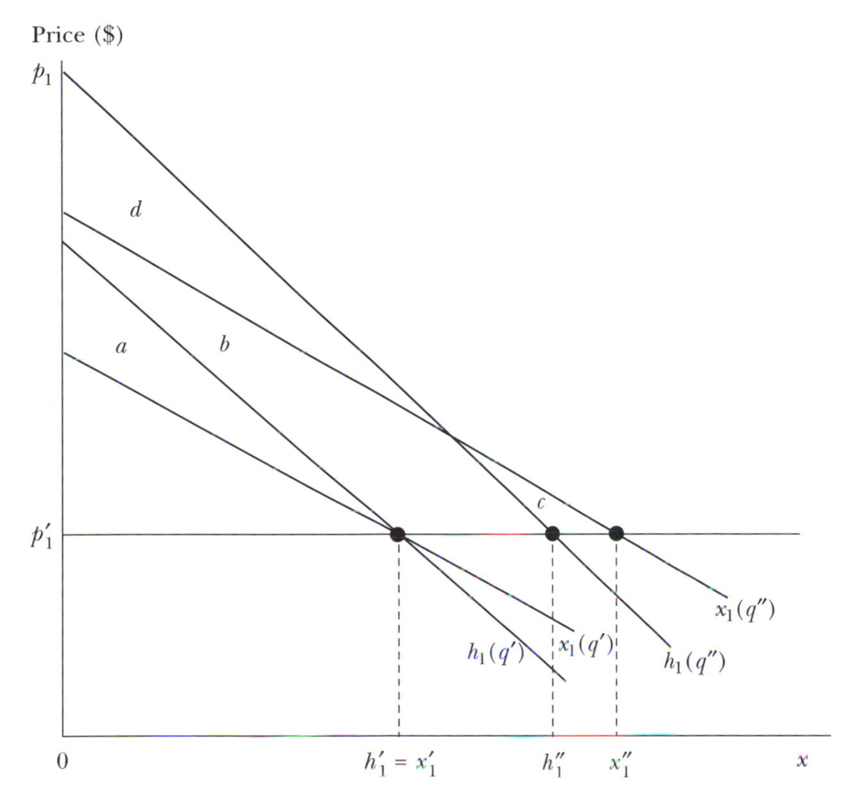

Figure 4-3. Approximating benefits using Marshallian demand curves with weak complementarity

under ordinary and compensated demand curves, they do not apply to areas between shifting ordinary and compensated demand curves.

This can best be shown graphically, as in Figure 4-3 (Bockstael et al. 1991). The compensated demand curves for the two levels of q are denoted as $h_1(q')$ and $h_1(q'')$. *CS* is measured by the area $b + d$. The ordinary demand curves are denoted by $x_1(q')$ and $x_1(q'')$. Observe that at the market price of $p_1{}'$ the increase in q causes the ordinary demand curve to shift out farther to the right than the compensated demand curve. This is because with the ordinary demand curve, there is no compensating reduction in income to hold utility constant. Taking areas between the ordinary demand curves would yield a consumer's surplus measure of $a + b + c$. The percentage error arising when the consumer's surplus measure is used to approximate the compensating surplus is

$$\%\ \text{error} = \frac{a + c - d}{b + d} \tag{4-55}$$

As can be seen by inspection, this error could be positive, negative, or, by coincidence, zero. Thus the practice of using ordinary demand curves to estimate welfare changes with weak complementarity can lead to errors of unknown sign and magnitude.

Fortunately, recent analyses have shown that it is possible to use the exact welfare measurement methods described in Chapter 3 to recover weakly complementary preferences from market demands. For demonstrations and discussions, see Larson 1991 and Bockstael and McConnell 1993.

Weak Complementarity and Household Production. As Bockstael and McConnell (1983) showed, if the individual choice problem is modeled in the household production function framework, there are conditions on the production technology and preferences that yield results equivalent to the weak complementarity model. They showed that if these conditions are satisfied, the welfare value of a change in q can be calculated from knowledge of the market good's compensated demand function even without knowledge of the household production technology. The demand function for the market good is a derived demand. If the conditions established by Bockstael and McConnell are satisfied, the welfare value of the change in q can be calculated by the area between the two compensated demand curves for the market good at the two levels of q.

The individual's choice problem is to maximize

$$u(\mathbf{Z}, q) \tag{4-56}$$

subject to

$$\mathbf{Z} = \mathbf{Z}(\mathbf{X}, q) \tag{4-57}$$

and

$$\boldsymbol{P} \cdot \mathbf{X} = M \tag{4-58}$$

The expenditure function is

$$e(\boldsymbol{P}, q, u) = \min\left\{\boldsymbol{P} \cdot \mathbf{X} \middle| u^0 = u\left[\mathbf{Z}(\mathbf{X}, q), q\right]\right\} \tag{4-59}$$

Suppose that z_1 is produced by the household by combining q and x_1 and that q *is* used only in the household production of z_1. As in the standard model, the compensated demand for x_1 is the derivative of the expenditure function with respect to p_1. However, in this case it is a derived demand, so that it reflects features of both preferences and the household production technology.

An increase in q will lower the marginal cost of the household production of z_1 and therefore increase the quantity demanded of x_1. If the change in q is to increase the compensated demand for x_1, then the household production technology and preferences together must result in an increase in the demand for z_1 sufficiently large that more x_1 is required. In other words, restrictions on the household production technology alone are not sufficient to establish the complementarity or substitutability of the derived demand for x_1 with respect to q. A comparative static analysis of the household maximization problem shows that with

$$\frac{\partial z_1}{\partial x_1} \cdot \frac{\partial x_1}{\partial q} > 0 \tag{4-60}$$

it is possible for an increase in q to reduce both the ordinary and the compensated derived demands for x_1. See, for example, Courant and Porter 1981, 325–326.

The area between the two derived compensated demand curves for x_1 is given by

$$e\left(p_1^*, q'', u^0\right) - e\left(p_1', q'', u^0\right) - e\left(p_1^*, q', u^0\right) + e\left(p_1', q', u^0\right) \tag{4-61}$$

If the first and third terms of this expression sum to zero, then this gives the welfare value of the increase in q. The sufficient conditions for these terms to sum to zero are the following:

A. that x_1 be essential in the production of z_1 so that if $x_1 = 0$, $\partial z_1/\partial q = 0$; and
B. that q not affect utility independent of its contribution to the production of z_1, in other words, that $\partial u/\partial q = 0$ when $x_1 = 0$.

These two conditions are sufficient to make the expenditure function independent of the level of q. Thus they play a role similar to the second condition of the weak complementarity model. The household production framework requires restrictions on both the household production technology and preferences for final services. However, there is no restriction on the Hicksian demand analogous to the existence of a choke price for the market good.

Bounds on Willingness To Pay

In this section I approach the relationship between changes in spending on market goods and changes in q in a more general way. The result is a set of expressions that place bounds on and provide an exact measure of marginal WTP and that may prove useful in some cases. The limitation of this line of analysis is that the bounds may prove to be too wide to be useful for policy purposes; the lower-bound estimate of the welfare improvement can even be negative. Also, the exact measure requires a good deal of information on preferences for unrelated goods, which may be difficult to obtain in practice.

First, let us suppose that each of the goods an individual consumes can be classified as being either a complement to (indicated by subscripts C), independent of (indicated by subscripts I), or a substitute (indicated by subscripts S) for q, depending on the effect of changes in q on the compensated demand functions. The definitions and notation are as follows:

The set of complements is $$\mathbf{X}_C : \frac{\partial h_{Ci}}{\partial b} > 0 \qquad i = 1, \ldots, m$$

The set of independent goods is $$\mathbf{X}_I : \frac{\partial h_{Ij}}{\partial b} = 0 \qquad j = 1, \ldots, n$$

The set of substitutes is $$\mathbf{X}_S : \frac{\partial h_{Sk}}{\partial b} < 0 \qquad k = 1, \ldots, p \tag{4-62}$$

where b is the virtual price of q (see Chapter 3). Neill (1988) proves that given these definitions, the following relationships between the compensated quantities and q hold:

$$\begin{aligned} &\frac{\partial h_{Ci}}{\partial q} > 0 \qquad i = 1, \ldots, m \\ &\frac{\partial h_{Ij}}{\partial q} = 0 \qquad j = 1, \ldots, n \\ &\frac{\partial h_{Sk}}{\partial q} < 0 \qquad k = 1, \ldots, p \end{aligned} \tag{4-63}$$

The compensated demands and the ordinary demands for each of these goods are related according to the following lemma:

LEMMA: $\partial h_{Ji}/\partial q = (\partial x_{Ji}/\partial q) - [w_q \cdot (\partial x_{Ji}/\partial M)]$ where $J = C, I, S$, i subscripts are goods; x are the ordinary demands; and w_q is the marginal value of a change in q.

Proof: In equilibrium, $h_{Ji} = x_{Ji}$ for all J and i. Because M is equal to the expenditure required to attain the initial level of utility, we can write

$$h_{Ji} = x_{Ji}[\boldsymbol{P}, q, e(\boldsymbol{P}, q, u^0)] \tag{4-64}$$

Differentiating yields

$$\partial h_{Ji}/\partial q = (\partial x_{Ji}/\partial q) + [(\partial x_{Ji}/\partial M)(\partial e/\partial q)] \tag{4-65}$$

for all J and i. Because $-(\partial e/\partial q) = w_q$, the lemma is proved. This means that we can write w_q as a function of the compensated and uncompensated changes in demand for any good:

$$w_q = \frac{(\partial x_{Ji}/\partial q)}{(\partial x_{Ji}/\partial M)} - \frac{(\partial h_{Ji}/\partial q)}{(\partial x_{Ji}/\partial M)} \tag{4-66}$$

or as a similar function of changes in the expenditure on the good:

$$w_q = \frac{(\partial E_{Ji}/\partial q)}{(\partial E_{Ji}/\partial M)} - \frac{(\partial H_{Ji}/\partial q)}{(\partial E_{Ji}/\partial M)} \tag{4-67}$$

where

$$E_{Ji} = \sum^{i} p_i \cdot x_{Ji} \tag{4-68}$$

and

$$H_{Ji} = \sum^{i} p_i \cdot h_{Ji} \tag{4-69}$$

Given knowledge of how the expenditure on a set of goods changes with changes in income and q, the last term in equation 4-67 represents the error involved in using that data as an estimate of w_q. For independent goods, this error term is zero. Thus an exact measure of marginal WTP can be obtained from observable changes in the demands for and expenditures on independent goods. The trick in applying this measure is to know enough about the structure of preferences to identify those goods that are independent by our definition and to know how the ordinary demand functions shift with changes in q.

If equation 4-67 is applied to data on complements, the result is an overestimate of w_q because the error term in equation 4-67 is negative, assuming that the complements are normal goods. The first term represents an upper bound on the true w_q. Theory does not shed any light on how big this error is likely to be. In the case of substitute goods, the first term on the right-hand side of equation 4-67 will often be negative for individual substitutes and will always be negative for the set of substitutes taken as a whole. Although this expression is a lower bound, this is not a very useful result concerning a number that is presumed to be positive.

A Hopeless Case

As these analyses have shown, there are conditions under which the benefits of a change in q can be estimated from information on the demand functions for market goods. The specification of these market good demand functions must reflect the interaction between q and the market goods in the utility function or the household production function. However, there is one form of utility function in which there are no interactions between q and market goods and with which it is impossible to estimate the value of q from market data.

Suppose that the utility function is strongly separable, with q as the single argument in one of the subsets. In other words,

$$u = v[u^1(\boldsymbol{X}) + u^2(\boldsymbol{Y}) + u^3(q)] \tag{4-70}$$

where $\boldsymbol{X}$ and $\boldsymbol{Y}$ are subsets of marketable goods. Strong separability means that the marginal rates of substitution between any x_i and y_j are independent of q. Because changes in q have no effect on marginal rates of substitution of any of the marketable goods, q can be excluded as an argument in all of the market demand functions. Although changes in q affect utility, they leave no record of this effect in the data on market transactions. In principle, it is not possible to estimate the demand for q from observable market data on transactions in x_i or y_j when the utility function is strongly separable in q.

Strong separability is a property of two commonly used functional forms for utility functions: the Cobb–Douglas function:

$$u = a \cdot \left(\prod_i x_i^{\alpha_i} \right) \cdot q^{\beta} \tag{4-71}$$

and the constant elasticity of substitution (CES) function:

$$u = a \left[\sum_i \left(b_i \cdot x_i^{-\alpha} \right) + \left(c \cdot q^{-\alpha} \right) \right]^{-1/\alpha} \tag{4-72}$$

This can be seen by writing them in their log transformations. Separability may be a characteristic of an important class of benefits. For example, those amenities of the urban environment that are not directly associated with private goods consumption may be separable. Also, as discussed in Chapter 5, if use of the environment is defined to mean purchase of a marketed good such as travel to a resource site, then nonuse values flow from a separable component of the utility function and cannot be recovered from data on market demands.

Hedonic Prices and the Value of *q*

The techniques described so far have been developed for the case where the level of the resource service or environmental quality is fixed and is the same for all individuals. Although this represents the textbook version of the public-good problem, it is not descriptive of all possible cases involving public goods or environmental quality. In some circumstances, the level of q can be considered to be a qualitative characteristic of a differentiated market good. In these cases individuals have some freedom to choose their effective consumption of the public good or environmental quality through their selection of a private-good consumption bundle. Another way to look at it is that there is a kind of complementarity between the public good and the market good in that as the quantity of the public good embodied in the market good increases, the demand for the market good increases.

In effect, the market for the differentiated private good functions also as a market for the public good or environmental quality. For example, people can choose the level of consumption of local public goods through their choice of a jurisdiction in which to reside; thus the housing market functions also as a market for the purchase of local public goods.

Where these choices are possible, information on public-good demand is embedded in the prices and consumption levels for private goods. For example, if air quality varies across space in an urban area, individuals may choose their exposure to air pollution through their decisions about residential location. Residential housing prices may include premiums for locations in clean areas and discounts for locations in dirty areas. If they do, it may be possible to estimate the demand for public goods such as clean air from the price differentials revealed in private markets. These price differentials are implicit prices for different levels of the public good. A job can also be considered as a differentiated good. Different jobs have different bundles of characteristics and different wage rates. Wage differentials can be interpreted as the implicit prices of job characteristics. These wage and price differentials and the implicit prices they reflect are the subject matter of hedonic price theory.

The hedonic price technique is a method for estimating the implicit prices of the characteristics that differentiate closely related products in a product class. Colwell and Dilmore (1999) cite examples of the technique being applied to prices of farmland as early as 1922. The earliest modern example appears to be Griliches' (1961) application to the prices of automobiles. Rosen (1974) developed the formal theory of hedonic prices in the context of competitive markets. For an early development of the use of hedonic prices for estimating the demand for environmental quality characteristics, see Freeman 1974.

In principle, if the product class contains enough products with different combinations of characteristics, it should be possible to estimate an implicit price relationship that gives the price of any model as a function of the quantities of its various characteristics. This relationship is called the hedonic price function. The partial derivative of the hedonic price function with respect to any characteristic gives its marginal implicit price, that is, the additional expenditure required to purchase a unit of the product with a marginally larger quantity of that characteristic.

More formally, let Y represent a product class. Any model of Y can be completely described by a vector of its characteristics. Let $\boldsymbol{Q} = q_1, \ldots, q_j, \ldots, q_n$ represent the vector of characteristics of Y. Then any model of Y, say y_i, can be described by its characteristics, that is $y_i = y_i(q_{i1}, \ldots, q_{ij}, \ldots, q_{in})$, where q_{ij} is the quantity of the jth characteristic provided by model i of good Y. The hedonic price function for Y gives the price of any model as a function of its characteristics. Specifically, for y_i,

$$p_y = p_y(q_{i1}, \ldots, q_{ij}, \ldots, q_{in}) \tag{4-73}$$

If $p_y(\cdot)$ can be estimated from observations of the prices and characteristics of different models, the price of any model can be calculated from knowledge of its characteristics.

Before turning to the interpretation of the hedonic price function, it will be useful to describe briefly how the hedonic price function is generated in a competitive market for a differentiated product. Following the analysis of Rosen (1974), assume that each individual purchases only one unit of Y in the relevant time period. An individual's utility depends upon that person's consumption of the numeraire, X, and the vector of characteristics provided by the unit of Y purchased:

$$u = u(X, \boldsymbol{Q}) \tag{4-74}$$

or

$$u = u(M - p_{y_i}, q_{i1}, \ldots, q_{ij}, \ldots, q_{in}) \tag{4-75}$$

if the individual purchases model i at p_{y_i}. To maximize equation 4-74 subject to the budget constraint $M - p_{y_j} - X = 0$, the individual must choose levels of each characteristic to satisfy

$$\frac{(\partial u / \partial q_j)}{(\partial u / \partial X)} = \frac{\partial p_y}{\partial q_j} \tag{4-76}$$

Alternatively, the marginal willingness to pay for q_j must just equal the marginal cost of purchasing more of q_j, other things being equal.

Inverting equation 4-75 and holding all but characteristic j constant, we obtain an indifference curve or bid curve that gives the maximum amount the individual would pay to obtain a model as a function of q_j, holding other things constant. It takes the form

$$B_j = B_j\left(M - p_{y_i}, q_j, \boldsymbol{Q}^*, u^*\right) \tag{4-77}$$

where u^* is the solution to the constrained utility-maximization problem and $\boldsymbol{Q}^*$ represents the optimally chosen quantities of the other characteristics.

Because of differences in preferences, incomes, or both, individuals can have different bid functions. Two such functions for individuals a and b are shown in Figure 4-4. They both show diminishing willingness to pay for q_j or a diminishing marginal rate of substitution between q_j and $\boldsymbol{X}$. Given the hedonic price function, these two individuals choose q_j^a and q_j^b, respectively.

Turning to the supply side of the market, firms' costs of production depend upon the levels of the characteristics of the models they produce. Assume that firms are heterogeneous so that their cost functions are different. Inverting a firm's profit function yields an offer curve for the characteristic of the form

$$C_j = C_j(q_j, \boldsymbol{Q}^*, \pi^*) \tag{4-78}$$

where π^* is the maximum attainable profit. Offer curves and the optimal quantities of q_j supplied for firms α and β are shown in Figure 4-5 for a given hedonic price function.

For all firms and individuals to be in equilibrium, all bid and offer curves for characteristics for each participant in the market must be tangent to the hedonic price function. Thus the hedonic price function is a double envelope of the two families of bid curves and offer curves (Rosen 1974). As a double envelope, the hedonic price function depends on the determinants of both the supply side and the demand side of the characteristics markets. It is a locus of equilibria between bids for and offers of characteristics.

The marginal implicit price of a characteristic can be found by differentiating the hedonic price function with respect to that characteristic. That is,

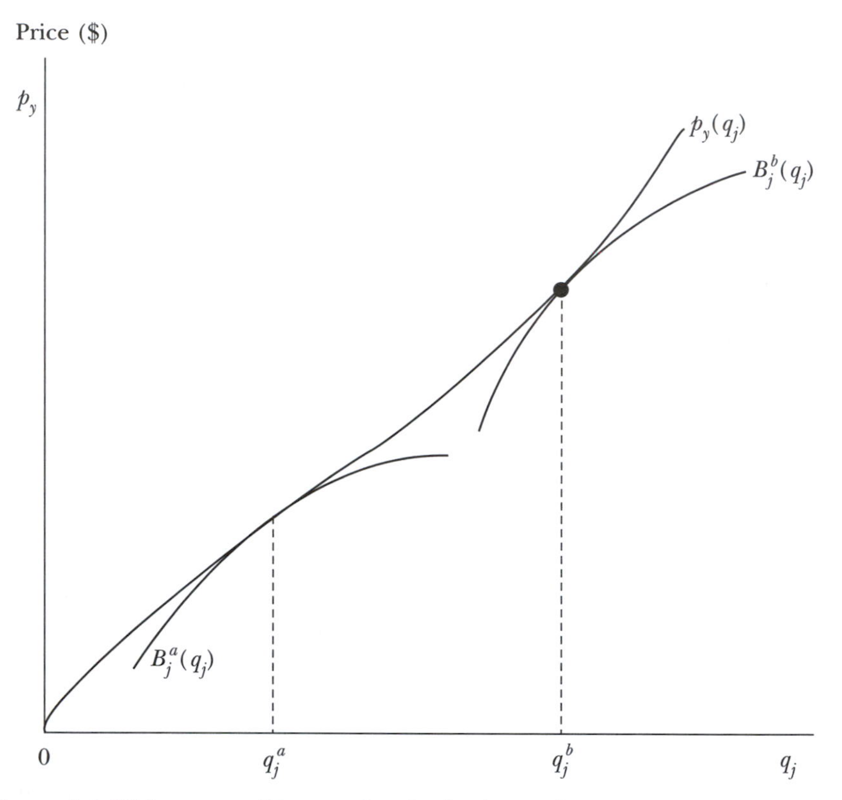

Figure 4-4. Bid curves of buyers in a hedonic market

$$\frac{\partial p_y}{\partial q_j} = p_{y_i}\left(q_{i1}, \ldots, q_{ij}, \ldots, q_{in}\right) \tag{4-79}$$

This gives the increase in expenditure on ***Y*** that is required to obtain a model with one more unit of q_j, other things being equal. If equation 4-73 is linear in the characteristics, then the implicit prices are constants for individuals. But if equation 4-73 is nonlinear, then the implicit price of an additional unit of a characteristic depends on the quantity of the characteristic being purchased.

Equation 4-73 need not be linear. Linearity will occur only if consumers can "arbitrage" attributes by untying and repackaging bundles of attributes (Rosen 1974, 37–38). For example, if individuals are indifferent between owning two two-door cars and one four-door car, other things being equal, they can create equivalents of four-door cars by repackaging smaller units. If both sizes exist on the market, the larger size must sell at twice the price of the smaller one, and the hedonic price of a car will be a linear function

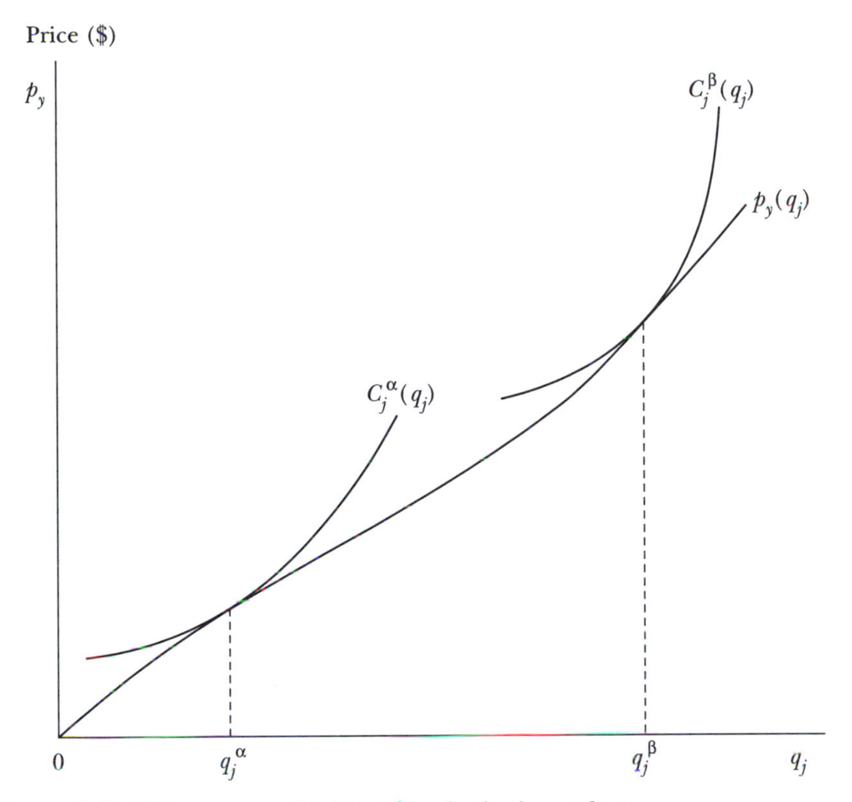

Figure 4-5. Offer curves of sellers in a hedonic market

of the number of doors. The example suggests that nonlinearity will be a common feature of hedonic price functions.

Another way of looking at the market equilibrium for a characteristic helps to make clearer the welfare implications of the hedonic price model. First, individual a's choice problem can be solved to obtain that person's uncompensated inverse demand function for q_j:

$$b_j^{*a} = b_j^{*a}\left(q_j, \boldsymbol{P}, M - p_y\right) \tag{4-80}$$

where a indexes the individual. This is shown in Figure 4-6, along with the marginal implicit price function for q_j, $p_{q_j}(\cdot)$. For each individual, the quantity of q_j purchased is known by observation, and its implicit price is known from equation 4-79. This point (q_j', p_{q_j}') can be interpreted as a utility-maximizing equilibrium for this individual resulting from the intersection of the individual's inverse demand curve and the locus of opportunities to purchase q_j, as defined by the marginal implicit price function. Thus from

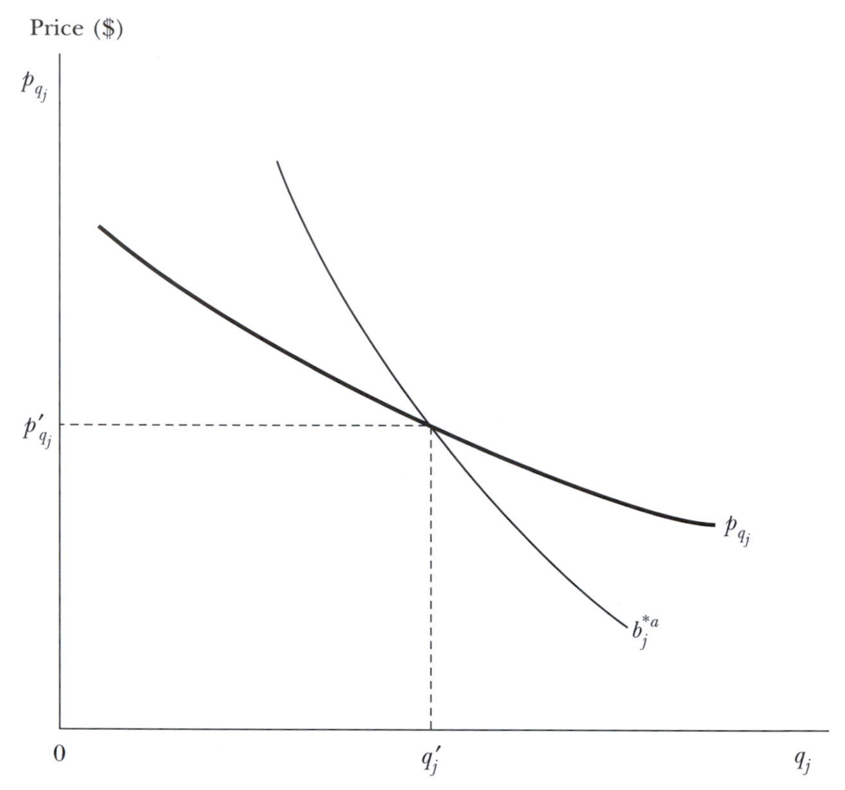

Figure 4-6. The marginal implicit price and inverse demand curves for *q*

the first-order conditions of equation 4-76, p_{q_j}' can be taken as a measure of the individual's equilibrium marginal willingness to pay for q_j.

Individuals can be viewed as moving out along p_{q_j} as long as their WTP, as reflected in their inverse demand curves, exceeds the marginal implicit price. Thus p_{q_j} is a locus of individuals' equilibrium marginal WTP.

It is also possible to specify a compensated marginal WTP function for q_j,

$$b_j^a = b_j^a\left(q_j, \boldsymbol{P}, u^*\right) \tag{4-81}$$

This function gives individual a's marginal willingness to pay for q_j, holding utility constant. The value of a nonmarginal change in q_j can be easily calculated if the compensated marginal WTP function (equation 4-81) is known. It is given by the integral of equation 4-81 over the range of the change in q_j.

This leads us naturally to the question of whether individuals' marginal WTP functions, their inverse demand functions, or both can be identified from observations of marginal implicit prices and quantities. The answer depends on the circumstances of the case. If the implicit price function is linear in q_j, then it is not possible to identify a demand curve for q_j. The price observation is the same for all individuals. However, p_{q_j}' can be interpreted as the marginal WTP or marginal benefit for small changes in q_j for each individual.

If the hedonic price function is nonlinear, different individuals selecting different bundles of characteristics will have different marginal implicit prices for q_j. There is one situation where the inverse demand function can be immediately identified, and that is if all individuals have identical incomes and utility functions. Then the marginal implicit price function is itself the inverse demand function. Recall that the marginal implicit price curve is a locus of equilibrium points on individuals' inverse demand curves. With identical incomes and preferences, all individuals have the same inverse demand curve. Because all the equilibrium points fall on the same inverse demand curve, they fully identify it.

In the case where differences in incomes, preferences, or other variables result in individuals having different inverse demand functions, Rosen (1974) argued that implicit price and quantity data from a single market could be used to estimate this inverse demand function, provided that the standard identification problem of econometrics could be solved. It is now clear that this analysis is incorrect. The problem is that the data from a single hedonic market are insufficient to identify how the same individuals would respond to different implicit prices and incomes.

There are at least two ways in which estimates of inverse demand functions for q_j can be obtained from hedonic analysis. The first is to increase the quantity of information obtained from marginal implicit prices by estimating hedonic price functions for several separate markets, and then pooling the cross-sectional data on the assumption that the underlying structure of demand is the same in all markets (Freeman 1974; Brown and Rosen 1982; Palmquist 1984). The second approach is to impose additional structure on the problem by invoking a priori assumptions about the form of the underlying utility function. The nature of the identification problem in hedonic price models and alternative approaches to dealing with it are major topics of Chapter 11.

Discrete Choice Models and Measures of Value

The models described in this chapter so far have all exploited the marginal equalities revealed when individuals optimize over choice variables that are continuously variable. For example, in the averting behavior model it was assumed that the individual could choose any level of the activity that was consistent with the budget constraint. A marginal rate of substitution could be inferred from knowledge of the slope of the budget constraint. This is not always a realistic way to model the individual choice problem. Some problems are better viewed as involving the choice of one option from a range of discrete alternatives. For example, the choice might be whether or not to take a once-in-a-lifetime cruise around the world, or whether to travel to work by private auto, bus, or on foot. The solutions of discrete choice problems of this sort are essentially corner solutions. As a consequence, there are no tangencies from which a marginal rate of substitution can be inferred. Discrete choice models have been developed both to predict individuals' behavior in these choice contexts and to draw inferences about welfare change on the basis of observed choices.

In this section, I describe a simple discrete choice model and show how measures of welfare change and value can be derived from it. In subsequent chapters I present more detailed discussions of applications of the model to several specific problems, including voting yes or no on a referendum question or accepting or rejecting a hypothetical offer for an environmental commodity (see Chapter 6), the choice of which of several alternative houses to live in (see Chapter 11), and the choices of whether or not to undertake a specific recreation activity or to visit a specific recreation site (see Chapter 13). For expositions of the specification, estimation, and interpretation of discrete choice models generally, see Ben-Akiva and Lerman 1985 and Train 1986. Hanemann (1999) gives a more advanced exposition in the context of valuing environmental changes; see also Johansson et al. 1989 and Hanemann 1989. Cameron has developed an alternative specification of the discrete choice model for deriving welfare measures from binary choice data (see Cameron and James 1987; Cameron 1988; and McConnell 1990 for more details). This alternative specification is described in Chapter 5 in the context of referendum contingent valuation questions.

Consider an individual's decision regarding which one of several alternative goods to purchase. The set of alternatives is $\boldsymbol{Y}$, and $\boldsymbol{P}_y$ denotes the vector of prices attached to the elements of $\boldsymbol{Y}$. Let y_i $(i = 1, \ldots, n)$ indicate the alternative chosen and p_i be its price, where p_i is an element in $\boldsymbol{P}_y$. Let the utility of the individual be a function of y_i and the consumption of a numeraire good $\boldsymbol{X}$ purchased at a normalized price of 1. Assume that only one unit of y_i is purchased, so that $\boldsymbol{X} = M - p_i$. There is a set of conditional utility functions

$$u_i = u(M - p_i, y_i, \boldsymbol{C}) \text{ if } y_i \text{ is purchased} \tag{4-82}$$

where $\boldsymbol{C}$ is a vector of individual characteristics that affect preferences. If the individual purchases y_i, that person reveals that $u_i > u_j$ for all $j \neq i$.

Now assume that we as researchers cannot fully observe utility. Rather, we can measure an observable component of utility, $v(M - p_i, y_i, C)$. Thus the utilities of all the possible outcomes are given by

$$u_i = v(M - p_i, y_i, C) + \varepsilon_i \tag{4-83}$$

where ε_i is the random unobservable component of utility. This stochastic specification of the utility function means that the probability that the individual chooses to purchase y_i can be expressed as the probability that the utility associated with i is greater than the utilities associated with all the other alternatives:

$$Pr(y_i) = Pr[v(M - p_i, y_i, C) + \varepsilon_i > v(M - p_j, y_j, C) + \varepsilon_j] \text{ for all } j \neq i \tag{4-84}$$

The logit model of McFadden (1973) shows that if the error terms are independently and identically distributed with a Type I extreme value distribution, this probability can be written as a function of the utilities, as follows:

$$Pr(y_i) = \frac{e^{v(M-p_i,y_i,C)}}{e^{v(M-p_i,y_i,C)} + \sum_{j \neq i}^{n} e^{v(M-p_j,y_j,C)}} = \left(1 + \sum_{j \neq i}^{n} e^{-\Delta v_{ij}}\right)^{-1} \tag{4-85}$$

where $\Delta v_{ij} = v_i - v_j$.

The logit model of choice implies certain restrictions on individuals' choices and preferences. The most notable restriction is that choices must have the property of the independence of irrelevant alternatives (IIA). Consider two alternatives from the choice set, i and k. From equation 4-85 we can see that the ratio of the probabilities of choosing y_i and y_k is independent of the characteristics of any other alternative, that is,

$$\frac{Pr(y_i)}{Pr(y_k)} = \frac{e^{v(M-p_i,y_i,C)}}{e^{v(M-p_k,y_k,C)}} \tag{4-86}$$

One implication of the property of the IIA is that the addition or deletion of options from the choice set does not affect the ratios of the probabilities associated with any other pair of alternatives. Yet it is easy to think of examples of choice problems in which the property of IIA is implausible. For example, suppose that the individual is choosing from among three sites for a day of fishing. The sites are lake A, lake B, and river C. Suppose that the two lakes are quite similar but that river fishing is a much different experience. Finally, suppose that the estimated probabilities are $Pr(A) = 0.25$; $Pr(B) = 0.25$; and $Pr(C) = 0.5$. It seems plausible that if lake A were eliminated from the choice set, the probability of visiting lake B would rise relative to the probability of visiting the river, but this would violate IIA. The logit model predicts that $Pr(B) = 0.33$ and $Pr(C) = 0.67$.

One approach to dealing with the property of IIA is to redefine the choice set so that two or more close substitutes are modeled as one alternative. In the example of recreational choice, the problem disappears if individuals are modeled as choosing between the alternatives of river fishing and lake fishing.

Now, suppose that we wish to estimate the WTP associated with the choice of y_i over some other option, say y_j, for a group of similar individuals. If all individuals face the same price set $\boldsymbol{P}_y$, and we observe that different individuals make different choices, it must be because of differences in income, differences in the vector of characteristics $\boldsymbol{C}$, or both. If we make some assumptions about the functional form of the observable component of utility in equation 4-83, the parameters of $v(\cdot)$ can be estimated from equation 4-85 using maximum likelihood methods. Once the relevant parameters of the utility function are known, it is possible to calculate welfare changes directly (Small and Rosen 1981; Hanemann 1999). For example, for the individual who chooses y_i when it is available but chooses y_j otherwise, the welfare value of y_i compared to y_j is given implicitly by

$$u(M - p_i, y_i, \boldsymbol{C}) = u(M - p_j + CS_i, y_j, \boldsymbol{C}) \tag{4-87}$$

or

$$v(M - p_i, y_i, \boldsymbol{C}) + \varepsilon_i - \varepsilon_j = v(M - p_j + CS_i, y_j, \boldsymbol{C}) \tag{4-88}$$

In other words, if y_i is removed from the choice set, CS_i compensates for this restriction in choice. The inclusion of the error terms makes CS_i a random variable with a known probability distribution. It can be calculated if the parameters of the utility function are known. Alternatively, the willingness to pay to have y_i added to the available choices is

$$u(M - p_j - CS_i^*, y_j, \boldsymbol{C}) - u(M - p_i, y_i, \boldsymbol{C}) \tag{4-89}$$

Similar equivalent variation measures can also be defined.

Hanemann (1984) describes three alternative approaches to obtaining a welfare measure from estimates of the parameters of $v(\cdot)$. The first is simply the researcher's expected value of CS_i derived from equation 4-89. The second is based on the expectation of $v(\cdot)$ and gives the sum of money that equates the researcher's expectation of the individual's utility with and without the good. The third is that payment which if imposed as a price would make the individual indifferent as to the two alternatives (that is, equally likely to accept or reject the good); it is equivalent to using the median of the distribution of CS_i. Hanemann argues for using the median-based measure because it will not be as sensitive to outliers and extreme responses in the data.

Summary

We have reviewed the conceptual basis for empirical techniques that attempt to take advantage of market information. Where q is an input into the production of marketed goods and services, our techniques focus on the effects of changes in q on output and factor markets, not on the utility of q per se. We have shown that there is a valid theoretical basis for examining changes in factor incomes such as land rents, cost savings in production, and changes in consumer's surplus associated with the private-good outputs.

When q is a consumption good and enters directly into the utility function, the problems are more severe. One of the more promising techniques is based on the concept of weak complementarity. Benefits are measured in terms of shifts in the demand curve for the private complementary good. The approach may be applicable to estimating recreation benefits due to water-quality changes.

Where the public good is either implicitly or explicitly an input, along with one or more market goods, in the household production of a final service flow, benefit estimation is straightforward if the substitution relationship or marginal rate of technical substitution is known and the problem has a sufficiently simple structure so that unobservable utility terms cancel out. Defensive expenditures and measures of additional costs (for example, for household cleaning, or for medical care and drugs in the case of health) are examples of estimates that are based on approximations of this approach.

Another promising case is where q varies across space, as in air pollution, or as a characteristic embodied in some private good. Then individuals can choose different quantities of q by varying residential locations or by choosing different private-good models. The hedonic price approach can be used to measure the implicit price of q, and under some circumstances the demand curve for q can be identified. Finally, where individuals make choices from among discrete alternatives, the fact that the actual choice is revealed to yield higher utility than the alternatives can be exploited to draw inferences about the utility function and value.

In most of the models described in this chapter, it makes intuitive sense to speak of the individual as using the environmental or resource service in question. For example, in the models of substitution and averting behavior, the individual can be interpreted as using clean air to produce a clean household or good health. Thus the values estimated with the models described here are sometimes referred to as use values. It is also possible for individuals to value environmental services independently of any use they might make of those services. The concept of so-called *nonuse values* is the topic of the next chapter. As we will see there, when nonuse values are defined and modeled in a plausible way, the result is that transactions in market goods reveal nothing about these nonuse values. Therefore, the revealed preference methods described in this chapter can shed no light on the possible magnitude of these values; instead, we must rely on stated preference methods. These are the topic of Chapter 6.

References

Bartik, Timothy J. 1988. Evaluating the Benefits of Non-Marginal Reductions in Pollution Using Information on Defensive Expenditures. *Journal of Environmental Economics and Management* 15(1): 111–127.

Ben-Akiva, Moshe, and Steven R. Lerman. 1985. *Discrete Choice Analysis: Theory and Application to Travel Demand.* Cambridge, MA: MIT Press.

Bockstael, Nancy E., and Catherine L. Kling. 1988. Valuing Environmental Quality: Weak Complementarity with Sets of Goods. *American Journal of Agricultural Economics* 70(3): 654–662.

Bockstael, Nancy E., and Kenneth E. McConnell. 1983. Welfare Measurement in the Household Production Framework. *American Economic Review* 73(4): 806–814.

———. 1993. Public Goods as Characteristics of Non-Market Commodities. *Economic Journal* 103(3): 1244–1257.

Bockstael, Nancy E., Kenneth E. McConnell, and Ivar [E.] Strand. 1991. Recreation. In *Measuring the Demand for Environmental Quality,* edited by John B. Braden and Charles D. Kolstad. Amsterdam: North-Holland.

Bradford, David F., and Gregory G. Hildebrandt. 1977. Observable Preferences for Public Goods. *Journal of Public Economics* 8(2): 111–121.

Brown, James N., and Harvey S. Rosen. 1982. On the Estimation of Structural Hedonic Price Models. *Econometrica* 50(3): 765–768.

Cameron, Trudy A. 1988. A New Paradigm for Valuing Non-Market Goods Using Referendum Data: Maximum Likelihood Estimation by Censored Logistic Regression. *Journal of Environmental Economics and Management* 15(3): 355–379.

Cameron, Trudy A., and Michelle D. James. 1987. Efficient Estimation Methods for Contingent Valuation Surveys. *Review of Economics and Statistics* 69(2): 269–276.

Collwell, Peter F., and Gene Dilmore. 1999. Who Was First? An Examination of an Early Hedonic Study. *Land Economics* 75(4): 620–626.

Courant, Paul N., and Richard Porter. 1981. Averting Expenditure and the Cost of Pollution. *Journal of Environmental Economics and Management* 8(4): 321–329.

Deaton, Angus, and John Muellbauer. 1980. *Economics and Consumer Behavior.* New York: Cambridge University Press.

Freeman, A. Myrick, III. 1974. On Estimating Air Pollution Control Benefits from Land Value Studies. *Journal of Environmental Economics and Management* 1(1): 74–83.

Goldman, S.M., and Hirofumi Uzawa. 1964. A Note on Separability in Demand Analysis. *Econometrica* 32(3): 387–398.

Griliches, Zvi. 1961. Hedonic Prices for Automobiles: An Econometric Analysis of Quality Change. In *The Price Statistics of the Federal* Government. General Series No. 73. New York: Columbia University Press for the National Bureau of Economic Research.

Hahn, Robert W., Paul C. Tetlock, and Jason K. Burnett. 2000. Should You Be Allowed To Use Your Cellular Phone While Driving? *Regulation* 23(3): 46–55.

Hanemann, W. Michael. 1984. Welfare Evaluations in Contingent Valuation Experiments with Discrete Responses. *American Journal of Agricultural Economics* 66(3): 332–341.

———. 1989. Welfare Evaluations in Contingent Valuation Experiments with Discrete Response Data: Reply. *American Journal of Agricultural Economics* 71(4): 1057–1061.

———. 1999. Welfare Evaluations with Discrete Choice Models. In *Valuing Recreation and the Environment: Revealed Preference Methods in Theory and Practice*, edited by Joseph A. Herriges and Catherine L. Kling. Cheltenham, U.K.: Edward Elgar.

Harford, Jon D. 1984. Averting Behavior and the Benefits of Reduced Soiling. *Journal of Environmental Economics and Management* 11(3): 296–302.

Harrington, Winston, and Paul R. Portney. 1987. Valuing the Benefits of Health and Safety Regulations. *Journal of Urban Economics* 22(1): 101–112.

Johansson, Per-Olov, Bengt Kristrom, and Karl-Göran Mäler. 1989. Welfare Evaluations in Contingent Valuation Experiments with Discrete Response Data: Comment. *American Journal of Agricultural Economics* 71(4): 1054–1056.

Katzner, Donald W. 1970. *Static Demand Theory.* New York: Macmillan.

Larson, Douglas M. 1991. Recovering Weakly Separable Preferences. *Journal of Environmental Economics and Management* 21(2): 97–108.

Mäler, Karl-Göran. 1974. *Environmental Economics: A Theoretical Inquiry.* Baltimore, MD: The Johns Hopkins University Press for Resources for the Future.

McConnell, Kenneth E. 1990. Models for Referendum Data: The Structure of Discrete Choice Models for Contingent Valuation. *Journal of Environmental Economics and Management* 18(1): 19–34.

McFadden, Daniel. 1973. Conditional Logit Analysis of Discrete Choice Behavior. In *Frontiers of Econometrics*, edited by Paul Zarembka. New York: Academic Press.

Mjelde, James W., Richard M. Adams, Bruce L. Dixon, and Phillip Garcia. 1984. Using Farmers' Actions To Measure Crop Loss Due to Air Pollution. *Journal of the Air Pollution Control Association* 34(4): 360–364.

Neill, Jon R. 1988. Another Theorem on Using Market Demands To Determine Willingness To Pay for Non-Traded Goods. *Journal of Environmental Economics and Management* 15(2): 224–232.

Palmquist, Raymond B. 1984. Estimating the Demand for Characteristics of Housing. *Review of Economics and Statistics* 64(3): 394–404.

Rosen, Sherwin. 1974. Hedonic Prices and Implicit Markets: Product Differentiation in Perfect Competition. *Journal of Political Economy* 82(1): 34–55.

Shibata, H., and J.S. Winrich. 1983. Control of Pollution when the Offended Defend Themselves. *Economica* 50(200): 425–438.

Small, Kenneth A., and Harvey S. Rosen. 1981. Applied Welfare Economics with Discrete Choice Models. *Econometrics* 49(1): 105–130.

Smith, V. Kerry. 1991. Household Production Functions and Environmental Benefit Estimation. In *Measuring the Demand for Environmental Quality*, edited by John B. Braden and Charles D. Kolstad. Amsterdam: North-Holland.

Train, Kenneth. 1986. *Qualitative Choice Analysis: Theory, Econometrics and an Application to Automobile Demand.* Cambridge, MA: MIT Press.

Willig, Robert D. 1976. Consumer's Surplus without Apology. *American Economic Review* 66(4): 589–597.

CHAPTER 5

Nonuse Values

This chapter is about the hypothesis that people place monetary values on natural resources and environmental characteristics that are independent of any present or future use these people might make of those resources. For example, people may gain utility from their knowledge that the Grand Canyon is preserved even though they expect never to visit the Grand Canyon. People may be willing to pay to ensure the survival of whales, eagles, and other endangered species even though they never expect to see them. Lying behind this hypothesis is the assumption that there is a meaningful way to define use so that values that arise from use can be distinguished from those that are independent of use. This assumption is discussed below.

In the economics literature, natural resource values that are independent of people's present use of the resource have been variously termed "existence," "intrinsic," "nonuse," and, more recently, "passive-use" values. This last term was coined by the D.C. Circuit Court of Appeals in its ruling in *Ohio v. U.S. Department of Interior* (880 F.2d 432 [1989]), which legitimized the inclusion of these values in natural resource damage cases brought by the federal government. These values are said to arise from a variety of motives, including a desire to bequeath certain environmental resources to one's heirs or future generations, a sense of stewardship or responsibility for preserving certain features of natural resources, and a desire to preserve options for future use.

The hypothesis of nonuse values has gained wide acceptance among economists working in the field of environmental and resource economics, at least partly as a result of findings of positive nonuse values in a variety of settings. For example, people have expressed willingness to pay to preserve the Grand Canyon even though they do not expect to visit it (Schulze and Brookshire 1983) and to ensure the survival of blue whales (Samples et al. 1986), eagles and striped shiners (Boyle and Bishop 1987), and other endangered species even though they never expect to see them. For references to other studies providing estimates of nonuse values, see Fisher and Raucher 1984; Randall 1991; and Freeman 1993a. Many economists believe that nonuse values can be large in the aggregate, at least in some circumstances. For example, a research team working for the state of Alaska estimated that the median lost nonuse value for U.S. residents due to the Exxon *Valdez* oil spill was $31 per household in 1991 dollars or about $2.8 billion for the United States as a whole (Carson et al. 1992). If nonuse values are large, ignoring them in natural resource policymaking could lead to serious errors and resource misallocations.

On the other hand, some economists argue that nonuse or existence values should play no role in economic assessments such as benefit–cost analyses and natural resource damage assessments. These critics generally emphasize three concerns:

- nonuse values should be seen as reflecting social and ethical norms that lie outside the realm of economic efficiency (e.g., Rosenthal and Nelson 1992);
- to the extent that nonuse values reflect altruism toward either other users or future generations, they are irrelevant to any economic assessment (Milgrom 1993; Diamond and Hausman 1994); and
- nonuse values cannot be measured reliably (e.g., Rosenthal and Nelson 1992; Diamond and Hausman 1993, 1994).

As for the first point, I will show that nonuse value can be defined within a utility theoretic framework and that it has the characteristics of a public good. Thus because nonuse value contributes to the well-being of individuals, it is relevant to economic efficiency (Kopp 1992). I will also show that in the case of nonpaternalistic or pure altruism, the critics' argument is correct. However, other forms of altruism can lead to policy-relevant measures of value (McConnell 1998). And there are other reasons besides altruism that might lead to a person's having nonuse values. Finally, because it is generally agreed that nonuse values can only be measured by using stated preference methods, the question of the reliability of measures of nonuse values will be left to the next chapter.

In the next section of this chapter, I review some of the early literature in which various forms of nonuse values are discussed. The most restrictive definition links use to the purchase of some complementary good, usually visits to the site of a resource. This definition has been convenient in that it provides a way of measuring use value—for example, through weak complementarity or hedonic price analysis. However, there are other ways to define use. We must also consider the question of whether the distinction between use and nonuse values is itself useful.

In the second section of this chapter I develop a consistent theoretical framework for defining nonuse value—a framework based on defining *use* in terms of the purchase of a related market good. I then consider an alternative definition of nonuse value that avoids the ambiguities in the concept of the use of the resource.

In the third section of this chapter, I discuss empirical techniques for the measurement of whatever we choose to call nonuse values. I consider the possibility of deriving measures of nonuse values from observed behavior of individuals. However, this does not appear to be a fruitful approach; thus we are left with the stated preference methods that are the focus of Chapter 6. I briefly discuss some of the issues involved in applying stated preference valuation techniques to the measurement of nonuse values.

Background: Motivation and Definitions

John Krutilla (1967) introduced the concept of existence or nonuse values into the mainstream economics literature. In his classic article "Conservation Reconsidered," he argued that individuals do not have to be active consumers of a resource, whose willingness to pay (WTP) can be captured by a price-discriminating monopoly owner, to derive value from the continuing existence of unique, irreplaceable environmental resources. He wrote that "when the existence of a grand scenic wonder or a unique and fragile ecosystem is involved, its preservation and continued availability are a significant part of the real income of many individuals" (Krutilla 1967, 779). In an accompanying footnote he added that "these [individuals] would be the spiritual descendants of John Muir, the present members of the Sierra Club, the Wilderness Society, the National Wildlife Federation, the Audubon Society and others to whom the loss of a species or the disfigurement of a scenic area causes acute distress and a sense of genuine relative impoverishment" (Krutilla 1967, 779n).

Krutilla went on to suggest at least two reasons why people might hold values unrelated to their current use of a resource. These reasons were related to bequeathing natural resources to one's heirs and preserving options for future use. In a footnote he suggested that the "phenomenon discussed may have an exclusive sentimental basis, but if we consider the bequest motivation in economic behavior ... it may be explained by an interest in preserving an option for one's heirs to view or use the object in question" (Krutilla 1967, 781n). He also wrote, "An option demand may exist not only among persons currently and prospectively active in the market for the object of the demand, but among others who place a value on the mere existence of biological and/or geomorphological variety and its widespread distribution" (Krutilla 1967, 781). However, more recent literature has recognized that the value of preserving options arises out of uncertainty about future demand and/or supply and has different implications for environmental valuation. For a brief discussion of this issue and references to the literature, see Chapter 8 of this book.

Later, Krutilla and Fisher wrote

> In the case of existence value, we conceived of individuals valuing an environment regardless of the fact that they feel certain they will never demand *in situ* the services it provides ... however, if we acknowledge that a bequest motivation operates in individual utility-maximizing behavior ... the existence value may be simply the value of preserving a peculiarly remarkable environment for benefit of heirs. (Krutilla and Fisher 1975, 124)

Whereas Krutilla and Fisher offer a bequest motivation as one of several possible explanations for a pure existence value, McConnell (1983) takes a different point of view:

> The notion that a good is valued only for its existence, that it provides no *in situ* services, is far fetched. In most cases, resources are valued for their use. Existence value occurs only insofar as bequest or altruistic notions prevail. We want resources there because they are valued by others of our own generation or by our heirs. Thus use value is the ultimate goal of preferences that yield existence demand, though the existence and use may be experienced by different individuals. (McConnell 1983, 258)

Thus, in McConnell's view, an altruistic attitude toward other people's use of a resource is the primary source of existence value. Although the source of existence value is related to someone's use, it is independent of any use made of the resource by the person holding the existence value.

Some have argued that people can have what are essentially existence values out of an ethical or altruistic concern for the status of nonhuman species or proper rules of human conduct (Kopp 1992). Whereas philosophers of ethics are not in agreement as to the validity and proper form of such concerns (e.g., see one of the collections on environmental ethics such as Van De Veer and Pierce 1994; DesJardins 1999; or Jamieson 2000), it is possible that some people hold such values and are willing to commit resources on the basis of those values. Such values could be entirely independent of any use of the environment.

There has also been a lot of discussion about the various forms that nonuse values might take. A typical approach in the literature is to define use value as the economic value associated with the in situ use of a resource, as, for example, through visiting a recreation site or observing a natural wildlife population. Total value is viewed as an individual's willingness to pay to preserve or maintain a resource in its present state. If total value exceeds use value, the difference is a nonuse value, or as some have called it, an "existence value," an "intrinsic value," or a "preservation value." Several authors have chosen one of these terms to represent the total difference and then identified various possible components of this total, often based on assumed motivations for holding these values. For example, Fisher and Raucher (1984) use the term "intrinsic value" to refer to the aggregate and state that the total intrinsic value is the sum of option value, aesthetic value, existence value, and bequest value. Sutherland and Walsh (1985) use "preservation value" to refer to the aggregate and state that it is the sum of option, existence, and bequest values.

I think that the attention given to questions of classification and motivation is misdirected. Motivations do not play an important role in the empirical analysis of the demands for market goods. There is little talk of "prestige value" and "speed value" in the literature on the demand for automobiles. So why should motivations be important in the case of nonuse values? Arguments about motivations seem to be offered primarily to persuade the reader of the plausibility of the hypothesis that nonuse values are positive. However, the real test of this hypothesis will come from the data. Rather than further debating definitions and possible motivations, it would be more useful to proceed with a test of the hypothesis that nonuse values (defined in a way that makes testing of the hypothesis feasible) are positive. If the evidence supports this hypothesis, then further research might be devoted to testing hypotheses about the determinants of the size of nonuse values in different cases. At that point, investigation of motivations might be useful in formulating hypotheses for testing. Then the choice of terms and explanatory variables would be governed by what are empirically meaningful distinctions.

One of the major issues in the literature on nonuse values is how to define the use that lies behind use value. The most common approach is to identify some market good or service that is a complement to the resource, as regards consumption, and to define and measure use in terms of the purchased quantity of this complementary good. If the resource is a park, the complementary good is the purchase of travel services to the park. Use of the resource is measured by the number of trips purchased. However, this approach to defining use is clearly a simplification of a more complex reality. The physical proximity that one normally thinks of as being an essential part of use can occur independently of the purchase of a complementary good such as travel. For example, people who live within the natural range of an endangered species such as the bald eagle or peregrine falcon may be able to view one of these birds and experience the utility and value associated with the sightings as an incidental part of their daily routine. In that case, there is no connection between that kind of use and any market good.

Some have argued that use does not require the physical proximity of the user and the resource. In their view, use can be defined to include the purchase of a complementary market good that embodies some visual or literary representation of the resource. Randall and Stoll (1983), for example, argue that there can be off-site uses, which they label as "vicarious consumption." "Thus," they explain, "we consider the values generated by reading about *Q* [the resource] in a book or magazine, looking at it in photographic representations, for example, to be use values. Clearly our definition of use includes vicarious consumption" (Randall and Stoll 1983, 267). This is essentially what Boyle and Bishop (1987) choose to call indirect use value.

Defining use in this way creates some problems for the measurement of value. One problem with such vicarious or indirect forms of use is that the observable market transaction, say the purchase of a nature magazine, often entails the simultaneous use of many environmental resources, so that allocation of the market transaction to specific resources is not possible. Furthermore, where vicarious uses involve information conveyed by photographs, films, and the like, the public-good dimension of information seems likely to virtually destroy any meaningful relationship between observed market behavior and underlying values. Vicarious use also has the odd feature that it can occur even though the resource itself no longer exists (as in the viewing of films and photographs).

Another unresolved question in the literature is whether the relevant distinction is between use and nonuse or whether it is more meaningful to distinguish between values to users and nonusers. I prefer the use-versus-nonuse distinction because it focuses on the presence or absence of activities involving the resource directly, rather than on the characteristics of the individuals holding the values. There is no logical reason why a user of a resource could not also hold values that are independent of that use and related to preservation, existence, or bequest motivations. Nonusers of a resource by definition can hold only nonuse values (if any), but users may hold both use and nonuse values for a resource.

This discussion of the possible definitions of and motivations for various types of use and nonuse values is inconclusive. Definitions can be considered a matter of taste. A set of definitions can be considered useful if it furthers research objectives and leads to useful answers to meaningful questions, and if the definitions are based on operationally meaningful distinctions. If use values are limited by definition to those associated with in situ use as measured by the purchase of a complementary good, then the definition has the virtue of distinguishing between cases where use of a resource generates observable market data and cases where no meaningful data can be obtained by observing market transactions. However, this definition leaves out vicarious or indirect uses as well as what I would call incidental uses. Perhaps it would be more useful to drop the distinction between use and nonuse values and instead to distinguish between values that can be estimated with revealed preference data and all other values. In the next section, I first present the standard model based on use of a complementary market good. I then consider the alternative formulation that distinguishes between those cases where changes in the level of q lead to changes in the behavior of individuals and those where they do not.

A Theoretical Framework

Assume that an individual has a preference ordering over a vector of market goods $\boldsymbol{X}$ and some nonmarketed resource q. The individual has no control over the level of q, but takes it as given. Here q is taken to be a scalar measure of some characteristic of the environment, such as the population of some species or the value of some parameter of water quality. In the abstract, q can represent a measure of either a quantity or a quality. The choice of a unit for measuring q has important implications for measurement in practice, but that question is not addressed here. The assumption that the environmental resource can be described by a single attribute is clearly a simplification. A more realistic model would allow for simultaneous changes in two or more quantitative or qualitative characteristics, or both, of the resource.

Following the analysis in the section on welfare measures for quantity changes in Chapter 3, and assuming that r (the price of q) is zero, the compensating surplus (CS) measure of welfare change for an increase in q is

$$v(\boldsymbol{P}, M - CS, q^1) = v(\boldsymbol{P}, M, q^0) \tag{5-1}$$

or alternatively

$$CS = e(\boldsymbol{P}, q^0, u^0) - e(\boldsymbol{P}, q^1, u^0) \tag{5-1'}$$

where u^0 is the solution to

$$\max: u(\boldsymbol{X}, q^0) \text{ subject to } \boldsymbol{P} \cdot \boldsymbol{X} \leq M \tag{5-2}$$

The derivative of the expenditure function with respect to q gives the compensated inverse demand function for q. If the inverse demand function is known, the compensating and equivalent surpluses can be measured by integrating over the range of change for q.

These are measures of total value, which can be interpreted as the sum of use value and nonuse value, that is:

$$CS = CS^U + CS^{NU} \tag{5-3}$$

To distinguish between use and nonuse value, it is necessary to add more structure to the model through assumptions about either preferences or the structure of household production. We turn now to this task.

Weak Complementarity and Use Value

One way to give additional structure to the model is to assume that use can be represented by the purchase of some market good, say x_1, and to define use to be zero when the individual's purchase of x_1 is zero. For examples of this modeling strategy, see McConnell 1983 and Smith 1987a, 1987b. Market good x_1 could be travel to a site for recreation, rental of a boat for fishing, or the use of the services of a guide for hunting or fishing. It is further assumed that q is an argument in the demand function for x_1 such that an increase in q increases the demand for x_1, other things being equal. Although q might be weakly complementary with several goods (see Chapter 4), for simplicity q is assumed here not to be an argument in the demand function for any other goods.

If the conditions for weak complementarity are satisfied, then the total value of an increase in q is reflected in the compensated demand function for x_1; the value of an increase in q that shifts the individual's Hicks-compensated demand curve out to the right can be measured by the geometric areas between the two demand curves. The resource value measured in this way is a use value because it is associated with the use of the resource as measured by the consumption of x_1. The conditions of weak complementarity mean that the individual derives no utility from an increase in q when use as measured by x_1 is zero. Therefore, nonuse value is zero. If the second condition of weak complementarity does not hold, the area between the two demand curves is still a measure of use value. However, because the individual derives additional utility from an increase in q when x_1 (use) is zero, there is an additional nonuse value that is not reflected in the demand curve for x_1.

The first condition of weak complementarity is that there is some price for x_1, say $p_1{}^*$, that chokes off the demand for x_1. This provides a basis for distinguishing use value from nonuse value. Nonuse value (CS^{NU}) can be calculated by evaluating the expenditure function at the price that chokes off the purchase of the private good that gives rise to use value:

$$CS^{NU} = e(p_1{}^*, q^0, u^0) - e(p_1{}^*, q^1, u^0) \tag{5-4}$$

where the prices of other goods are suppressed for convenience. The *CS* measure of use value associated with q^0 is given by $e(p_1^*, q^0, u^0) - e(p_1, q^0, u^0)$, while the use value associated with q^1 is given by $e(p_1^*, q^1, u^0) - e(p_1, q^1, u^0)$. Thus the increase in use value due to the increase in q is

$$CS^U = e(p_1^*, q^1, u^0) - e(p_1, q^1, u^0) - [e(p_1^*, q^0, u^0) - e(p_1, q^0, u^0)] \qquad (5\text{-}5)$$

Substituting equations 5-4 and 5-5 into equation 5-3 yields equation 5-1′:

$$\begin{aligned} CS &= CS^U + CS^{NU} \\ &= e(p_1^*, q^1, u^0) - e(p_1, q^1, u^0) + e(p_1, q^0, u^0) - e(p_1^*, q^0, u^0) \\ &\quad + e(p_1^*, q^0, u^0) - e(p_1^*, q^1, u^0) \\ &= e(p_1, q^0, u^0) - e(p_1, q^1, u^0) \qquad (5\text{-}6) \end{aligned}$$

However, if the first condition does not hold (that x_1 be nonessential), there is no finite price that reduces use to zero, and there is no economically meaningful distinction between use and nonuse values.

This modeling approach shows that it is possible for a user of the resource to have both use and nonuse values. It also suggests a way of defining a nonuser. A nonuser is one for whom the current price of the complementary good is greater than his or her choke price, that is, $p_1 > p_1^*$.

Separable Preferences and Use versus Nonuse Values

As noted above, the physical proximity that one normally thinks of as being an essential part of use can occur independently of the purchase of a complementary good. An example of this would be a person who sees members of an endangered species as an incidental part of his or her daily activities. Another example is the person who drives by a nature refuge or scenic outlook while commuting to work. In both cases, the presence or absence of the resource may have no effect on the behavior of the individual, yet this person could have a willingness to pay to preserve the resource. However, nothing can be inferred about this value from the demand curves for market goods. Thus estimation of this type of use value by revealed preference methods will not be possible.

Given the ambiguity of the concept of use in these cases, an alternative approach, apparently first suggested by Mäler and colleagues (1994), might be more appropriate. From the perspective of the measurement of value, the most meaningful distinction is between those cases where changes in the level of q lead to changes in the behavior of individuals and those where they do not. Mäler and colleagues propose to do away entirely with the "use value" and "nonuse value" terminology and to talk instead of "values revealed from market behavior" and "values *not* revealed by market behavior." I think that the distinction identified by Mäler and colleagues is the most useful one for dealing with the issues raised in this chapter. However, I suspect that the "use value/nonuse value" terminology is too deeply embedded in the literature now to make Mäler and colleagues' suggestion to do away with it feasible. So I will use "use value" to refer to "values revealed from market behavior" and "nonuse value" to refer to "values not revealed by market behavior." In this, I follow the practice of Carson and colleagues (1999).

What sort of preference structure implies no changes in market-related behavior? Suppose that the utility function is strongly separable, with q being the single argument in one of the separable subsets. This is the "hopeless case" described in Chapter 4. Changes in q have no effect on the demand functions for any market goods. The utility associated with q is independent of the level of use of any market good. Changes in q have no effect on purchases of market goods. Values revealed from market behavior are therefore zero.

An alternative formulation that allows individuals to simultaneously hold both types of values is to assume weakly separable preferences as follows:

$$u = u[g(\mathbf{X}, q), h(q)] \tag{5-7}$$

The approach of the preceding section based on a complementary relationship between q and a market good is a special case of this formulation. Maximizing either equation 5-7 or $g(\mathbf{X}, q)$ subject to the budget constraint and the exogenous q yields a set of demand functions for $\mathbf{X}$:

$$x_i(\mathbf{P}, M, q) \tag{5-8}$$

By substituting equation 5-8 in equation 5-7, we get the indirect utility function:

$$u = u[v(\mathbf{P}, M, q), h(q)] \tag{5-9}$$

The use value of a change in q flows from the expression $v(\cdot)$ and can be recovered from the data on market demands for $\mathbf{X}$, while nonuse value flows from $h(\cdot)$. But does this way of looking at things yield a unique nonuse value? Modifying equation 5-1 to account for the assumed structure of preferences gives us the following expression to solve for CS:

$$u[v(\mathbf{P}, M - CS, q^1), h(q^1)] = u[v(\mathbf{P}, M, q^0), h(q^0)] \tag{5-10}$$

Given the separability assumption, use value, or CS^U, can be defined as the solution to:

$$v(\mathbf{P}, M - CS^U, q^1) = v(\mathbf{P}, M, q^0) \tag{5-11}$$

Then, because we must have $CS = CS^U + CS^{NU}$, nonuse value is the solution to the following:

$$u[v(\mathbf{P}, M - CS^U - CS^{NU}, q^1), h(q^1)] = u[v(\mathbf{P}, M - CS^U, q^1), h(q^0)] \tag{5-12}$$

In other words, the expression for nonuse value must be evaluated assuming the compensating payment for use value has already been paid.

However, this is not the only way we could make use of this framework to define use and nonuse values. Alternatively, we could evaluate nonuse value first. Thus CS^{NU*} could also be defined as the solution to

$$u[v(\mathbf{P}, M - CS^{NU*}, q^0), h(q^1)] = u[v(\mathbf{P}, M, q^0), h(q^0)] \tag{5-13}$$

and CS^{U*} as the solution to

$$v(\mathbf{P}, M - CS^{NU*} - CS^{U*}, q^1) = v(\mathbf{P}, M - CS^{NU*}, q^0) \tag{5-14}$$

Because equation 5-14 is evaluated at a different income level than equation 5-11, in general it will give a different value for CS^U.

This is analogous to the way that multiple price changes are sequenced in calculating compensating variations. By path independence, the sum, CS, is independent of the sequence of evaluation. However, in the case of multiple price changes, the individual components of the sum are not independent of the sequence of evaluation. Mäler and colleagues (1994) make this clear by defining values in terms of differences between expenditure functions evaluated at different positions, as in equation 5-1′. They show that there is no unique way of partitioning total value into its use and nonuse components.

The Household Production Framework and Use Values

The theory of household production provides an alternative framework for thinking about how resources affect utility and welfare and what it means to assume that x_1 is a measure of the use of a resource. Recall that in the household production framework, the demand functions for market goods are derived demands. Assume that q is an input in the production of only one final service flow, z_1. If x_1 is also an input in the production of z_1, then (as shown in Chapter 4) q will be an argument in the ordinary and compensated demand functions for x_1. If x_1 and q are the only inputs in z_1, then the value of q in producing z_1 will be reflected in the derived demand for x_1 (Bockstael and McConnell 1983). If x_1 and q are both essential inputs, then the marginal product of each is zero if the other input is absent. A positive level of input for x_1 is required to use q in household production. This is equivalent to saying that the use of q and hence its use value is reflected in the level of x_1.

Another advantage of thinking about resource values in the context of the household production framework is the light that it sheds on the nature of what have been called indirect or vicarious use values. It appears that there are two categories of indirect or vicarious values—those that derive from household production and those that derive from market production. If the vicarious use involves the viewing of home-produced films and photos, then the theory suggests that the place to look for measures of indirect use values is in the influence of q on the demands for other inputs in the production of films and photos. If a market good is an essential input to this indirect use, then in principle the use value of changes in q can be measured by areas between the demand curves for the input in home production.

On the other hand, if indirect or vicarious use is the viewing of commercially produced films and television programs or the reading of books and magazines, the source of value for q is in its influence either on the production function or the quality-differentiated demand functions for these marketed goods. The place to look for measures of this type of value is either in changes in costs and prices or in the hedonic price functions for these goods. Thus the role of q is analytically no different from the role of, say, air quality in the production of agricultural crops. There is a well-developed methodology for measuring values in this case, as shown in Chapter 9. However, it seems doubtful that a reduction in the availability of a specific resource would increase the cost of producing the editorial content of films, books, and magazines, so the empirical significance of this approach to measuring values of changes is in question.

It seems reasonable to assume that the demand for a specific magazine or television program will depend in part on the quality and the information content of the material presented. However, the link between changes in q and changes in the demand for information about a specific resource is complex and may not meet the requirements of the standard models for estimating welfare values from changes in demand. For example, a major pollution event that damages a unique resource system may increase the demand for information about that resource, and producers are likely to respond with more articles and programs about that resource. This would be an increase in vicarious use, but it could hardly be called a benefit resulting from damage to the resource.

Distinguishing Nonuse and Existence Values

Because most discussion of nonuse values occurs in the context of proposals for irreversible development of a natural resource or the destruction or degradation of some natural amenity, it seems most useful here to talk in terms of the payment required to compensate for a reduction in q. In other words, welfare measures are defined for decreases in q using the *CS* definition of welfare change. With appropriate modification, similar conclusions can be reached for increases in q and/or equivalent surplus (*ES*) measures of welfare change. In Freeman (1993a, 1993b), I showed that there can be a meaningful distinction between the loss of nonuse values associated with the degradation of a resource that continues to exist and the loss associated with the destruction of a resource (the extinction of a species, the flooding of the Grand Canyon, or the clear-cutting of an old-growth redwood forest), the latter being a pure existence value. I also showed that these values can be held by both users and nonusers at current prices and qualities. What I called *pure nonuse value* could occur for an individual when current prices precluded use but there was some lower price for x_1 at which some use as represented by the purchase of a positive quantity for x_1 would occur. A resource was said to cease to exist if q fell below some level q^{min}, at which x_1 fell to zero even at a zero price. This change would result in a loss of existence value. In that analysis I showed that it is possible to define use, existence, and nonuse values in a consistent manner such that total value satisfies an adding-up condition. Where there is a threshold value for q, the total value of a decrease in q that crosses that threshold is the sum of three components:

- a loss of use value because of crossing the threshold that makes use no longer possible;
- the loss of user's nonuse value associated with degradation of q down to the threshold; and
- the loss of pure existence value associated with falling below the threshold of existence.

However, as shown above, this partitioning of total value into its components is not unique but depends on the path followed in identifying the components of total value.

The Temporal Dimension of Nonuse Values

The focus of this section so far has been on the structure of preferences in a static, timeless framework. I now take time explicitly into account by looking at the implications of different temporal patterns of changes in q.

First consider a permanent reduction in q from q^0 to q^1 such that $q_t^1 < q_t^0$ for all future t, where t indexes future time. An example would be the irreversible destruction of a unique natural asset. As developed more fully in Chapter 7, the proper measure of the welfare loss is the compensating increase in wealth (*CW*) required to maintain lifetime utility at its original level. This is approximated by the present value of the *CS*s for all future years. This measure could, in principle, reflect either use or nonuse values, or both.

Now consider a temporary reduction in q, that is, assume that q returns to its original level at, say, t^* ($q_t^1 = q_t^0$ for those t, $t^* < t \leq \infty$). This could be the result of natural recovery or of investment of resources in cleanup and restoration. The temporary reduction in q would likely result in lost use values. But would there also be lost nonuse values? There is nothing in the logic of the structure we have imposed on preferences up to this point that would preclude there being lost nonuse values. However, the early literature on nonuse values emphasized irreversible change and the irreplaceability of natural assets as justifications for the presence of nonuse values (see Krutilla 1967; Krutilla and Fisher 1975). It seems to me that whether irreversibility or the irreplaceable character of a natural asset is a prerequisite for nonuse values is an open question that can only be settled on the basis of empirical evidence. However, the available evidence suggests that irreversibility is not a prerequisite. In the state of Alaska stated preference study of the losses associated with the Exxon *Valdez* oil spill in Prince William Sound, respondents were told that the damages to natural resources and living species were not expected to be permanent. Yet the study showed significant willingness to pay to avoid future similar accidents (Carson et al. 1992).

Altruism and the Relevance of Nonuse Values

As noted in the first section of this chapter, one possible reason that nonusers might value the preservation of a resource is an altruistic feeling toward those who do use it. There is a literature on altruism and how the presence of altruistic feelings toward others affect the Pareto optimum or efficient allocations of resources. One conclusion in this literature is that if those feelings take a form known as *nonpaternalistic altruism*, they have no effect on the Pareto optimum allocation, that is, they are irrelevant for benefit–cost analysis (see Lazo et al. 1997 and references cited therein). If nonusers' willingness to pay for the preservation or improvement of a resource is due to nonpaternalistic altruism, the nonuse values should not be counted in a benefit–cost analysis (Milgrom 1993).

Nonpaternalistic altruism refers to a case where one individual cares about the general level of well-being of others but does not have any preferences regarding the composition of consumption bundles of others. If individual A had such preferences, they could be represented by

$$u^A = u^A[X^A, u^B(X^B, q^B)] \qquad (5\text{-}15)$$

An increase in q would result in an increase in u^B; therefore A would also be better off. A's willingness to pay for this improvement could be defined in the usual fashion, but it would not be proper to add this to B's WTP in an economic assessment of the policy, at least if the increase in q was not costless. This is because someone has to bear the cost of the increase in q. If B bears the cost, this decreases B's utility and therefore A's WTP for the increase. As long as altruism takes the form shown in equation 5-15, the terms representing altruism cancel out of the conditions for Pareto optimality. However, if A's altruism stems from a concern for the level of q that B experiences, then A's willingness to pay for B's improvement in q is relevant for economic assessments (Lazo et al. 1997; McConnell 1998).

In summary, if the sole source of nonuse values is nonpaternalistic altruism, these values are irrelevant for policy purposes. However, if altruism takes the paternalistic form, then the resulting nonuse values are relevant for policy analysis. Also, nonuse values can arise for other reasons besides or in addition to altruism.

Conclusions

In this section I have presented some results regarding nonuse values derived from several ways of defining and modeling the use of a resource. I have shown that where there is a complementary market good that can be chosen to represent use, something about the value of the resource is revealed in the observable market demand for the complementary good. If the assumption of weak complementarity holds, then market demand information can reveal everything there is to know about the value of the resource; that is, value measures derived from market data reflect total value. If the conditions of weak complementarity do not hold, then there is another component of value that is not reflected in market demands. There is also a consistent way of distinguishing between pure nonuse and existence values within this framework. However, whether preferences actually take the form that leads to this distinction is an empirical question.

If we adopt a broader concept of use based on physical proximity of the individual to the resource, we must confront the question of how to model those uses that do not require purchase of a complementary good. One possibility is that the requisite proximity is a characteristic of some of the houses in a residential housing market. If that is the case, then hedonic property value models can be used to learn something about that type of use value. Similarly, if some jobs provide opportunities for incidental viewing of peregrine falcons out of office windows, hedonic wage models might be used to estimate this type of use value. Many (but not all) these incidental uses require time. In principle, it would be possible to infer the "time value" of incidental use from data on individuals' allocation of time across activities. But it is possible that some types of incidental use do not leave any traces in market behavior or in the allocation of time. Thus as a practical matter, it may be best to define use value as the value that can be estimated using revealed preference methods based on observed market behaviors and nonuse value as the value that is not revealed through market behavior.

Toward Measurement

If we accept the argument that for practical purposes nonuse values should be defined as those values not revealed by market behavior, then by definition, they can be measured only with the stated preference methods discussed in Chapter 6. However, before turning to the stated preference methods, I want to ask whether there are any methods based on observed behavior that might lead to measurement of those values that are not directly linked to in situ use of the resource. Because my conclusions on this question are pessimistic, I then consider how stated preference methods might be used as part of a research strategy for measuring nonuse values as a separate component of total value. I conclude with a discussion of other issues that must be resolved in any effort to measure nonuse values.

Because preservation of a resource for those who do not make in situ use of it has the properties of nonexcludability and nondepletability, we would expect markets to fail to provide these preservation services, or at least to provide them in suboptimal quantities. Unless there is a market for preservation of a resource or for enhancing its quality for nonusers, there will be no market transactions to reflect the preservation values of individuals, and only stated preference methods of estimating nonuse values will be feasible.

We observe environmental organizations undertaking a variety of activities to protect and preserve natural environments, and people support these activities through voluntary contributions of time and money. The question is whether this revealed behavior provides an adequate basis for measuring WTP for preservation. For several reasons, I do not think that economic data on either the activities of these organizations or individuals' contributions to them can be relied upon as measures of the value of preservation for policymaking purposes. Here I discuss a range of these activities and why I think that data on these activities are not appropriate for the task of valuation.

The activities of environmental organizations can be placed in one of two categories: direct provision of preservation through acquisition, and advocacy in an effort to influence public-sector provision of preservation.

Organizations such as the Nature Conservancy accept private donations and use the funds to purchase lands with special ecological, geological, or scenic characteristics for the purpose of protection and preservation. Individuals' donations and dues paid to such organizations are manifestations of willingness to pay for preservation. However, if "free-rider" behavior is significant, these donations would be only a lower bound of true aggregate WTP. Furthermore, at least in some instances the lands acquired by such organizations are accessible to individual use. Therefore, individual donations could reflect a combination of use and nonuse values.

Many environmental organizations devote a substantial portion of their budgets to advocacy activities on behalf of environmental preservation in general and for policy actions to protect specific natural resources. Again, individuals' membership dues and donations in support of these activities reflect individuals' willingness to pay for preservation. However, because of free-rider behavior, aggregate donations are likely to represent less than total WTP. Also, the organizations undertaking these activities frequently have multiple purposes; they provide such services to members as magazines, other publications, and field trips. This means that only that portion of dues and donations supporting the incremental cost of advocacy to the organization is relevant for estimating preservation values.

In the policymaking arena, another factor weakens the relationship between individual donations to support advocacy and preservation values. This factor is the uncertainty concerning the outcome of the policy process and the contribution of advocacy activities to the desired outcome. A rational organization with limited resources would estimate the probabilities of successful advocacy as a function of the resource commitment for each specific issue and would allocate resources to maximize the expected value of the outcome. The observed allocation of advocacy resources across specific issues would reflect the interaction of the probabilities of a successful advocacy, the marginal productivity of advocacy in increasing the probability of success, and the value of success to the organization (Freeman 1969). It would probably prove difficult, if not impossible, to model this complex policy process to identify individuals' WTP on the basis of observed contributions to advocacy on specific issues.

It is reasonable to conclude that individuals' donations to environmental organizations involved in acquisition and advocacy reflect willingness to pay for preservation and nonuse values. These activities provide evidence in support of the hypothesis of significant nonuse values. However, for the reasons outlined here, observed acquisition and advocacy expenditures are likely to be an underestimate of the values held by people who are not in situ users of the resource.

Some people have argued that it is unrealistic to think that q can matter to an individual and still have no effect on his or her behavior. Individuals normally absorb some costs—if only in acquiring information—to possess values for the existence of q (McConnell 1983). Larson (1993, 381) argues that either changes in q affect behavior in ways that can in principle be detected and used to measure values, or such values are likely to be vanishingly small. He points out that a more realistic model of choice would include a time constraint and that even if q is not in the demand functions for market goods, changes in q are likely to affect the allocation of time to different activities. However, I am unaware of any successful efforts to measure values for environmental goods based on Larson's insights.

The stated preference valuation methods described in the next chapter are likely to offer the only feasible approaches to estimating nonuse values. Of course, what matters for policy purposes is total value, regardless of how it is divided between use and nonuse value. For individuals who are nonusers (or who do not alter observable behavior in response to changes in q), stated preference questions provide estimates of total value—which consists entirely of nonuse value. For individuals who are users and have separable preferences, such as those represented by equation 5-7, stated preference methods still yield estimates of total value; revealed preference methods yield underestimates of total value.

However, for research purposes there is some interest in learning more about the magnitudes of nonuse values and what factors determine their size. Broadly speaking, there are two approaches to estimating nonuse values. The first is to use a stated preference question to obtain an estimate of total value and then to deduct from this a separately obtained estimate of use value. The latter could be obtained, for example, by the travel cost method for estimating the demand for visits to a recreation site. As should be apparent, a problem with this approach to imputing nonuse values is that the imputation is based on the difference between two other values, both of which are measured with some unknown error. So without some understanding of the error properties of the other measures, one cannot know whether the imputed value is simply the result of measurement error or is a true nonuse value.

The second approach to estimating nonuse values is to ask people explicitly about their nonuse values, either by questioning people who are known to be nonusers or by asking people to assume that their own use is zero. Any value revealed by a nonuser is, by definition, a nonuse value. But if present users are asked to assume that they are not using the resource, their responses to stated preference questions may not be valid indicators of nonuse value. This is because the situation they are being asked to see themselves in is unfamiliar to them, and they might misunderstand it or reject it outright.

Three additional issues concerning research study design deserve at least brief mention. The first issue concerns the definition and description of the resource to be valued. Resources have both quantitative and qualitative dimensions that can be affected by policy decisions or damaged by pollution events. Yet in the theoretical discussion above, I assumed that a resource was measured in a single dimension. Respondents must be given a clear description of the changes in all relevant dimensions of resource quantity and quality. The determination of what is relevant must come in part from the judgement of experienced researchers in this field and in part from research specifically designed to determine what characteristics of the resource are important to people.

The second issue concerns the relevant population for sampling when nonuse values are involved. If the resource to be valued is in California, for example, should the sample include East Coast residents? Or should it be limited only to westerners or to California residents? Casting the sampling net too wide wastes scarce research resources, but important values may be missed if the geographic scope of the sample is too narrow. Even small per capita values can loom large when aggregated over a large population. Again, experience and the results of research designed specifically to shed light on this set of issues can help to guide research study design. It is unclear what more can be said about this issue until we know a lot more about what characteristics of resources are likely to give rise to significant nonuse values. Some resources such as the Amazonian rain forest and African elephants may have worldwide significance, implying that the relevant population is the world population. Also, one cannot rule out the possibility that where there are regionally significant resources, important nonuse values are held only by people within that region. This is an important research question.

The third issue concerns the choice of the temporal dimension of the value measure being elicited. In the preceding section, I showed that the correct welfare measure was the compensating change in wealth, that is, a "one-shot" payment. This amount would be elicited by asking individuals a lump-sum question. Some studies, on the other hand, have asked respondents about their willingness to pay in equal annual payments over some period of time. For this form of question, it is necessary to specify the time period involved. Individuals are likely to find this form of question difficult to answer because they are being asked implicitly to carry out an annuitization calculation. This seems likely to reduce the reliability of responses.

Summary

As this discussion of the concept of nonuse values and approaches to their measurement has shown, several questions remain unresolved. One of the most fundamental is how to distinguish between use and nonuse values.

The present practice is a consequence of how the field of natural resource valuation has evolved over time. Economists have devoted their energies to developing techniques for measuring particular pieces of the total value puzzle. The success of such a piecemeal strategy depends on how we choose to break up the pieces. The easy pieces have been those associated with identifiable uses such as recreation. As more specific uses have been identified, modeled, and measured, the use-versus-nonuse distinction has seemed to take on some validity. Some attention has also been devoted to the separate development of a theory of nonuse values.

However, the question of how to model nonuse values cannot be separated from that of modeling use values. The theory of economic welfare gives unambiguous guidance only on defining total values as compensating income changes for changes in a resource. The question of whether nonuse values, however defined, are positive takes on meaning only after some decision has been made about what use values measure because nonuse value is simply total value minus whatever has been called use value.

Most theorizing on nonuse value starts by defining use in terms of observable market-related behavior, that is, the purchase of a complementary market good. This has been a useful strategy because of the availability of revealed preference methods for estimating values from observed market data. However, it is not unfair to characterize this strategy as one of shaping the definitions to fit the available empirical methods and data. This problem cannot be avoided by declaring that the relevant distinction is between users and nonusers because users cannot be identified until we have decided what activity defines *use*. I have also shown that there is no a priori reason why users cannot also hold nonuse values.

I urge that we not accept uncritically the present conventions for distinguishing between use and nonuse values. Ultimately we want to be able to measure total value. Any distinction between use and nonuse values is itself useful only if it helps in the task of measuring total values. The present convention has been useful because it makes possible the measurement of certain kinds of use values by invoking the weak complementarity model, but this is not the only way to look at the problem.

Another important question is, when are nonuse values likely to be important? The long literature on nonuse values emphasizes the uniqueness or specialness of the resource in question and the irreversibility of loss or injury. For example, economists have suggested that there are important nonuse values in preserving the Grand Canyon in its natural state and in preventing the global or local extinction of species and the destruction of unique ecological communities. In contrast, resources such as ordinary streams and lakes or a subpopulation of a widely dispersed wildlife species are not likely to generate significant nonuse values because of the availability of close substitutes. Moreover, the literature does not suggest that nonuse values are likely to be important where recovery from an injury is quick and complete, either through natural processes or restoration. However, there are problems in giving operational meaning to the idea of uniqueness. In economic terms, uniqueness would be reflected in the absence of substitutes and a low price elasticity of demand. This idea comes from the literature on industrial economics, but there is no threshold on price elasticity that distinguishes between the presence or absence of close substitutes. Similarly, long-term injury with slow recovery could give rise to nonuse values that are of the same order of magnitude as those associated with irreversible injury.

It would be desirable to give empirical researchers some guidance about how to determine when nonuse values are likely to be significant. However, at the present time there is no general method for determining whether a resource is sufficiently unique or a resource change is of sufficient duration to generate important nonuse values. This important question should be the subject of continued research.

References

Bockstael, Nancy E., and Kenneth E. McConnell. 1983. Welfare Measurement in the Household Production Framework. *American Economic Review* 73(4): 806–814.

Boyle, Kevin J., and Richard C. Bishop. 1987. Valuing Wildlife in Benefit–Cost Analyses: A Case Study Involving Endangered Species. *Water Resources Research* 23(5): 943–950.

Carson, Richard T., Nicholas E. Flores, and Robert C. Mitchell. 1999. The Theory and Measurement of Passive-Use Value. In *Valuing Environmental Preferences: Theory and Practice of the Contingent Valuation Method in the US, EU, and Developing Countries*, edited by Ian J. Bateman and Kenneth G. Willis. Oxford, U.K.: Oxford University Press.

Carson, Richard T., Robert C. Mitchell, W. Michael Hanemann, Raymond J. Kopp, Stanley Presser, and Paul A. Ruud. 1992. *A Contingent Valuation Study of Lost Passive Use Values Resulting from the Exxon Valdez Oil Spill.* Report to the Attorney General of the State of Alaska, La Jolla, CA: Natural Resource Damage Assessment (NRDA), Inc.

DesJardins, Joseph. 1999. *Environmental Ethics: Concepts, Policy, and Theory.* London: Mayfield.

Diamond, Peter A., and Jerry A. Hausman. 1993. On Contingent Valuation Measurement of Nonuse Values. In *Contingent Valuation: A Critical Assessment*, edited by Jerry A. Hausman. Amsterdam: North-Holland.

———. 1994. Contingent Valuation: Is Some Number Better Than No Number? *Journal of Economic Perspectives* 8(4): 455–464.

Fisher, Ann, and Robert Raucher. 1984. Intrinsic Benefits of Improved Water Quality: Conceptual and Empirical Perspectives. In *Advances in Applied Microeconomics*, edited by V. Kerry Smith and Ann Dryden Witte: Greenwich, CT: JAI Press.

Freeman, A. Myrick, III. 1969. Advocacy and Resource Allocation Decisions in the Public Sector. *Natural Resources Journal* 9(2): 166–175.

———. 1993a. Nonuse Values in Natural Damage Assessment. In *Valuing Natural Assets: The Economics of Natural Resource Damage Assessment*, edited by Raymond J. Kopp and V. Kerry Smith. Washington, DC: Resources for the Future.

———. 1993b. *The Measurement of Environmental and Resource Values: Theory and Methods*. Washington, DC: Resources for the Future.

Jamieson, Dale. 2000. *A Companion to Environmental Philosophy*. Maldon, MA: Blackwell.

Kopp, Raymond J. 1992. Why Existence Value *Should* Be Used in Cost–Benefit Analysis. *Journal of Policy Analysis and Management* 11(1): 123–130.

Krutilla, John V. 1967. Conservation Reconsidered. *American Economic Review* 57(4): 777–786.

Krutilla, John V., and Anthony C. Fisher. 1975. *The Economics of Natural Environments: Studies in the Valuation of Commodity and Amenity Resources*. Baltimore, MD: The Johns Hopkins University Press.

Larson, Douglas M. 1993. On Measuring Existence Value. *Land Economics* 69(4): 377–388.

Lazo, Jeffrey K., Gary H. McClelland, and William D. Schulze. 1997. Economic Theory and Psychology of Non-Use Values. *Land Economics* 73(3): 358–371.

Mäler, K.-G., I.-M. Gren, and C. Folke. 1994. Multiple Use of Environmental Resources: A Household Production Function Approach to Valuing Natural Capital. In *Investing in Natural Capital: The Ecological Economics Approach to Sustainability*, edited by A.-M. Jansson, M. Hammer, C. Folke, and R. Costanza. Washington, DC: Island Press.

McConnell, Kenneth E. 1983. Existence and Bequest Value. In *Managing Air Quality and Scenic Resources at National Parks and Wilderness Areas*, edited by Robert D. Rowe and Lauraine G. Chestnut. Boulder, CO: Westview Press.

———. 1998. Does Altruism Undermine Existence Value? *Journal of Environmental Economics and Management* 32(1): 22–37.

Milgrom, Paul. 1993. Is Sympathy an Economic Value? Philosophy, Economics, and the Contingent Valuation Method. In *Contingent Valuation: A Critical Assessment*, edited by Jerry A. Hausman. Amsterdam: North-Holland.

Randall, Alan. 1991. Total and Nonuse Values. In *Measuring the Demand for Environmental Quality*, edited by John B. Braden and Charles D. Kolstad. Amsterdam: North-Holland.

Randall, Alan, and John R. Stoll. 1983. Existence Value in a Total Valuation Framework. In *Managing Air Quality and Scenic Resources at National Parks and Wilderness Areas*, edited by Robert D. Rowe and Lauraine G. Chestnut. Boulder, CO: Westview Press.

Rosenthal, Donald H., and Robert Nelson. 1992. Why Existence Value Should *Not* Be Used in Cost–Benefit Analysis. *Journal of Policy Analysis and Management* 11(1): 116–122.

Samples, K.C., J.A. Dickson, and M.M. Gowen. 1986. Information Disclosure and Endangered Species Valuation. *Land Economics* 62(3): 306–312.

Schulze, W.D., and D.S. Brookshire. 1983. The Economic Benefits of Preserving Visibility in the National Park Lands of the Southwest. *Natural Resources Journal* 23(1): 149–173.

Smith, V. Kerry. 1987a. Non-Use Values in Benefit–Cost Analysis. *Southern Economic Journal* 54(1): 19–26.

———. 1987b. Uncertainty, Benefit–Cost Analysis, and the Treatment of Option Value. *Journal of Environmental Economics and Management* 14(3): 283–292.

Sutherland, Ronald J., and Richard G. Walsh. 1985. Effect of Distance on the Preservation Value of Water Quality. *Land Economics* 61(3): 281–291.

Van De Veer, Donald, and Christine Pierce. 1994. *The Environmental Ethics and Policy Book: Philosophy, Ecology, Economics.* Belmont, CA: Wadsworth.

CHAPTER

6

Stated Preference Methods for Valuation

In Chapter 4, I described several methods for inferring individuals' values for environmental amenities from their observed choices of quantities of market goods and bundles of amenities when they are assumed to make optimal choices subject to income and other constraints. However, discussions in Chapters 4 and 5 also showed that there are many circumstances under which value measures cannot be derived from observations of individuals' choices. In this chapter I describe a variety of methods that have in common their source of data for analysis: individuals' responses to questions about hypothetical situations such as: "Would you pay $\$X$ for ...?", "What is the most that you would be willing to pay for ...?", "What would you do if ...?", or "Which of the following alternatives do you prefer ...?" Because values are inferred from stated responses to such questions, these methods are now commonly referred to as stated preference (SP) methods.

Although not all authors use the same terminology, the term *stated preference methods* has come to refer to any survey-based study in which respondents are asked questions that are designed to reveal information about their preferences or values. The term encompasses three broad types of questions. The first type involves questions that ask directly about monetary values for a specified commodity or environmental change. These are usually called *contingent valuation methods* (CVM). The most commonly used CVM questions simply ask people what value they place on a specified change in an environmental amenity or the maximum amount they would be willing to pay to have it occur. In contrast to the discrete choice questions described next, these are usually open-ended in that the individual has to state a number rather than respond to a number offered by the researcher. The responses to these questions, if truthful, are direct expressions of value and would be interpreted as measures of compensating surplus (*CS*). The other major type of CVM question asks for a yes or no answer to the question, "Would you be willing to pay $\$X$ ...?" Each individual's response reveals only an upper bound (for a no) or a lower bound (for a yes) on the relevant welfare measure. Questions of this sort are termed *discrete choice questions.* Responses to discrete choice questions can be used to estimate willingness-to-pay (WTP) functions or indirect utility functions.

The second and third major types of SP methods do not reveal monetary measures directly. Rather, they require an analytical model to derive welfare measures from responses to questions. In the second approach to questioning, respondents are given a set of hypothetical alternatives, each depicting a different bundle of environmental attributes. Respondents are asked to choose the most preferred alternative, to rank the alternatives in order of preference, or to rate them on some scale. Responses to these questions can then be analyzed to determine, in effect, the marginal rates of substitution between any pair of attributes that differentiate the alternatives. If one of the other characteristics has a monetary price, then it is possible to compute the respondent's willingness to pay for the attribute on the basis of the responses. Studies based on this form of question are usually referred to as *stated choice* or sometimes *attribute-based* methods (Holmes and Adamowicz 2003).

In the third type of SP question, individuals are asked how they would change the level of some activity in response to a change in an environmental amenity. If the activity can be interpreted in the context of some behavioral model such as an averting behavior model or a recreation travel cost demand model, the appropriate indirect valuation method can be used to obtain a measure of WTP. These are known as *contingent behavior* or sometimes *contingent activity* questions.

Use of the set of SP methods for environmental valuation has been controversial. This controversy became especially heated after the Exxon *Valdez* oil spill in March 1989, when it became known that a major component of the legal claims for damages was likely to be based on CVM estimates of lost nonuse or existence values. The controversy had the salutary effect of stimulating a substantial body of new research on both SP practice and on the credibility or validity of SP-based estimates of value. A good overview of the issues raised in this controversy is contained in the three essays published as a symposium in the *Journal of Economic Perspectives* (Portney 1994; Hanemann 1994; Diamond and Hausman 1994). To get a sense of the intensity of the controversy, see the collection of essays sponsored by Exxon Corporation and especially the transcripts of the discussions by audience members and authors at the public presentation of these studies (Hausman 1993).

In the next section, I describe in more detail the major types of SP question formats and, where appropriate, I explain how responses can be analyzed to obtain measures of welfare for changes in the environmental amenity. A major question regarding all SP methods concerns the validity and reliability of the data, that is, whether the hypothetical nature of the questions asked inevitably leads to some kind of bias or results in so much "noise" that the data are not useful for drawing inferences. In the second section of this chapter I consider approaches to assessing the validity of measures of value obtained with SP methods, where by *validity* I mean the degree of correspondence between the measure obtained and the theoretical concept of value. I conclude with an assessment of the current state of the art of the SP methods.

My treatment of the issues surrounding SP methods is selective. I will not discuss a number of design and implementation problems, such as the type of contact with respondents (mail, telephone, personal interview) and sampling design. I will not review the extensive literature reporting empirical results of SP surveys. Mitchell and Carson's (1989) pioneering treatise is still the primary reference on CVM, especially for design and implementation questions (see also Carson 1991). Two new works that focus on best practice and empirical estimation for CVM and stated choice studies are Boyle (2003) and Holmes and Adamowicz (2003), respectively. The so-called NOAA Blue Ribbon Panel (U.S. National Oceanographic and Atmospheric Administration 1993) reviewed CVM in the context of assessing damages to natural resources in support of litigation and provided its guidelines for best practice. Other important references are Bjornstad and Kahn 1996 for a review of theoretical and empirical issues that includes assessments by both proponents and critics of SP methods; see Kopp et al. 1997 and Bateman and Willis 1999.

The Welfare Economics of Stated Preference Questions

Open-Ended Contingent Valuation Questions

Two Types of Open-Ended Questions. Data gained from open-ended value questions are the simplest to interpret. Each respondent is typically asked to state his or her maximum willingness to pay for an environmental improvement (compensating surplus, *CS*) or to avoid a loss (equivalent surplus, *ES*). There are several ways to elicit this number. Many of the earliest studies used an iterative technique, which has come to be called *the bidding game.* In the bidding game, individuals are first asked whether they would be willing to pay \$*X*. If the individual answers yes, the question is repeated with a higher price. The procedure is repeated until the individual answers no. The highest price with a yes response is interpreted as the maximum WTP. If the original response is no, the iteration proceeds downward until a yes response is received. When researchers divided their samples and tested for the effects of different starting points for the bidding game, they often found that the starting point influenced the outcomes. See Whitehead (2002) for a recent example. This has been called the problem of *starting point bias.*

An alternative elicitation technique is simply to ask open-ended questions of the form, how much would you be willing to pay? The major problem with this approach is that it confronts people with an unfamiliar problem. In most real market settings, individuals are faced with choices among sets of goods with listed prices. Seldom are they asked to offer a one-shot bid that may be either accepted or rejected by the seller. People appear to have difficulty dealing with open-ended direct questions. As a consequence, surveys using this form of elicitation method typically result in high rates of nonresponse to the valuation question, or high proportions of implausibly high or low stated values, or both.

A variation on the open-ended approach is to show respondents a card with a range of alternative payment values on it and ask them either to pick a number from the card or to state their own value if that is not to be found on the card. Some authors have also experimented with payment cards that indicate the amounts that typical respondents are paying in taxes for such public programs as police protection, health care, and national defense. Rowe and others (1996) conducted a study to see if changes in the ranges covered by payment cards affected the distribution of expressed maximum WTP. They found no effect as long as the payment card did not truncate the upper range of values.

Analysis of Open-Ended Responses. A direct valuation question produces a set of welfare measures W_i ($i = 1, \ldots, i, \ldots, n$) for the n respondents in the sample. As noted above, W_i is either a *CS* or *ES* measure, depending on the format of the question. An estimate of the total value of the welfare change for the population from which the sample is drawn can be obtained by calculating the sample mean W^* and multiplying by the total population. Alternatively, the responses can be regressed on income and other socioeconomic characteristics to obtain a bid function for a given change in q:

$$B_i^* = B^*(M, \mathbf{S}_i) \tag{6-1}$$

where $\mathbf{S}$ is a vector of socioeconomic characteristics that affect individuals' valuations. Then data on the characteristics of the relevant population can be used to calculate B^* for every member of the population. An alternative approach is to include variation in the size of the change in q across the sample as part of the survey design. Then the bid function

$$B_i^* = B^*(\Delta q_i, M, \boldsymbol{S}_i) \tag{6-2}$$

can be estimated and used to calculate values for alternative scenarios of environmental or resource change.

Mean values, and therefore aggregate values for the population, are sensitive to the presence of large individual bids. Many surveys obtain at least a few bids that are so large relative to the sample mean as to be obviously invalid for valuation purposes. It would be desirable to have an unambiguous rule for identifying such outliers so that they could be deleted from the sample before aggregate values or valuation functions are estimated. Several procedures for identifying outliers have been used in the literature, but each has a degree of arbitrariness associated with it.

The simplest and most commonly used procedure is to adopt a rule of thumb concerning the relationship between the stated bid and the respondent's income. For example, the rule could be to delete all responses representing more than X% of reported income. The chosen value for X could depend on the analyst's assessment of the likely importance to people of the environmental or resource change being valued. As an alternative to this procedure, Mitchell and Carson (1989, 226–227) advocate the calculation of the α-trimmed mean, where the analyst chooses the value of α. The procedure is to delete a proportion of responses equal to the selected value of α from the top and bottom of the rank-ordered responses. For $\alpha = 0$, the α-trimmed mean is the true mean of the sample. For $\alpha = 0.5$, the α-trimmed mean is equal to the true median. Regression analysis of the data would be done after the α-trimming was carried out. A third alternative is to use a set of regression diagnostic procedures (those of Belsley et al. 1980, for example) to delete observations with unduly large influences on regression coefficients. The deletion of extreme values from the sample by any procedure may be creating a different sort of bias problem—that due to item nonresponse. Item nonresponse bias and strategies for dealing with it are discussed extensively in Mitchell and Carson (1989, 267–278).

In addition to the problem of invalid large responses, a sample could contain invalid zero responses, so-called *protest zeros.* Protest zeros occur when respondents reject some aspect of the constructed market scenario by reporting a zero value even though they place a positive value on the amenity or resource being valued. The process of α-trimming the sample is not a solution to the problem of protest zeros because α could be greater than or less than the true proportion of protest zeros. Rather, some means must be found to identify protest zeros for deletion before applying the procedure for deleting outliers. One approach is to ask every respondent who gives a zero value to indicate a reason for doing so. For example, respondents could be asked the following question:

Which statement best expresses your reason for giving a zero response?

I. I can't afford to pay for the good.
II. The good is not important to me.
III. I don't think that I should have to pay for the good.

Responses of those choosing the third statement would be classified as protest zeros and deleted from the sample, while responses of those choosing the first and second would be considered valid zeros. The deletion of protest zeros can also lead to item nonresponse bias if those who protest are systematically different in some respect from those who give proper responses.

Discrete Choice Contingent Valuation Questions

Discrete Choice Question Format. In its simplest form, the discrete choice question format asks each respondent whether he or she would be willing to pay a specified amount of money to obtain the environmental change in question. If a respondent answers yes, that person has indicated a WTP that is greater than or equal to the specified sum. If the response is no, then that sum of money can be taken as an upper bound on true WTP. Respondents should be assigned randomly to different subsamples, with each subsample asked to respond to a different dollar amount. It is then possible to test the hypothesis that the proportion of yes responses decreases with an increase in the price of the environmental good. These data can then be analyzed with a model of discrete choice to obtain estimates of indirect utility functions or bid functions.

Whereas the discrete choice question makes relatively inefficient use of a sample, the information from a given sample size can be increased by asking each respondent a follow-up question. If the initial answer is yes, the respondent can be asked a second discrete choice question on a higher amount, typically twice the original amount. If the initial response is no, the respondent can be asked a second question about a lower amount, for example, half the original amount. This so-called *double-bounded format* increases the information obtained from each respondent and therefore increases the statistical efficiency of welfare estimation. However, it also raises some important statistical issues that I will briefly discuss below.

The design of the set of first and second prices offered to subjects is an important consideration. If the range of offers is too low, the estimated mean WTP will be biased downward because of the lack of information from respondents who would likely answer yes to the first question but no to the second one. See Alberini 1995 and Hanemann and Kaninen 1999 for discussions of the optimal design of bids for double-bounded questions.

The discrete choice format has at least three advantages relative to the bidding game and WTP formats. First, it places people in a relatively familiar social context. Many private market transactions involve goods offered on a take-it-or-leave-it basis, in which the individual decides whether or not to purchase the good at the offered price. If the payment vehicle is a tax, the discrete choice question simulates a referendum of the sort found everywhere from small New England town meetings to statewide votes on highway bond issues. The second advantage is that because only a yes or no answer is required, the discrete choice question poses a relatively simple decision for individuals. This may result in lower levels of item nonresponse and fewer refusals to participate in the survey. Third, in at least some circumstances, it is incentive-compatible, that is, respondents' best strategy is to be truthful in answering the question. I will return to this point later.

The hypothetical nature of discrete choice surveys makes it possible to gain more information than would be available by observing individual choices in most real-world markets or discrete choice settings. In real-world choice settings, usually all individuals face the same set of prices for the alternatives from which they must choose. For example, in choosing a mode of transportation, all individuals face the same fares. Differences in responses must then be due to differences in individual characteristics. However, in a discrete choice setting it is possible to present respondents in different randomly chosen subsamples with different prices.

Analysis of Discrete Choice Responses. To convert data on yes or no responses to a discrete choice question into a monetary measure, it is necessary to use an explicit utility theoretic model of choice. The discrete choice model introduced in Chapter 4 is well suited to the task. As Cameron (1988) and Cameron and James (1987) pointed out, the variation in prices across the sample makes it possible to explain individuals' choices in terms of a WTP function rather than in terms of differences in indirect utility. McConnell (1990) compared the Cameron model with the model based on utility functions outlined by Hanemann (1984) and by Sellar and others (1986). He showed that the two models can be derived from the same underlying utility theoretic framework. In a deterministic formulation, the two models yield the same predictions about behavior and choice. Because the stochastic forms of the models introduce random components in different ways, the two models are not in general equivalent, although there are special cases in which the two models are dual to each other. Moreover, as Cameron (1988) showed, the WTP function model permits the straightforward calculation of marginal values for all the arguments in the WTP function, whereas this is not possible with the utility function model.

In what follows, I first model the individual's response to a single dichotomous choice question in the utility difference framework of Hanemann

(1984) and show how welfare measures can be derived. I then show the parallel development of the WTP function model. Finally, I identify some of the statistical issues raised by the double-bounded format.

Consider an individual who must decide whether to answer yes or no to the following question: "Would you vote for a program to permanently increase environmental quality from q^0 to q^1 if it would increase your taxes by \$$T$ for this year?" Let the indirect utility function be $u(M, q, \mathbf{C})$, where $\mathbf{C}$ is a vector of individual characteristics and the vector of market prices, $\mathbf{P}$, is omitted because prices are assumed to be constant. The individual responds yes if

$$u(M - T, q^1, \mathbf{C}) - u(M, q^0, \mathbf{C}) \geq 0 \tag{6-3}$$

and no otherwise.

Letting $v(\cdot)$ represent the observable component of utility, the probability of a yes response is given by

$$Pr(Y) = Pr[v(M - T, q^1, \mathbf{C}) + \varepsilon_1 > v(M, q^0, \mathbf{C}) + \varepsilon_0] \tag{6-4}$$

where the ε_i $(i = 0, 1)$ are the random, unobserved components of utility. As discussed in Chapter 4, if the random terms are independently and identically distributed with a Type I extreme value distribution, then this probability can be expressed as

$$Pr(Y) = (1 + e^{-\Delta v})^{-1} \tag{6-5}$$

where $\Delta v = v^1 - v^0$. Also, reversing the sign on the probability difference gives the expression for the probability of rejecting the offer:

$$Pr(N) = (1 + e^{\Delta v})^{-1} \tag{6-6}$$

The willingness to pay for q^1 (CS) is defined implicitly by

$$u(M - CS, q^1, \mathbf{C}) = u(M, q^0, \mathbf{C}) \tag{6-7}$$

and

$$v(M - CS, q^1, \mathbf{C}) + \varepsilon_1 - \varepsilon_0 = v(M, q^0, \mathbf{C}) \tag{6-8}$$

In terms of the observable utility function, CS is thus a random variable because of the term $\varepsilon_1 - \varepsilon_0$. The probability of accepting the offer is also, then, the probability that $CS \geq T$, and the probability of rejecting the offer is the probability that $CS < T$. This is a cumulative density function (c.d.f.), denoted here as $F(T)$, and shown in Figure 6-1, which plots the probability

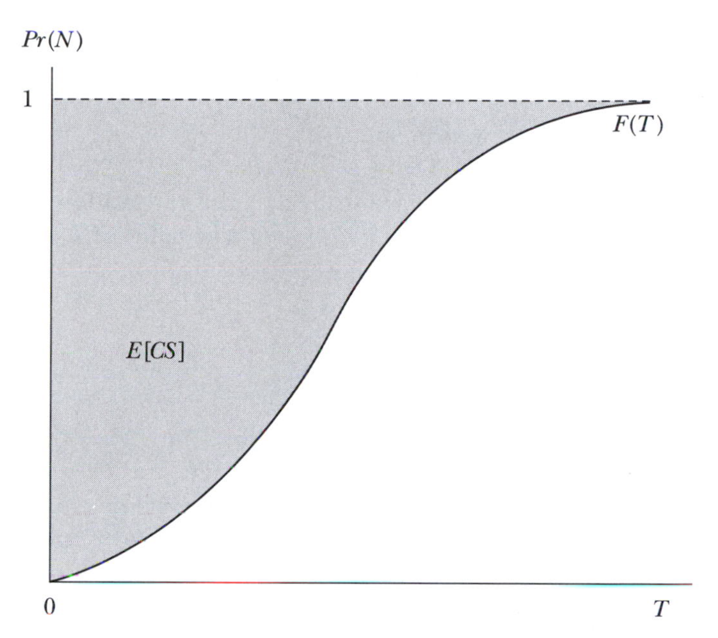

Figure 6-1. The cumulative density function for rejecting the offer of q^1 at T and the expected value of CS

of a "no response" ($Pr(N)$) as a function of T. As Hanemann (1984) pointed out, the expected value of the random variable CS can be found from the c.d.f. as follows:

$$E[CS] = \int_0^\infty \left[1 - F(T)\right] dT \tag{6-9}$$

Graphically, in Figure 6-1, the expected value of CS is the shaded area above the c.d.f. and below $Pr(N) = 1$. For example, for any given T, an individual with a lower willingness to pay for the change in q would have a higher probability of rejecting the offer, and the shaded area would be smaller. Alternatively, an offer of $q^2 > q^1$ at any T would decrease the probability of an individual rejecting the offer. So $F(T)$ would be shifted down and $E[CS]$ would be larger.

Specifying a functional form for the observable component of utility makes it possible to estimate the parameters of the utility difference in equations 6-5 and 6-6. For example (following Hanemann 1984), if $u = a + b \cdot \ln M + c \cdot \ln q$ (ignoring the $\boldsymbol{C}$ vector for simplicity), then

$$\Delta u = (a^1 - a^0) + \left[b \cdot \ln\left(1 - \frac{T}{M}\right)\right] + \left[c \cdot \ln\left(\frac{q^1}{q^0}\right)\right] \tag{6-10}$$

Then with the parameters from equation 6-10, equations 6-5 (or 6-6) and 6-9 can be used to calculate $E[CS]$ (for example, Hanemann 1984; Seller et al. 1986).

As equation 6-9 shows, in principle $F(T)$ is integrated over the range to infinity. To avoid the implausibly high estimates of $E[CS]$ that can sometimes result, some researchers have truncated the integration of $F(T)$ at some "reasonable" finite value for T. However, estimates of $E[CS]$ can be highly sensitive to the value of T_{max} chosen for truncation, at least in some cases (Bishop and Heberlein 1979; Hanemann 1984; Mitchell and Carson 1989, 103, 196–197).

There is an alternative approach to dealing with the sensitivity of estimates of *CS* to outliers in the distribution. As Hanemann (1984) suggested, the median WTP for the sample can be found by setting $Pr(Y)$ in equation 6-5 equal to 0.5 and solving for T. This gives the value for T that makes the representative individual indifferent as to accepting or rejecting the offer. Generalizing to the sample, one would expect half the respondents to accept (and half to reject) the offer, other things (including income and $\mathbf{C}$) held constant. This procedure is less sensitive to extreme values in the data.

The alternative model for analyzing discrete choice data is based on the bid or WTP function. This function can be derived from the expenditure function as follows:

$$B(q^0, q^1, u^0, \mathbf{C}) = e(q^0, u^0, \mathbf{C}) - e(q^1, u^0, \mathbf{C}) \qquad (6\text{-}11)$$

Cameron (1988) calls this a valuation function; McConnell (1990) calls it a variation function because it can be defined for either compensating or equivalent welfare measures. The formulation here gives the compensating surplus measure of value for an improvement. Other versions can easily be specified for losses and for *ES* measures. See Carson 1991, especially 143–144, for further discussion of this model.

The individual will respond yes if

$$B(q^0, q^1, u^0, \mathbf{C}) \geq T \qquad (6\text{-}12)$$

and no otherwise. The probability of accepting the offer of q^1 at T can be expressed in terms of the function

$$Pr(Y) = Pr[B^*(q^0, q^1, u^0, \mathbf{C}) - T > \eta] \qquad (6\text{-}13)$$

where B^* is the observable component of the bid function and η is the unobserved random component of WTP. In other words,

$$B(q^0, q^1, u^0, \mathbf{C}) = B^*(q^0, q^1, u^0, \mathbf{C}) + \eta \qquad (6\text{-}14)$$

The next step is to make some assumptions about the distribution of the random component of the bid. Cameron and James (1987) assume that η is normally distributed with zero mean and standard deviation of σ. The result is a form of probit model in which

$$Pr(Y) = 1 - \Phi\left(T - \frac{\mathbf{X}'\beta}{\sigma}\right) \tag{6-15}$$

where $\mathbf{X}'\beta$ is the set of explanatory variables and coefficients and Φ is the standard normal cumulative density. The bid is given by

$$B^* = \mathbf{X}'\beta + \sigma \tag{6-16}$$

Once the parameters of equation 6-16 have been estimated, calculations of total and marginal bids for individuals and in the aggregate are straightforward. Alternatively, if η is assumed to be distributed as in the logistic model, then (following Cameron 1988)

$$Pr(Y) = [1 + e^{-(B^*-T)}]^{-1} \tag{6-17}$$

Again, estimation of the parameters of the bid function makes it possible to calculate individual total and marginal WTPs as well as the probabilities of accepting an offer. For a careful, detailed, and thorough discussion of the statistical analysis of discrete choice responses and the implications of the lack of independence for statistical estimation, see Hanemann and Kaninen 1999.

Just as in the case of direct value questions, respondents might express rejection of the discrete choice market scenario by saying no to all offers even though their WTP exceeds the lowest offered price. Respondents who refuse the lowest offer should be queried about their reasons, as discussed above, and respondents who signal that their no vote is a protest should be deleted from the sample.

As noted above, the double-bounded question format doubles the information that is obtained from a given sample. However, this added information comes at a cost. In early applications of the double-bounded format, researchers treated the responses to the first and second questions as independent draws from the same distribution (for example, Hanemann et al. 1991). Subsequently, though, others subjected this assumption to statistical tests and found that it was not consistent with the data (Cameron and Quiggin 1994). This result can lead to biased welfare estimates. In general, the WTP calculated from both questions together is often less than the WTP based on responses to the first question alone. Carson and others

(2000) offer an explanation based on an examination of the incentive properties of this question format. I will return to this point below.

Stated Choice Questions

Preferred Rankings and Ordered Alternatives. In Chapter 4, I described the use of the discrete choice model to estimate the parameters of a utility function from individuals' choices of one alternative from a set of *n* alternatives. The discrete choice contingent valuation format described above is a simplified version of this approach in that individuals make their selection of one alternative from a set of two (the alternatives of yes or no). The stated choice or conjoint analysis method of estimating values asks individuals to provide more information about their preferences by giving them more alternatives than the discrete choice approach and by asking them either to select their most preferred option or to rank the alternatives in order of preference.* Each alternative could have several different attributes. Normally one attribute would have a monetary dimension—for example, a price—to facilitate calculation of monetary values.

In this respect, the stated choice method bears more than a passing resemblance to the random utility models used in the analysis of recreation demand and value (see Chapter 13). In both cases, the objects of choice are differentiated by embodying different levels of a set of attributes. By focusing on the trade-offs among attributes, both methods yield estimates of the marginal rates of substitution between pairs of attributes and, where price is one of the attributes, the marginal willingness to pay for the attribute. Furthermore, with the stated choice method the analyst has experimental control through the design of the attributes presented in the choice set.

One of the earliest applications of the stated choice method in economics was that of Beggs and others (1981). They used an ordered logit model to estimate the values of characteristics of alternative models of cars, including electric-powered vehicles. Their respondents were asked to rank 16 alternative vehicle designs; each design had nine attributes, including purchase price and fuel costs per mile. The method has subsequently been used in the environmental realm to value rural visibility (Rae 1983), water quality (Smith and Desvousges 1986), and the avoidance of diesel odor (Lareau and Rae 1989). The Smith and Desvousges study was perhaps the

*Some researchers have asked respondents to rate the alternatives on some scale. See for example, Roe et al. 1996 and Boyle et al. 2001. Holmes and Adamowicz (2003) also discuss this approach but conclude that the pick-one or ranking formats are preferable because people may not be able to translate differences of strength of preference into numerical ratings.

simplest for respondents: they were asked to rank only four alternatives, each of which involved only two attributes, a description of water quality and an annual fee. Other studies have posed the simpler task of asking respondents to indicate only their most preferred alternative. Examples include Adamowicz et al. 1994 and Morrison et al. 1999.

In comparing the stated choice method with other methods of obtaining hypothetical data, one question is whether individuals find it easier to rank alternatives ordinally than to establish a monetary (cardinal) value for a commodity. Intuitively, it would seem that ordinal ranking is easier than establishing a value under the open-ended and bidding game formats. On the other hand, ranking several alternatives, each with several attributes, seems likely to be more difficult and certainly more time-consuming than ranking only two alternatives, each with only two attributes, as in the case of the discrete choice contingent valuation method format.

Analysis of Stated Choice Responses. The analytical model used to extract information about preferences from the stated choice responses is a straightforward extension of the discrete choice model. This model provides a set of parameter weights on the attributes that maximizes the likelihood of realizing the observed choice (the most preferred alternative or the complete rank ordering). Provided one of the attributes is a money measure, parameter weights can then be used to calculate the marginal willingness to pay for an attribute (the marginal rate of substitution between income and the attribute) or the willingness to pay for a nonmarginal change in its level.

Suppose that an individual is asked to pick the most preferred alternative or to rank a set of alternatives $\boldsymbol{Y}$ ($\boldsymbol{Y} = y_1, \ldots, y_i, \ldots, y_n$) and that each element in $\boldsymbol{Y}$ can be described by its price p_i and a vector of nonprice attributes $\boldsymbol{Q}$ ($\boldsymbol{Q} = q_1, \ldots, q_j, \ldots, q_m$). The individual's indirect utility associated with alternative i consists of two components. The observable component is $v(M, \boldsymbol{Q}_i, p_i, \boldsymbol{C})$ where $\boldsymbol{C}$ is a vector of individual taste parameters. The unobservable component is ε_i, which is assumed to be random. From Chapter 4, we know that if the ε_i are independently and identically distributed with a Type I extreme value distribution, the probability of the individual picking alternative 1 or ranking it first is

$$Pr\left[u_1 > u_j; \quad \text{for all } j \neq 1\right] = \frac{e^{v_1}}{\sum_{i=1}^{n} e^{v_i}} \tag{6-18}$$

In the case of contingent ranking, suppose that the individual values the alternatives in the following order from most preferred: $y_1, y_2, \ldots, y_n$. The probability of this ranking is the probability of ranking y_1 above the remaining $n - 1$ alternatives times the probability of ranking y_2 above the remaining $n - 2$ alternatives, and so forth. Or

$$Pr\left[u_1 > u_2 > \ldots u_n\right] = \frac{e^{v_1}}{\sum_{j=1}^{n} e^{v_j}} \cdot \frac{e^{v_2}}{\sum_{j=2}^{n-1} e^{v_j}} \ldots = \prod_{i=1}^{n}\left[\frac{e^{v_r}}{\sum_{j=r}^{n+1-r} e^{v_j}}\right] \qquad (6\text{-}19)$$

where r is the rank ordering of the alternatives.

As in the other applications of the discrete choice model, once a functional form for the observable component of the indirect utility function has been specified, its parameters can be estimated from data on the rankings of a sample of individuals, using maximum likelihood methods. Welfare measures can then be derived by computing the marginal rate of substitution between income and any of the elements of $\boldsymbol{Q}$.

Contingent Behavior

This term refers to the use of hypothetical questions about activities to obtain data for use in one of the behavioral models of the sort described in Chapter 4. For example, suppose one had used travel cost data to estimate a demand function for visits to a recreation site, but one wanted to know the value of a change in one of the environmental attributes of that site. The relevant behavioral model would be weak complementarity. However, in the absence of an observed change in the environmental attribute, it might not be possible to predict the shift in the demand curve for visits to the site. Visitors could be asked how their visitation behavior would change if the environmental attribute were to change in a specified way. This contingent change in visitation rates could be used to estimate the shift in the demand curve for visits. McConnell (1986) used this approach to estimate the benefits to visitors of local beaches if the pollution of New Bedford Harbor, Massachusetts, by polychlorinated biphenyls could be eliminated. For other recent applications, see Thayer 1981 and Cameron et al. 1996.

Assessing the Validity of Stated Preference Welfare Measures

As Mitchell and Carson (1989, 190) put it, "The validity of a measure is the degree to which it measures the theoretical construct under investigation." The theoretical construct of interest here is the individual's true value, her true probability of accepting an offer or selecting one of the alternatives offered, his true ranking of alternatives, or her true change in the level of an activity. Ideally, one would like to assess the validity of an SP measure by comparing it with the true measure. However, at least in the cases of monetary values, probabilities, and rankings, the true measure cannot be known, so this option is not available. In the case of contingent behavior questions, we will not often be able to observe "true" changes in activity levels under precisely the same conditions that prevailed when the question was asked. So, much of the discussion of validity in the literature focuses on less direct approaches to validity assessment, such as asking whether the results are consistent with economic theory (construct validity) or whether the study was conducted in accordance with generally accepted principles.

In this section, I briefly describe four types of validity and approaches to assessment for each type. These are not mutually exclusive alternatives but should be seen as focusing on different aspects of the way a study is conducted and of the information generated by the study. I also summarize what can be said at present about the validity of SP methods in general, as distinct from the validity of particular studies. This is not a full and complete review of either the methods for assessment or the evidence about validity. For a more detailed discussion of validity and a review of the evidence available as of their writing, see Mitchell and Carson 1989. The reader should also consult Carson et al. 2001, Hausman 1993, and some of the essays included in Bjornstad and Kahn 1996 and Kopp et al. 1997. Bishop 2003 also includes an insightful discussion of concepts of validity.

The Several Forms of Validity

Criterion Validity. Assessing criterion validity involves comparing the value as measured by the SP method with some alternative measure that can be taken as the criterion for assessment. It is tempting to think of the "true" value as the appropriate criterion. However, the "true" value is a theoretical construct: that sum that creates a state of indifference between two alternative states of the world. As such, it is not directly observable (Bishop 2003).

Some authors have created simulated or experimental market settings in which individuals can engage in actual transactions. In simulated or experimental markets, transactions actually take place and respondents live with the consequences of their choices. These experiments allow the comparison of responses to hypothetical questions with responses in similarly designed experimental or simulated markets. These authors have used the observed values from these transactions as the criterion for comparison with SP responses. However, Carson and others (1996) have argued that because of sampling variability and the role of model specification in obtaining value measures from both the real and hypothetical transactions, these studies do not really assess criterion validity.

One of the pioneering studies of this type was conducted by Bishop and Heberlein (1979), who made hypothetical closed-ended offers to purchase special goose-hunting permits from a sample of hunters. At the same time, they made actual cash offers for a limited number of permits from a different sample, so that they could compare the values revealed. The mean value of a permit calculated from the responses to the hypothetical offer (a willingness-to-accept (WTA) compensation measure) was about 60% higher than the mean revealed by the real cash offers.

List and Gallet (2001) reviewed the results of 29 studies dealing with the provision of both private and public goods, including that of Bishop and Heberlein. They used a meta-analysis to explain the reported calibration ratios, that is, the ratios of minimum, median, and maximum hypothetical values to the corresponding values from the real transactions. Although the median calibration ratio for their sample was about 3.0, they found that using a WTP rather than WTA question and using a first-price sealed-bid auction mechanism both had statistically significant effects in lowering the calibration ratio. Evidence such as that summarized in List and Gallet has led some to suggest that compensating variation (*CV*) values be reduced by a calibration factor (for example, Diamond and Hausman 1994, 54).

Convergent Validity. A related form of validity assessment is the comparison of SP values with measures derived from some revealed preference (RP) method. This provides an assessment of convergent validity. The problem with this measure of validity is that if a discrepancy between the SP value and the alternative is discovered, it could be because of the lack of validity of the SP measure, the lack of validity of the alternative measure derived from observation, or both. This strategy is as old as the contingent valuation method itself (see Knetsch and Davis 1966). For earlier reviews of this evidence, see Cummings et al. 1986 and Mitchell and Carson 1989.

Carson and others (1996) conducted a large study of convergent validity of contingent valuation estimates of willingness to pay for quasi-public goods (goods provided by the government for which exclusion of users is possible). The estimates covered were for three types of goods: access to outdoor recreation activities, environmental amenities, and changes in health risks. They found that the ratio of *CV* to *RP* values averaged 0.89 for the whole sample with a 95% confidence interval of 0.81 to 0.96. Based on their evidence, the authors argue against systematic calibration of *CV* values.

Construct Validity. Construct validity asks whether *CV* responses are related to variables that economic theory suggests should be predictors of WTP. Construct validity can be assessed by regressing expressed values on characteristics of the good being valued and characteristics of the respondent. One test of construct validity is to see whether expressed values vary systematically and in the expected manner with the explanatory variables that theory suggests should be important.

For example, it is usually thought that values should be an increasing function of income, other things being equal, and that the estimated income elasticity of WTP or price flexibility of income should be "reasonable." However, Flores and Carson (1997) showed that the relationship between the income elasticity of demand for a quantity-constrained good and the corresponding income elasticity of WTP is not as simple one might think. They showed that in the case of more than one quantity-constrained good, the income elasticity of willingness to pay for a good depends not only on the income elasticities of demand for all the quantity-constrained goods but also on the cross price substitution effects for those goods and on the expenditures for market goods as a percentage of virtual income (see Chapter 3). This latter term will always be less than one, leading to a likelihood that the income elasticity of WTP will be less than the income elasticity of demand. However, they also show by example that plausible values for the three determinants of the income elasticity of demand for WTP can lead to values greater than the income elasticity of demand or even negative values for so-called *normal goods*. Therefore, very small or even negative values for the income elasticity of WTP for a nonmarket good are not prima facie evidence of lack of construct validity.

When an experimental design involves facing different individuals with different quantities, implied prices, or both, the choices should be consistent with the axioms of revealed preference. There are nonparametric tests for consistency with these axioms. For an example of an application of this strategy, see Adamowicz and Graham-Tomasi 1991.

Another form of construct validity assessment is the development of different scenarios to test hypotheses about the effects of certain features of a scenario on the mean value of the sample. These tests can involve either comparing mean responses across sample groups given different scenarios (between-subject comparisons) or comparing responses of the same sample group to different scenarios (within-subject comparisons). The hypotheses can take two forms. One form is the hypothesis that individual or sample mean values vary in a consistent fashion with relevant and meaningful variations in the scenario. For example, an increase in the quantity of the good being valued should produce an increase in the magnitude of expressed values. Tests of this hypothesis have come to be known as *scope tests*. The NOAA Panel (U.S. National Oceanographic and Atmospheric Administration 1993) recommended including a scope test in its guidelines.

Inadequate response of expressed WTP to changes in scope is often cited as a reason to reject *CV* values by critics of the methods; see, for example, Diamond and Hausman 1994. However, several authors have reviewed the evidence from scope tests and found that they can reject the hypothesis of inadequate responsiveness on expressed values to changes in scope. See Smith and Osborne 1996; Kopp and Smith 1997; and Carson 1997.

The other form of hypothesis is that bids are invariant to irrelevant changes in the scenario. Experiments in settings different from contingent valuation have produced a large body of evidence that responses depend on irrelevant changes in descriptions or contexts. For some examples, see Tversky et al. 1990. However, at least one study (Boyle 1989) has shown that contingent valuation responses can be invariant to irrelevant changes in the scenario.

SP questions can be asked either in WTP or WTA formats. A large body of evidence suggests that WTP and WTA formats for SP values lead to quite different results. WTA responses are typically several times larger (and, in at least one case, two orders of magnitude larger) than WTP responses for the same change. See Cummings et al. 1986, especially 35, for a review of earlier studies comparing WTP and WTA. Brookshire and Coursey (1987) also find large differences. Some have argued that the disparity between WTP and WTA responses is evidence that SP methods lack construct validity. However, as discussed in Chapter 3, there are theoretically plausible reasons why there could be large differences between WTP and WTA, at least in the case of environmental resources and amenities. Thus the observed disparities between WTP and WTA are not prima facie evidence of lack of construct validity.

Content Validity. Content validity refers to the extent to which the design and implementation of the survey conform to the generally recognized best practice or state of the art. It should involve an examination of the survey instrument, including the scenario specification, the elicitation question—especially its incentive properties—and the payment vehicle, as well as procedural matters, such as sample size and design, and the analysis of data. For a more complete discussion of the assessment of content validity, see Bishop and McCollum 1997.

Practitioners in the field of SP valuation methods have identified a number of problem areas and issues that can affect the validity of responses. There is general agreement, for example, that the form of the valuation question is important. Discrete choice questions work better than the bidding game format. WTP questions work better than WTA questions. Several sources of potential bias have been identified. Thus, much can be learned from a systematic examination of the survey instrument.

Mitchell and Carson (1989, especially Chapter 11) provided a classification of ways in which scenarios can affect individuals' responses and introduce systematic error or bias into the value measures that are derived from these responses. Systematic errors arise from several sources. One is scenario misspecification. Were the question and the information provided with it to establish its context understood by the respondent in the way intended by the investigator? Another source is implied value cues. To what extent did the information provided to the respondent predispose him or her toward a particular response? A third source is the choice of a payment vehicle. Did the payment vehicle tap into attitudes about government and taxation or business pricing policies that might bias responses? And finally, were there incentives for misrepresentation of values? Part of the art of asking SP questions involves avoiding problems such as these. I now take up each of these sources of error in turn.

The various types of scenario misspecification discussed by Mitchell and Carson (1989, 246–258) all have the effect of creating a divergence between what the respondent understands about the choice situation and what the investigator intends the respondent to understand. The investigator can increase the likelihood that the questions will be understood as intended by developing an awareness of how people think about the issues at hand and what language and terminology they use in talking about it. This can be accomplished through the use of focus groups, pilot surveys, pretests, and the like. Focus groups involve structured interviews with small groups of individuals from the population to be surveyed. Focus groups can be used to investigate perceptions of environmental problems, to determine which terminologies have particular salience, and to try out various descriptions and question formats. Pretests and pilot surveys may be useful in testing different scenarios and question formats, especially if respondents are queried afterward to identify questions and problems they may have had with specific wording.

If a respondent has well-formed preferences regarding the hypothetical choice problem posed to him, he should find the task of formulating a response to the question fairly easy. However, if the choice problem posed to the individual is unfamiliar and the individual is not clear about her preferences, she might, quite unconsciously, seek clues regarding the "correct" choice or value from the information supplied as part of the scenario specification. If such clues are present, they may systematically bias individuals' choices and values. Such clues are termed *implied value cues*. The best-studied implied value cue is the starting point in bidding game versions of direct value questions. Other forms of direct value question may avoid starting-point biases.

Implied value cues can be found outside of the scenario itself. For example, respondents who are asked about the value of access to a recreation site or to a fishing experience may anchor their responses to the implied value provided by the normal admission fee or fishing license fee. The task for the investigator is to find ways to thwart this tendency, perhaps by specifying a payment vehicle that focuses attention away from the customary admission fee or license fee. For example, recreationists might be asked about their willingness to pay higher travel costs to reach a site.

Stated choice and contingent behavior questions appear to be less susceptible to implied value cues. However, there could be a tendency for the results of a stated choice question to depend upon the order in which options are initially presented for ranking. This kind of order effect can be prevented by randomizing the order of presentation of alternatives across respondents.

Another feature of scenario design that may be important is the specification of a payment vehicle for collecting the respondents' WTP. To make contingent market scenarios plausible to respondents, it is necessary to inform them how their stated value will be collected. Examples of payment vehicles used in past studies include an increase in the prices of all goods resulting in a higher expenditure of $\$X$, an increase in taxes of $\$X$, and a surcharge on utility bills of $\$X$. If the vehicle involves a higher price for a market good, the question must be clear about whether it refers to a change in expenditure or a change in price. If the question is about a change in price, then the welfare calculation must take account of the reduction in quantity that a price increase will cause. In some studies where respondents were asked for a maximum increase in a price per unit, total WTP was calculated by multiplying the increase in price by the observed quantity. However, assuming that respondents accurately stated the maximum price increase they would accept, the correct welfare measure is the area to the left of the demand curve between the two prices. This will be less than the original quantity times the increase in price.

There is evidence that the specification of a payment vehicle can exert an independent influence on bids, a phenomenon that has been termed "vehicle bias" in the literature. However, judging whether this influence is a bias in the usual sense of the word is a more difficult matter because the respondent's "true" value is probably not known. The influence of the payment vehicle on bids is one example of the importance of what Fischhoff and Furby (1988) called the social context of the hypothetical transaction. They argued that in both designing and interpreting surveys that elicit hypothetical values, researchers must be sensitive to how individuals understand and interpret the environmental good being valued, the payment they are being asked to consider, and the social context in which the exchange takes place.

Biased responses can occur if the scenario specification and the framework for eliciting values are not incentive-compatible. In an incentive-compatible framework there is no reason for an individual to respond to any valuation question other than in terms of that person's true preferences. If an individual has any reason to believe that his or her response will have any effect on either the provision of public goods or environmental quality or on the individual's taxes or other payment responsibilities, then the respondent may try to use the response to influence the outcome of the public policy process.

A classic form of bias in *CV* responses arises from the efforts of respondents to "free ride," that is, to reduce their repayment obligation by stating low values. If a respondent believes that the level of provision of the public good is independent of his response, but that his tax liability does depend on the response, that person has an incentive to understate his true value for the good. Alternatively, if a respondent believes that her response can have some effect on the level of provision of the good but will not affect her payment obligation, then she has an incentive to bid high if she prefers more of the public good or bid low (or zero) if she prefers less of the good.

Strategic bias has been discussed extensively in the literature. For earlier discussions of the topic, see Cummings et al. 1986 and Mitchell and Carson 1989. Carson and others (2000) provide the most thorough treatment of the subject of which I am aware. The rest of this section is based on their paper.

The analysis of Carson and others (2000) is based on reasoning from first principles about what is in the best interest of respondents faced with a scenario, payment vehicle, and elicitation question. The first point is that unless the survey question is about something that matters to the respondent and respondents believe that their responses might affect the outcome of the policy issue that is the subject of the survey, there is no basis in economics for predicting individuals' responses. If both conditions hold, then the survey question is termed "consequential" to respondents. For consequential questions, it is possible to reason from an assumption of acting on rational self-interest to predict whether responses will be truthful and if not, then at least in some cases what the direction of bias will be.

For consequential questions, the only question format that can in principle be incentive-compatible is the single discrete choice question. This form requires the further condition that the government agency is perceived as able to compel payment of some amount from the respondent if the good is provided. For example, questions that ask about the willingness to make a voluntary contribution to support some government action fail this condition and provide incentives to respond "yes" even when the requested contribution is greater than the respondent's WTP.

In the cases of double-bounded discrete choice questions, the offer of the good at a second price destroys the incentive compatibility of the discrete choice question. The actual result depends on how respondents interpret the new information they have received. For example, the second offer could create uncertainty about what the actual price to be charged will be. The individual will respond differently because of this uncertainty. Alternatively, the individual might interpret the second price as signaling that the agency is willing to bargain about the price. Then the individual might provide a no in the hope that an even lower offer will be forthcoming.

Carson and colleagues confirm the lack of incentive compatibility of open-ended questions. The analysis of repeated paired comparisons (a form of stated choice) and ranking questions is more complex, but the conclusion is the same. In the end, only the single discrete choice question remains as fully incentive-compatible, and then only if certain conditions are met. With the double-bounded discrete choice question, there is a trade-off between bias and variance. However, the terms of the trade-off (that is, the direction and magnitude of the bias) have not been well characterized. This is a question for future research.

Conclusions

Economists and other social scientists have made a great deal of progress in the design and implementation of SP methods since the earliest contingent valuation studies of the mid-1960s and early 1970s. One major advance is the recognition that the original bidding game format is just one of a family of SP elicitation methods. Other advances include the development of discrete choice questions, choice experiments or stated choice questions, contingent ranking, and contingent behavior methods. Each of these methods has its strengths and limitations relative to the others, but together they provide a powerful array of tools for dealing with a variety of valuation problems.

A second advance has been the development of a greater awareness of and sensitivity to the ways in which scenario specification can influence responses and thereby affect the validity of the value measures. Practitioners in economics owe a debt to researchers from other fields, especially those in cognitive psychology, for their contributions to our understanding of how people respond to questions and of the importance of how questions are asked.

A third advance has been the development of more sophisticated approaches to the analysis of data and the calculation of welfare measures, especially with respect to discrete choice models. Fourth, there has developed a much greater sensitivity to a variety of sampling issues and related problems, such as the need to deal with item nonresponse and to identify and treat outliers and protest zeros. Finally, researchers are developing a variety of methods for combining stated preference and revealed preference data in one statistical model. I will have more to say about these models in Chapter 14.

All of these developments have helped to make the set of SP methods a powerful tool for dealing with amenity and resource valuation problems.

Yet, SP methods continue to be controversial, with many people remaining highly skeptical that they can generate valid estimates of economic value. The NOAA "Blue Ribbon Panel" concluded that "… under those conditions [that were discussed earlier in the report], CV studies convey useful information…. [and] can produce estimates reliable enough to be the starting point of a judicial process of damage assessment …" (U.S. National Oceanographic and Atmospheric Administration 1993, 4610). This conclusion is in sharp contrast to that of Diamond and Hausman (1994, 62), "… that contingent valuation is a deeply flawed methodology for measuring nonuse values, one that does not estimate what its proponents claim to be estimating."

Critics of SP methods often cite the many studies showing differences in estimates for a good when different elicitation formats are used. For example, Diamond and Hausman say, "This skepticism [about CVM] comes from the belief that the internal consistency problems [described in their paper] come from an absence of preferences, not a flaw in the survey methodology" (1994, 63). However, Hanemann and Kaninen say, "Some critics of CV have asserted that one should expect the same results regardless of the elicitation procedure…. the evidence in market research as well as environmental valuation overwhelmingly rejects procedural or task invariance" (1999, 408). After reviewing the incentive properties of the various elicitation formats, Carson and colleagues add, "Rather than be taken as evidence that respondents don't have well-defined preferences, differences between the estimates obtained using different elicitation formats, if predicted by economic theory, should be taken as evidence supporting the proposition that respondents are taking the scenario posed seriously" (2000, 38).

My own assessment of SP methods is cautiously optimistic. Some people seem to be attracted to the methodology because of a perception that it represents a relatively easy and inexpensive way to get usable values for environmental resources. However, I think that it is hard to do an SP study well. Reading this chapter is not sufficient to equip one to do an SP study well. There is no substitute for a careful reading in the now substantial and rapidly growing literature—including, but not limited to, the sources cited here.

References

Adamowicz, Wiktor L., and Theodore Graham-Tomasi. 1991. Revealed Preference Tests of Nonmarket Goods Valuation Methods. *Journal of Environmental Economics and Management* 20(1): 29–45.

Adamowicz, W., J. Louviere, and M. Williams. 1994. Combining Revealed and Stated Preference Methods for Valuing Environmental Amenities. *Journal of Environmental Economics and Management* 26(3): 271–292.

Alberini, Anna. 1995. Optimal Designs for Discrete Choice Contingent Valuation Surveys: Single-Bound, Double-Bound, and Bivariate Models. *Journal of Environmental Economics and Management* 28(3): 287–306.

Bateman, Ian J., and Kenneth G. Willis (eds.). 1999. *Valuing Environmental Preferences: Theory and Practice of the Contingent Valuation Method in the US, EU, and Developing Countries.* Oxford, U.K.: Oxford University Press.

Beggs, S., S. Cardell, and J. Hausman. 1981. Assessing the Potential Demand for Electric Cars. *Journal of Econometrics* 16(1): 1–19.

Belsley, David A., Edwin Kuh, and Roy E. Welsch. 1980. *Regression Diagnostics: Identifying Influential Data and Sources of Collinearity.* New York: John Wiley.

Bishop, Richard C. 2003. Where to from Here? In *A Primer on Non-Market Valuation,* edited by Kevin J. Boyle and Patricia A. Champ. Boston: Kluwer Academic Publishers.

Bishop, Richard C., and Thomas A. Heberlein. 1979. Measuring Values of Extra Market Goods: Are Indirect Measures Biased? *American Journal of Agricultural Economics* 61(5): 926–930.

Bishop, Richard C., and Daniel W. McCollum. 1997. Assessing the Content Validity of Contingent Valuation Studies. Unpublished (available from Bishop, Department of Agricultural and Applied Economics, University of Wisconsin, Madison, WI 53706).

Bjornstad, David J., and James R. Kahn (eds.). 1996. *The Contingent Valuation of Environmental Resources: Methodological Issues and Research Needs.* Cheltenham, U.K.: Edward Elgar.

Boyle, Kevin J. 1989. Commodity Specification and the Framing of Contingent-Valuation Questions. *Land Economics* 65(1): 57–63.

———. 2003. Contingent Valuation in Practice. In *A Primer on Non-Market Valuation,* edited by Kevin J. Boyle and Patricia A. Champ. Boston: Kluwer Academic Publishers.

Boyle, Kevin J., Thomas P. Holmes, Mario Teisl, and Brian Roe. 2001. A Comparison of Conjoint Analysis Response Formats. *American Journal of Agricultural Economics* 83(2): 441–454.

Brookshire, David S., and Don L. Coursey. 1987. Measuring the Value of a Public Good: An Empirical Comparison of Elicitation Procedures. *American Economic Review* 77(4): 554–566.

Cameron, Trudy A. 1988. A New Paradigm for Valuing Non-Market Goods Using Referendum Data: Maximum Likelihood Estimation by Censored Logistic Regression. *Journal of Environmental Economics and Management* 15(3): 355–379.

Cameron, Trudy A., and Michelle D. James. 1987. Efficient Estimation Methods for Contingent Valuation Surveys. *Review of Economics and Statistics* 69(2): 269–276.

Cameron, Trudy A., and John Quiggin. 1994. Estimation Using Contingent Valuation Data from a "Dichotomous Choice with Follow-Up" Questionnaire. *Journal of Environmental Economics and Management* 27(3): 218–234.

Cameron, Trudy A., W. Douglass Shaw, Shannon E. Ragland, J. Mac Callaway, and Sally Keefe. 1996. Using Actual and Contingent Behavior Data with Differing

Levels of Time Aggregation To Model Recreation Demand. *Journal of Agricultural and Resource Economics* 21(1): 130–149.

Carson, Richard T. 1991. Constructed Markets. In *Measuring the Demand for Environmental Quality*, edited by John Braden and Charles Kolstad. Amsterdam: North-Holland.

———. 1997. Contingent Valuation Surveys and Tests of Insensitivity to Scope. In *Determining the Value of Non-Marketed Goods: Economic, Psychological, and Policy Relevant Aspects of Contingent Valuation Methods*, edited by Raymond J. Kopp, Werner W. Pommerehne, and Norbert Schwarz. Boston: Kluwer Academic Publishers.

Carson, Richard T., Nicholas E. Flores, Kerry M. Martin, and Jennifer L. Wright. 1996. Contingent Valuation and Revealed Preference Methodologies: Comparing the Estimates for Quasi-Public Goods. *Land Economics* 72(1): 80–99.

Carson, Richard T., Nicholas E. Flores, and Norman F. Meade. 2001. Contingent Valuation: Controversies and Evidence. *Environmental and Resource Economics* 19(2): 173–210.

Carson, Richard T., Theodore Groves, and Mark J. Machina. 2000. Incentive and Informational Properties of Preference Questions. Unpublished. http://weber.ucsd.edu/~rcarson/ (accessed August 20, 2002).

Cummings, Ronald G., David S. Brookshire, and William D. Schulze (eds.). 1986. *Valuing Environmental Goods: An Assessment of the Contingent Valuation Method.* Totowa, NJ: Rowman and Allanheld.

Diamond, Peter A., and Jerry A. Hausman. 1994. Contingent Valuation: Is Some Number Better Than No Number? *Journal of Economic Perspectives* 8(4): 45–64.

Fischhoff, Baruch, and Lita Furby. 1988. Measuring Values: A Conceptual Framework for Interpreting Transactions with Special Reference to Contingent Valuation of Visibility. *Journal of Risk and Uncertainty* 1(2): 147–184.

Flores, Nicholas E., and Richard T. Carson. 1997. The Relationship between the Income Elasticities of Demand and Willingness To Pay. *Journal of Environmental Economics and Management* 33(3): 287–295.

Hanemann, W. Michael. 1984. Welfare Evaluations in Contingent Valuation Experiments with Discrete Responses. *American Journal of Agricultural Economics* 66(3): 332–341.

———. 1994. Valuing the Environment through Contingent Valuation. *Journal of Economic Perspectives* 8(4): 19–43.

Hanemann, W. Michael, and Barbara Kaninen. 1999. The Statistical Analysis of Discrete-Response CV Data. In *Valuing Environmental Preferences: Theory and Practice of the Contingent Valuation Method in the US, EU, and Developing Countries*, edited by Ian J. Bateman and Kenneth G. Wills. Oxford, U.K.: Oxford University Press.

Hanemann, W. Michael, John Loomis, and Barbara Kaninen. 1991. Statistical Efficiency of Double-Bounded Dichotomous Choice Contingent Valuation. *American Journal of Agricultural Economics* 73(4): 1255–1263.

Hausman, Jerry A. (ed.). 1993. *Contingent Valuation: A Critical Assessment.* Amsterdam: North-Holland.

Holmes, T., and W. Adamowicz. 2003. Attribute-Based Methods. In *A Primer on Non-Market Valuation,* edited by Kevin J. Boyle and Patricia A. Champ. Boston: Kluwer Academic Publishers.

Knetsch, Jack L., and Robert K. Davis. 1966. Comparisons of Methods for Recreation Evaluation. In *Water Research,* edited by Allen V. Kneese and Stephen C. Smith . Baltimore, MD: The Johns Hopkins University Press for Resources for the Future.

Kopp, Raymond J., Werner W. Pommerehne, and Norbert Schwarz (eds.). 1997. *Determining the Value of Non-Marketed Goods: Economic, Psychological, and Policy Relevant Aspects of Contingent Valuation Methods.* Boston: Kluwer Academic Publishers.

Kopp, Raymond J., and V. Kerry Smith. 1997. Constructing Measures of Economic Value. In *Determining the Value of Non-Marketed Goods: Economic, Psychological, and Policy Relevant Aspects of Contingent Valuation Methods,* edited by Raymond J. Kopp, Werner W. Pommerehne, and Norbert Schwarz. Boston: Kluwer Academic Publishers.

Lareau, Thomas J., and Douglas A. Rae. 1989. Valuing WTP for Diesel Odor Reductions: An Application of Contingent Ranking Technique. *Southern Economic Journal* 55(3): 728–742.

List, John A., and Craig A. Gallet. 2001. What Experimental Protocol Influence Disparities between Actual and Hypothetical Stated Values? *Environmental and Resource Economics* 20(3): 241–254.

McConnell, Kenneth E. 1986. *The Damages to Recreational Activities for PCBs in New Bedford Harbor.* Cambridge, MA: Industrial Economics.

———. 1990. Models for Referendum Data: The Structure of Discrete Choice Models for Contingent Valuation. *Journal of Environmental Economics and Management* 18(1): 19–34.

Mitchell, Robert Cameron, and Richard T. Carson. 1989. *Using Surveys To Value Public Goods: The Contingent Valuation Method.* Washington, DC: Resources for the Future.

Morrison, Mark, Jeff Bennett, and Russell Blamey. 1999. Valuing Improved Wetland Quality Using Choice Modeling. *Water Resources Research* 35(9): 2805–2914.

Portney, Paul R. 1994. The Contingent Valuation Debate: Why Economists Should Care. *Journal of Economic Perspectives* 8(4): 3–17.

Rae, Douglas A. 1983. The Value to Visitors of Improving Visibility at Mesa Verde and Great Smoky Mountain National Parks. In *Managing Air Quality and Scenic Resources and National Parks and Wilderness Areas,* edited by Robert D. Rowe and Lauraine G. Chestnut. Boulder, CO: Westview Press.

Roe, Brian, Kevin J. Boyle, and Mario Teisl. 1996. Using Conjoint Analysis To Derive Estimates of Compensating Variation. *Journal of Environmental Economics and Management* 31(2): 145–159.

Rowe, Robert D., William D. Schulze, and William S. Breffle. 1996. A Test for Payment Card Biases. *Journal of Environmental Economics and Management* 31(2): 178–185.

Sellar, Christine, Jean-Paul Chavas, and John R. Stoll. 1986. Specification of the Logit Model: The Case of Valuation of Nonmarket Goods. *Journal of Environmental Economics and Management* 13(4): 382–390.

Smith, V. Kerry, and William H. Desvousges. 1986. *Measuring Water Quality Benefits.* Norwell, MA: Kluwer-Nijhoff.

Smith, V. Kerry, and Laura L. Osborne. 1996. Do Contingent Valuation Estimates Pass a "Scope" Test? A Meta-Analysis. *Journal of Environmental Economics and Management* 31(3): 287–301.

Thayer, Mark A. 1981. Contingent Valuation Techniques for Assessing Environmental Impacts: Further Evidence. *Journal of Environmental Economics and Management* 8(1): 27–44.

Tversky, Amos, Paul Slovic, and Daniel Kahneman. 1990. The Causes of Preference Reversal. *American Economic Review* 80(1): 204–217.

U.S. National Oceanographic and Atmospheric Administration. 1993. *Report of the NOAA Panel on Contingent Valuation.* http: //web.lexis-nexis.com/congcomp/printdoc (accessed August 12, 2002).

Whitehead, John C. 2002. Incentive Incompatibility and Starting-Point Bias in Iterative Valuation Questions. *Land Economics* 78(2): 285–297.

CHAPTER

7

Aggregation of Values across Time

For purposes of economic planning, management, and policymaking, it may often be useful to view environmental and natural resources as assets that yield flows of services over time. We have now developed the theoretical framework necessary for defining and measuring the values of these service flows to the individuals who receive them. However, this theory of value has been developed without explicit consideration of the temporal dimension of these service flows. Specifically, the theory has been applied to the values of services only at a single point in time or over some relatively short period of time (a day, a week, or a year) during which it is reasonable to treat other things such as income, prices, and the consumption of other goods and services as fixed. The theory has defined values only at the time at which the service flow is received. Thus we are left with the question of how to aggregate values that are realized at different times and that might vary from one moment to another.

In this chapter I deal explicitly with the temporal dimension of value and welfare theory. In the first section, I review the standard theory of individual preferences and choice in an intertemporal setting. In the absence of taxes and other capital market imperfections, utility-maximizing individuals borrow or lend so as to equate their marginal rate of substitution between present and future consumption with the market rate of interest. Thus the interest rate can be taken as a revealed preference indicator of individuals' intertemporal marginal rates of substitution.

In the second section, I show that an individual's marginal rate of substitution can be used to convert a marginal welfare measure for one moment in time into its equivalent at any other point in time. Measures for several different points in time can be made commensurate by converting them all to one period, for example, the present. Then these measures can be added to obtain an intertemporal aggregate welfare measure.

After reviewing the theory of intertemporal choice and using this theory to develop measures of intertemporal welfare change, I turn in the third section to the question of choosing an interest rate in a world with many different market interest rates and other intertemporal prices. The multitude of market interest rates arises because of inflation, taxation, and risk, among other things. As a consequence, no single number can be identified as the "correct" interest rate for intertemporal welfare measurement. However, it is possible to suggest a reasonable range that will be appropriate in many circumstances.

One consequence of the multiplicity of interest rates is that environmental policies that divert resources from other investments with high rates of return may have intertemporal opportunity costs at the margin that are different from individuals' intertemporal marginal rates of substitution. In the fourth section, I briefly describe one approach to calculating these opportunity costs and making them commensurate with the measure of welfare gains developed in this chapter.

The theory of intertemporal welfare measurement presented here has been developed to examine the welfare implications of changes that affect one individual at different points in time. In the last section, I take up the question of interpersonal welfare comparisons across time and focus attention on those cases where costs and benefits accrue to different generations. The specific issue is intergenerational equity and discounting.

Individual Preferences and Intertemporal Choice

The theory of preferences that we have been working with up to this point models only single-period choices and makes single-period utility a function of only the single-period consumption of marketed goods and services and environmental quality. With time introduced explicitly, the single-period utility in time period t is given by

$$u_t = u_t(X_t, q_t) \tag{7-1}$$

where X_t measures the expenditure on consumption in t (X_t is a numeraire good with a unit price) and q_t is the level of the environmental or resource service flow at t. This model has obvious limitations in that it cannot be used to analyze intertemporal choices and such things as saving and lending.

The alternative is a model of lifetime utility. This general representation of preferences would have an individual at any point in time experiencing a level of "lifetime" utility from the goods and services and environmental quality that person anticipates over his or her remaining lifetime:

$$u^* = u^*(X_1, \ldots, X_t, \ldots, X_T, q_1, \ldots, q_t, \ldots, q_T) \tag{7-2}$$

where T is the number of years of remaining life. For simplicity, this formulation implies that the individual gains no satisfaction from making a bequest to others at the end of his or her life. However, it is a straightforward matter to add a bequest motivation if that is an important part of the problem being analyzed (see, for example, Cropper and Sussman 1988). Also in this analysis, we abstract from all considerations of uncertainty concerning the future.

The existence of single-period preferences as described by equation 7-1 implies that the lifetime utility function (equation 7-2) is separable. Brekke argues that this is a dubious assumption because the utility of X_t and q_t could plausibly depend on the past or future levels of either variable for a variety of reasons (1997, 93, 108–113). If we accept Brekke's argument, this means that the expressions for the marginal values of q_t derived later in this chapter would have to include as arguments all past and future levels of X_t and q_t. This would vastly complicate the empirical task of estimation. Despite Brekke's argument, I will continue in the tradition of standard welfare analysis and assume that as a pragmatic simplification lifetime utility can be written thus:

$$u^* = u[u_1(X_1, q_1), \ldots, u_t(X_t, q_t), \ldots, u_T(X_T, q_T)] \tag{7-3}$$

It is common to assume that this expression is additively separable and that the single-period utility function is the same for all periods. Thus we can write

$$u^* = \sum_{t=1}^{T} D_t \cdot u(X_t, q_t) \tag{7-4}$$

where $u(\cdot)$ is invariant in t, $\partial u^*/\partial X_t$ is positive and decreasing in X_t, and where D_t is meant to capture the individual's time preference.* For example, if an individual intrinsically preferred present consumption over future consumption, D_t would decrease with t and the marginal lifetime utility of X_t would be a decreasing function of t.

*See Trostel and Taylor 2001 for an interesting model of time preference that makes $u(\cdot)$ a function of age.

A useful way to characterize time preference is to let $D_t = 1/(1 + d)^t$ where d can be interpreted as a subjective rate of time preference analogous to an interest rate. This form of constant d and geometrically decreasing D_t over time is necessary for intertemporal consistency of individual choice (Strotz 1956; Page 1977). However, experiments in which people are offered either hypothetical or real choices between payments of different sizes and dates consistently reveal implicit discount rates that are both high for short periods of time and declining as the interval lengthens (Loewenstein and Thaler 1989; Ainslie 1991; Cropper et al. 1994). This evidence implies nongeometric discounting. For discussions of the implications of nongeometric or time-inconsistent discounting for individual behavior and the evaluation of public policies, see Harvey 1994; Laibson 1997; O'Donoghue and Rabin 1999; and Cropper and Laibson 1999. For a critical review of the literature on time preference and individual discounting, see Frederick et al. 2002.

Given exponential discounting, the relationship between d and time preference is as follows:

$d > 0$ represents positive time preference with the marginal utility of X_t decreasing with t;
$d = 0$ represents neutral time preference; and
$d < 0$ represents negative time preference with the marginal utility of X_t increasing with t.

Olson and Bailey (1981) argue that empirical evidence points to $d > 0$ as the most common pattern of individual preference. As they also usefully make clear, an individual's lifetime marginal utility of X_1 $(t > 1)$ could be less than the lifetime marginal utility of X_1 for one (or both) of two reasons: X_t is greater than X_1, or D_t is less than one $(d > 0)$.

Now assume that the individual knows with certainty that he or she will receive a stream of income payments, M_t, and a stream of environmental services, q_t, and that the individual can borrow or lend in a perfect capital market at an interest rate of r% per period. Assume that wealth is measured as of the beginning of the first period and that income and expenditures are made at the end of each period. Then, if the individual does not wish to leave a bequest and assuming that all accounts must be settled in year T, the individual's lifetime budget constraint is

$$\sum_{t=1}^{T} X_t \cdot (1+r)^{-t} = W^* = W_0 + \sum_{t=1}^{T} M_t \cdot (1+r)^{-t} \qquad (7\text{-}5)$$

where W^* is lifetime wealth and W_0 is initial wealth. Knowledge of the initial wealth, the streams of income and environmental services, and the interest rate is sufficient to determine the lifetime pattern of consumption of X that maximizes lifetime utility. In each period, if X_t is greater than (or less than) M_t, the individual must borrow (or lend) the difference.

The first-order conditions for maximizing equation 7-4 subject to the wealth constraint (equation 7-5) take the form

$$\frac{(\partial u^* / \partial X_t)}{(\partial u^* / \partial X_{t+1})} = 1 + r \tag{7-6}$$

or, because $\partial u^*/\partial X_t = D_t \cdot \partial u/\partial X_t = (1 + d)^{-t} \partial u/\partial X_t$,

$$(1+d)\frac{(\partial u / \partial X_t)}{(\partial u / \partial X_{t+1})} = 1 + r \quad \text{for all } t = 1, \ldots, t, \ldots, T-1 \tag{7-7}$$

The individual's intertemporal marginal rate of substitution must be equal to one plus the interest rate. Thus the individual's intertemporal marginal rate of substitution can be inferred by observing the interest rate that governs these intertemporal trade-offs. The first-order conditions imply that if both r and d equal 0, the individual will equate the single-period marginal utility of consumption across all periods, thus choosing a constant stream of consumption over time. If $r = 0$ and $d > 0$, the single-period marginal utility of consumption in t must be less than that in $t + 1$, implying that the individual chooses to consume relatively more in period t. An increase in r, other things being equal, requires adjustments to increase the marginal utility of consumption in earlier periods, thus implying a deferral of consumption to future time periods.

Measures of Welfare Change

We turn now to the development of measures of intertemporal welfare change. First, consider a marginal increase in q_t. The individual would be willing to accept a reduced quantity of the numeraire in that period to hold utility constant. This compensating surplus measure of welfare change is the marginal rate of substitution between q_t and X_t, or

$$w_{qt}^t = \frac{(\partial u / \partial q_t)}{(\partial u / \partial X_t)} \tag{7-8}$$

From equations 7-6 or 7-7, in intertemporal equilibrium the individual would be indifferent between paying w_{qt}^t in period t and paying

$$w_{qt}^{0} = \frac{w_{qt}^{t}}{(1+r)^{t}} \tag{7-9}$$

now. If the whole stream of future Q increases, the marginal willingness to pay now is the sum of the willingnesses to pay for each of the components of the increase, or

$$w_{Q}^{0} = \sum_{t=1}^{T} \frac{w_{qt}^{t}}{(1+r)^{t}} \tag{7-10}$$

If the changes in $\boldsymbol{Q}$ are nonmarginal, aggregation over time is not so simple. As in Chapter 3, we can define single-period compensating surplus (CS) and equivalent surplus (ES) welfare measures for changes in any q_t. These are changes in single-period income based on single-period utility. In an analogous fashion, we can use equations 7-2 or 7-3 combined with equation 7-5 to define compensating and equivalent changes in initial wealth that are based on lifetime utility. The relationship between these two alternative perspectives on the measurement of welfare change is of interest.

To keep it simple, consider a policy that would result in a change in q during only one period. Specifically, suppose that the policy results in $q_t'' > q_t'$. What is the welfare gain to the individual from this change? The single-period CS measure is the solution to

$$u(X_t - CS_t, q_t'') = u(X_t, q_t') \tag{7-11}$$

The present value of the single-period measure of gain is $(1 + r)^{-t}CS_t$.

To derive the compensating and equivalent wealth measures of lifetime welfare change, we first use the solution to the lifetime utility maximization problem to obtain the lifetime indirect utility function:

$$v^*(W^*, q_1, \ldots, q_t, \ldots, q_T) \tag{7-12}$$

From the perspective of lifetime utility, the individual is willing to make a payment at $t = 0$ that will equate the lifetime utilities with and without the change in q_t. Call this the compensating wealth, or CW. It is the solution to

$$v^*(W^* - CW, q_1, \ldots, q_t'', \ldots, q_T) = v^*(W^*, q_1, \ldots, q_t', \ldots, q_T) \tag{7-13}$$

Following the analysis by Blackorby and others (1984), it can be shown that the present-value welfare measures based on equations 7-11 and 7-13 are different except under certain special conditions. Specifically,

$$|CW| \geq |(1+r)^{-t} CS_t| \tag{7-14}$$

This result can be easily explained. If the individual pays CS in period t, then according to equation 7-11 single-period utility is restored to its initial level, and so is lifetime utility. However, if the single-period marginal utility of consumption in t is affected by the change in q_t, then the first-order conditions for an optimum allocation of consumption over time will not be satisfied. The individual will wish to reallocate consumption across time through some pattern of additional borrowing and lending. Such reallocation of consumption will increase utility. This means that the individual could have made a larger single-period payment to keep lifetime utility at its original level. This establishes the inequality of equation 7-14.

The equality will hold only if the marginal utility of consumption is invariant to the combined changes in q_t and its offsetting CS_t. For example, suppose that the increase in q_t causes a decrease in the marginal utility of consumption in that period. The payment of CS_t will increase the marginal utility of consumption in period t. These two forces might exactly offset each other, leaving the marginal utility of consumption and the optimal intertemporal pattern of consumption unchanged. However, if the single-period utility function is separable in q_t so that the marginal utility of consumption is unchanged by the change in q_t, or if the increase in q_t leads to an increase in the marginal utility of consumption, then the net effect of the change in q_t and the offsetting compensation payment will be to increase the marginal utility of consumption in period t and to cause a reallocation of consumption across time.

Similar inequalities can be derived for equivalent surplus and equivalent wealth (EW) measures of welfare gain and for both types of measures for welfare losses associated with a decrease in q_t. For example, suppose that q_t decreased. If the individual instead had to pay ES_t as the equivalent measure of welfare loss, and if this resulted in an increase in the marginal utility of consumption in t, then the individual could restore part of the loss of lifetime utility by reallocating spending toward period t, and ES_t would be an overestimate of the lifetime welfare loss.

These results can be summarized as follows:

For gains (q_t increases)	*For losses (q_t decreases)*
$CW \geq (1+r)^{-t} CS_t$	$\lvert CW \rvert \leq \lvert (1+r)^{-t} CS_t \rvert$
$EW \leq (1+r)^{-t} ES_t$	$\lvert EW \rvert \geq \lvert (1+r)^{-t} ES_t \rvert$

All of these expressions can be easily generalized to the case where there are changes in $\boldsymbol{Q}$ in many periods.

These results have somewhat disturbing implications for applied welfare analysis. The typical practice in evaluating environmental policies that affect welfare over many years is to estimate single-period welfare measures on the basis of some indirect technique and to compute the present value of these single-period changes using some interest rate. However, as shown here, if *CS* measures of single-period welfare change are used, this procedure will underestimate the true lifetime welfare benefit of the stream of improvements in Q and overstate the loss of decreases in Q. If the present value of the single-period CS_t is greater than zero, then the project unambiguously passes a Kaldor potential compensation test. However, projects with negative aggregate CS_t could still be efficient. Conversely, if *ES* single-period measures are used, the conventional practice leads to upward-biased estimates of total welfare gains and downward-biased estimates of lifetime losses. Thus if the sum of the single-period ES_t is less than zero, the project unambiguously fails a Hicks potential compensation test. But projects with positive aggregate present values of single-period ES_t might decrease aggregate welfare (Blackorby et al. 1984).

The important question is, how big is this bias? Keen (1990) shed some light on this question. He showed that the difference between the single-period and the lifetime welfare measures was a term that he called the *intertemporal compensating variation,* or the maximum amount the individual would be willing to pay to be able to reallocate expenditure and utility across periods. Keen then derived expressions for this term and showed that in most instances the term would be of only second-order importance. In general, the error will be significant only if high values for the elasticity of intertemporal substitution combine with large values of CS_t over time.

Although most of the indirect methods for measuring welfare change produce only single-period measures, sometimes it is possible to obtain direct unbiased estimates of *CW* or *EW*. If changes in Q affect the market price of an asset, the change in its price might provide a basis for estimating *CW* or *EW*. For example, the hedonic price method is based on differences in the prices of assets such as houses. Because differences in the market prices for the housing asset are observed, price differentials between low- and high-quality houses reflect the compensating wealth change associated with the difference in the expected stream of environmental quality associated with the house. It is common practice in applied welfare analysis to convert the *CW* measure to an annual equivalent by using some interest rate or discount factor. Provided that the interest rate properly represents the individual's opportunities for intertemporal substitution, this annual equivalent can be used as an unbiased indicator of lifetime welfare change.

To summarize, if an individual can reallocate his stream of consumption over time through transactions at some interest rate r, that person will equate his intertemporal marginal rate of substitution with a discount factor equal to $1 + r$. This means that r can be interpreted as an intertemporal price or measure of intertemporal value, at least for marginal changes.

For nonmarginal changes in Q, the correct measure of intertemporal welfare change is the compensating or equivalent change in initial wealth. However, many of the available methods for measuring welfare change yield estimates of single-period values rather than lifetime values; the present value of these single-period welfare measures is, at best, only an approximation (an upper or lower bound) on the lifetime welfare measure. Given the limited availability of data, the approximation may be all that is obtainable.

Which Interest Rate Is the Right Intertemporal Price?

How can we estimate an individual's rate of time preference? Standard economic theory predicts that in a simple economy with no taxes and with perfect capital markets in which all individuals can borrow or lend at the market rate of interest, an individual will arrange her time pattern of consumption by borrowing and lending so that at each point in time her rate of time preference is just equal to the observed market rate of interest (Lind 1982; Varian 1990). In such a simple economy, this is the interest rate that should be used for the intertemporal aggregation of individuals' single-period welfare measures. In fact, though, there are taxes (certainly), inflation (at least sometimes), and various market imperfections. We also observe a multiplicity of market interest rates. This section briefly discusses the choice of an interest rate for welfare evaluation when factors such as inflation and taxation are present and suggests some approaches to choosing an intertemporal price when faced with many rates of interest and tax rates.

Consider first the problem of the effects of taxes on income. Taxes can drive wedges between market prices and individuals' marginal rates of substitution. Suppose that an individual's interest income is taxed at the rate of $t\%$. Thus money lent at $r\%$ will return only $r \cdot (1 - t)\%$ net of tax. The optimizing lender will equate his intertemporal marginal rate of substitution to this after-tax rate of return. Similarly, if interest expense is treated as deductible in calculating tax liabilities, then the net cost of borrowing at a market rate of r is $r \cdot (1 - t)\%$. Again, this after-tax net borrowing cost will govern the individual's borrowing decisions. Discounting for welfare evaluation should be done at this after-tax interest rate.

A second problem is inflation. If there is price inflation, then the distinction between real and nominal interest rates becomes important. The individual would still arrange her time pattern of consumption so as to equate her rate of time preference with the expected real rate of interest. If inflation of $i\%$ is perfectly anticipated, then the observed market interest rate m and the real rate of interest r are related as follows:

$$\frac{(1+m)}{(1+i)} = 1 + r \quad \text{or} \quad m = r + i + (r \cdot i) \tag{7-15}$$

Because for reasonable rates of inflation and real interest rates, the term $r \cdot i$ in equation 7-15 will be small and can be ignored, the real rate is (approximately) the market interest rate less the rate of inflation. If the estimates of individuals' single-period CS_t are based on the expectation of actual prices in that period, then discounting should be done at the market rate of interest, which will also reflect the anticipated rate of inflation, i. However, the conventional practice in benefit measurement is to abstract from inflation by basing estimates of future values of CS_t on the present price level. In that case, discounting should be done at the real rate interest.

What is a reasonable value for this real rate of interest? One range of estimates comes from Lind (1982). After reviewing historical evidence on rates of return to a variety of investments, Lind concluded that rates of time preference could be as low as 0%, if historical real rates of return on U.S. Treasury bills were taken as the bench mark (Lind, 1982, 84). Over this same period, the real after-tax return on a broad portfolio of common stocks was about 4.6% (Lind 1982, 83). This could be taken as an upper bound on the appropriate discount rate and would be the correct discount rate if the payments or receipts being discounted had the same risk characteristics as the market portfolio of common stocks.

Given the increasing openness of financial markets and the ease of moving funds between nations, world average interest rates may be more relevant than rates from any one nation's financial market (Lind 1990). Barro and Martin (1990) presented estimates of an index of real expected and real actual short-term interest rates for nine major industrialized nations, including the United States. For more than two-thirds of the period studied, average pretax real rates of interest were below 2%, and they never exceeded 6%. At least in the United States, real rates have been in the lower half of this range during the past decade. These results are broadly consistent with Lind's conclusions (Lind 1982). They also demonstrate the variability of real rates of interest over time.

Rather than look at market rates of interest, an alternative approach is to estimate the implicit interest rate that must be guiding individuals' observed choices over alternative temporal patterns of benefits and costs. For example, suppose an individual who has a choice between receiving $10 today and $15 in one year chooses the smaller immediate payment. This implies that the individual discounts future consumption at a rate of at least 50%. Empirical studies on choices of this sort often reveal unusually high implicit interest rates—more than 20%, and up to and above 100%. Several people have estimated the discount rates implied by the choice to trade off price against lifetime operating costs when more energy-efficient appliances have higher purchase prices (Hausman 1979; Gately 1980; Ruderman et al. 1986). These implied rates must be far higher than the effective interest rates available to the people making the choices. See also Frederick et al. 2002 for a review of this evidence.

On the other hand, Moore and Viscusi (1990) estimated the discount rates implied by the way workers make trade-offs between higher wages now and loss of future years of life through accidental death on the job. They found implicit discount rates in the range of 2–12%, depending on the data source and the model specification. They concluded that *"Overall, we find no evidence of a significant discrepancy between rates of discount for health risks and financial rates of time preference"* (Moore and Viscusi 1990, 5–52; emphasis in the original).

Overall, the evidence from financial markets suggests that individuals' real after-tax rate of interest (which can vary across individuals) lies in the range of 1–4%. This is much lower than the 7% discount rate mandated for use by federal agencies (U.S. Office of Management and Budget 1992). However, EPA's recommended rate for conducting economic analyses of its regulations (2–3%) lies within this range (U.S. EPA 2000, 48). I am comfortable with the EPA recommendation.

Even if we have estimates of market rates of interest or implicit discount rates from revealed behavior, another problem arises in the calculation of welfare measures because of the fact that, for a variety of reasons, people face different interest rates. For example, individuals' real after-tax rates of interest will differ because individuals face different marginal tax rates and have different portfolios of investments. Differences in effective interest rates can also arise from market imperfections, transaction costs, and differences in the tax treatment of interest paid and interest received. For example, under current federal tax law most interest income is taxable, but interest paid on consumer loans is not tax-deductible. How should we deal with differences in effective interest rates across individuals? In principle, each individual's single-period welfare changes should be discounted at the interest rate that the person uses to make intertemporal allocation decisions. However, in practice, we typically have single-period welfare measures that are aggregated across individuals who face quite different rates of interest. Here I will sketch a procedure for calculating aggregate values in the second-best situation, where individuals face different effective interest rates at the margin.

Assume that no individual borrows and lends at the same time and that at any time each individual's transactions are entered into at a single interest rate that may vary across individuals. Consider an individual who at period t could borrow for one period at r_{bt} or could lend at another rate r_{st}, where both are real after-tax rates of interest. The individual, in effect, faces a kinked budget line. Whichever direction the individual chooses to move, the interest rate governing that form of transaction will be the one with which that person equates his or her intertemporal marginal rate of substitution. That interest rate is the one that should be used to discount that individual's welfare measure over that interval of time. So if the individual is a net lender between period t and period $(t + 1)$, and there is a change in q_{t+1}, the after-tax lending rate should be used to discount CS_{t+1} back to t. If the individual neither borrows nor lends, there is no transaction from which to infer a marginal rate of substitution.

Now consider aggregation across individuals who face different effective interest rates at any point in time, for example, because of differences in marginal income tax rates or differences in borrowing and lending behavior. Specifically, assume that the ith individual faces an effective real interest rate r_i that is constant over time. Assume then that we wish to calculate the sum of the CW payments that would leave each individual in his or her initial lifetime utility position. For individual i, the lifetime CW_i for a change in $\boldsymbol{Q}$ is approximated by the present value of the stream of single-period CSs:

$$CW_i \simeq \sum_{t=1}^{T} \frac{CS_{it}}{(1+r_i)^t} \tag{7-16}$$

The aggregate change for the n affected individuals is

$$\overline{CW} \simeq \sum_{i=1}^{n} \sum_{t=1}^{T} \frac{CS_{it}}{(1+r_i)^t} \tag{7-17}$$

There is a simple way of calculating a weighted-average discount factor that can be used with equation 7-17. First, each individual's share of each period's aggregate compensating surplus is calculated:

$$s_{it} = \frac{CS_{it}}{\left(\sum_{i=1}^{n} CS_{it}\right)} \tag{7-18}$$

Then this expression is used to substitute for CS_{it} in equation 7-17 to obtain

$$\overline{CW} \simeq \sum_{i=1}^{n} \sum_{t=1}^{T} s_{it} \frac{\sum CS_{it}}{(1+r_i)^t} \tag{7-19}$$

The preceding analysis is based on the assumption that each individual enters into transactions at only one interest rate in any one period. However, introspection and anecdotal evidence suggest that this assumption is often violated in practice. Many people simultaneously save by contributing to pension funds and borrow to buy new houses and consumer durables. They may also earn after-tax rates of return on savings instruments that are only a third or a quarter of their effective borrowing costs on outstanding credit card balances. This behavior raises two questions. First, how can this behavior be reconciled with the assumption of economic rationality? Second, what does this behavior imply for inferring intertemporal marginal rates of substitution and making intertemporal welfare measurements?

Lind (1990) has suggested that recent developments in behavioral economics can help to explain these practices. One key paper in this literature (Shefrin and Thaler 1988) hypothesizes that people place different types of income, expenditures, and assets into different mental accounts and that they make different intertemporal allocation decisions for different sets of accounts. If it should prove possible to identify unique marginal rates of substitution or implicit discount rates for each mental account, and if it is possible to discover to which mental accounts people assign changes in environmental quality, then it may prove possible to develop a consistent discounting framework for intertemporal welfare measurement. However, this possibility should be considered only as speculation and a possible area for future research.

In summary, the existence of multiple interest rates in the marketplace does not preclude the consistent intertemporal aggregation of individuals' welfare measures. In principle, each individual's set of welfare measures is aggregated across time, using that individual's intertemporal marginal rate of substitution as revealed by the interest rate at which that person transacts. In practice, there is a weighted-average discount factor that reproduces this calculation. The case of individuals who simultaneously transact at different interest rates is more problematic, but the concept of mental accounts may help to explain this behavior and may also provide a basis for calculating the relevant intertemporal marginal rate of substitution.

The Capital Costs of Environmental Policies

A major issue in the economic evaluation of public investment projects has been how to take account of the fact that financing public investment through some combination of taxes and borrowing is likely to displace some private investment that has a higher rate of return than the effective interest rate governing individuals' intertemporal substitutions. The divergence between the rate of return on private investment and the social discount rate can be attributed at least in part to the taxes on corporate and personal income. In this section, I briefly review the "shadow-cost-of-capital" approach to measuring the cost of public investment projects. I then discuss the implications of the shadow cost of capital for the economic evaluation of public policies that require private investments that may displace other private investments.

Consider a public investment project with an initial capital cost of K_0. Assume that no further operating or maintenance costs are associated with the project. Let the project yield benefits of B per year in perpetuity. As before, let r represent the effective after-tax rate of interest governing all individuals' borrowing and lending decisions. Because r will reflect individuals' willingness to trade off present consumption for future consumption, it should be used as the intertemporal price for welfare evaluation, given the individualist welfare perspective adopted in this book.* Let s represent the real marginal rate of return on investments in the private sector. Because of taxes on capital income at both the corporate and personal level, $s > r$.

Using r, we can calculate the present value of the perpetual stream of benefits as a lower bound to the aggregate compensating wealth payment that is our desired lifetime welfare indicator. However, if we find $B/r > K_0$, this does not necessarily signal that the project will improve welfare because the relevant welfare comparison is not between the present value of the stream of benefits and the capital cost of the project. Rather, the relevant comparison is between the present value of the benefits of the project and the opportunity cost of the project as reflected in the stream of consumption forgone. In other words, the capital cost must be converted to its consumption equivalent for comparison with the benefit stream.

If K_0 were invested at the marginal rate of return in the private sector, it would produce a perpetual stream of $s \cdot K_0$ per year of future consumption. This is what is lost by diverting K_0 of capital resources from private investment to public investment, and this is the opportunity cost of the public investment. The present value of the stream of future consumption forgone is

$$\frac{s \cdot K_0}{r} \tag{7-20}$$

Thus in this simple example the shadow price or true social opportunity cost of one dollar of capital diverted from private investment is s/r, which is greater than one. In practice, the shadow price of capital will depend on a variety of factors, reflecting the extent to which private investment is dis-

*For an alternative perspective on the social rate of discount, see Marglin 1963a.

placed, on net. If no private investment is displaced, the shadow price is one. More realistic formulations of the shadow price expression would take into account the facts that public investment will not necessarily displace private investment on a dollar-for-dollar basis, that public and private investments have less than infinite lifetimes, and that some portion of the future returns from displaced private investment could be reinvested for future consumption rather than being consumed immediately. For a review of some of these issues and some estimates of the magnitude of the shadow price, see Lind 1982. Earlier important references include Bradford 1975 and Marglin 1963b.

To summarize, the shadow price of capital is the present value (discounted at the individual rate of time preference, r) of the stream of future consumption forgone from one dollar of public investment. The proper procedure for project evaluation is to discount benefits (and operating and maintenance costs, if any) at the social rate of discount. The present value of the benefit stream would then be compared with the social cost of the capital investment. The latter would be calculated by multiplying the dollar cost of the investment by the appropriately calculated shadow price of capital.

Now suppose there is a pollution control regulation that, instead of requiring public investment, requires private firms to invest capital in pollution control. As in the case of public investment, the relevant opportunity cost is the present value of displaced private consumption. The private investment can affect consumption through a variety of channels. Here I present a very simple model to illustrate the basic principles. For more detailed analyses, see Kolb and Scheraga 1990 and Lesser and Zerbe 1994.

Suppose also that this regulation will yield environmental improvements with an aggregate value of B per year in perpetuity. Once again, the present value of the benefit stream from the regulation is B/r. Assume that it is estimated that firms can comply with the regulation by installing equipment with an aggregate capital cost of K_0, that there are operating, maintenance, and repair costs of R per year, and that the equipment will last forever. And suppose further that firms face infinitely elastic factor supply curves for all factor inputs and have constant returns-to-scale production functions. This assumption ensures that all the costs of complying with the regulation will be passed on to consumers as higher prices.

Because under these assumptions the aggregate supply of capital is infinitely elastic, the regulation does not result in any displacement of investment in the economy. The cost of the regulation comes in the form of higher prices to consumers. Prices must be raised sufficiently to amortize the investment of K_0 and to cover the operating costs of R per year. Ignoring the effects of higher prices on demand, the required price increase must generate additional revenues of $s \cdot K_0 + R$. This is the annual cost of the regulation to consumers. To calculate the present value of this stream of costs, the effective consumers' interest rate, r, should be used. The present value of the social cost to consumers is approximated by $(s/r) \cdot K_0 + R/r$. In effect, the capital cost is first annualized using the pretax marginal rate of return on private investment (s) and then converted back to a present value using the consumption rate of interest (r).

To take account of the effects of higher prices on demand and output, it is necessary to solve for the K_0 that is required to comply with the regulation after taking account of the effect of higher prices on output and the need for capital. For example, if some firms are expected to exit the industry, then fewer firms would have to make investments. The aggregate capital requirement would be smaller than if output and the number of firms were assumed to be unchanged. Also, the conceptually correct measure of cost is the *CV* of the consumers who must pay the higher prices.

Discounting and Aggregation across Generations

We have seen that the aggregation of an individual's single-period welfare measures over time can be done using a discount factor that represents the individual's marginal rate of intertemporal substitution or rate of time preference. For rational individuals this rate will be equal to the relevant after-tax real rate of interest. The ethical justification for discounting benefits and costs that accrue within one generation's lifetime lies in the observation that this discount rate is a reflection of the individuals' preferences for their own present consumption relative to future consumption.

However, as is well known, discounting future benefits or costs over the lifetimes of two or more generations is controversial. Applying any discount rate to a future sum can make it look very small in present-value terms when the interval is 100 years or more. For example, at 2%, the present value of $1 in 100 years is about 14 cents; if the $1 comes in 200 years, its present value is only ½ cent at 2%.

The principal argument for discounting for economic welfare analysis also provides some insight into why one might choose not to make discounting calculations across generations. Recall that what is being discounted is individuals' willingness to pay or required compensation, either for a change (compensating measure) or to avoid a change (equivalent measure). The ethical justification for making these welfare calculations is their use in applying either the Kaldor or Hicks form of compensation test to determine whether it is possible to compensate the losers so that no one is made worse off. When the gainers and losers are part of the same generation, actual compensation is feasible, and in any event both groups can participate in the decision. If a proposed policy would impose costs on a future generation, it is hard to imagine mechanisms for ensuring that the compensation would be paid if it were thought desirable or necessary to obtain the consent of future generations. Also, the future generation has no voice in present decisions. Similar problems arise in the case of a policy that would yield benefits to the future but impose costs on the present. The real problem with discounting across generations is not, then, that it results in very small numbers for the present values of future effects; rather, it is that it is hard to imagine a meaningful role for measures of compensation in the context of intergenerational resource reallocations.

So what is to be done? One place to start is with the recognition that, just as in the case of interpersonal welfare comparisons within a generation, a social welfare function is required to make comparisons across generations. One approach to defining a social welfare function has its origins in work by Frank Ramsey nearly 75 years ago (Ramsey 1928). For a more recent restatement and discussion, see Arrow et al. 1996. In this formulation, the social rate of discount, r_s, emerges as the sum of two components. The first is the social rate of time preference, ρ, which reflects society's judgement regarding the relative intrinsic deservingness of different generations. The most ethically appealing judgement is that ρ is zero, meaning that all generations are treated as equally deserving. The second component reflects the expected relative economic position of the present and future generations. It is the product of the elasticity of the marginal utility of consumption, θ, and the expected rate of growth of future per capita consumption, g.

Combining these terms gives the following equation:

$$r_s = \rho + (\theta \cdot g) \tag{7-21}$$

The justification for the second term, $(\theta \cdot g)$, is that if per capita consumption is higher in the future, an extra dollar of consumption in the future will have a lower marginal utility to that generation and therefore should be discounted accordingly. Following this approach, with $\rho = 0$, a growth rate of consumption of 1% and an elasticity of 1.5 would yield a social discount rate of 1.5%. If growth in consumption were expected to be negative, equation 7-21 would produce a negative discount rate to give increments of consumption to future generations a greater social value than equal increments to the present generation.

Another approach to the problem of intergenerational discounting is to argue for a discount rate that is a decreasing function of the time interval over which discounting takes place, for example, hyperbolic discounting. For interesting rationales for a discount rate that declines with time, see Harvey 1994; Weitzman 1998; and Newell and Pizer 2001. Some of the questions surrounding intergenerational discounting were the subject of a workshop at Resources for the Future that has been reported in Portney and Weyant 1999. As the chapters in that volume show, this is a challenging set of issues, and economists have not yet reached a consensus on the appropriate answers.

References

Ainslie, George. 1991. Derivation of "Rational" Economic Behavior from Hyperbolic Discount Curves. *American Economic Review* 81(2): 334–340.

Arrow, K.J., W.R. Kline, K.-G. Mäler, M. Munasinghe, R. Squitieri, and J.E. Stiglitz. 1996. Intertemporal Equity, Discounting, and Economic Efficiency. In *Climate Change 1995: Economic and Social Dimensions of Climate Change,* edited by James P. Bruce, Hoesung Lee, and Erik F. Haites for Intergovernmental Panel on Climate Change. Cambridge, U.K.: Cambridge University Press.

Barro, Robert J., and Xavier Sala i Martin. 1990. World Real Interest Rates. *NBER Macroeconomics Annual* 5: 15–59.

Blackorby, Charles, David Donaldson, and David Moloney. 1984. Consumer's Surplus and Welfare Change in a Simple Dynamic Model. *Review of Economic Studies* 51(164): 171–176.

Bradford, David F. 1975. Constraints on Government Investment Opportunities and the Choice of Discount Rate. *American Economic Review* 65(5): 887–899.

Brekke, Kjell Arne. 1997. *Economic Growth and the Environment: On the Measurement of Income and Welfare.* Cheltenham, U.K.: Edward Elgar.

Cropper, Maureen L., Sema K. Aydede, and Paul R. Portney. 1994. Preferences for Life Saving Programs: How the Public Discounts Time and Age. *Journal of Risk and Uncertainty* 8(3): 243–265.

Cropper, Maureen L., and David Laibson. 1999. The Implications of Hyperbolic Discounting for Project Evaluation. In *Discounting and Intergenerational Equity,* edited by Paul R. Portney and John P. Weyant. Washington, DC: Resources for the Future.

Cropper, Maureen L., and Frances G. Sussman. 1988. Families and the Economics of Risks to Life. *American Economic Review* 78(1): 255–260.

Frederick, Shane, George Loewenstein, and Ted O'Donoghue. 2002. Time Discounting and Time Preference: A Critical Review. *Journal of Economic Literature* 40(2): 351–401.

Gately, Dermot. 1980. Individual Discount Rates and the Purchase and Utilization of Energy-Using Durables: Comment. *Bell Journal of Economics* 11(1): 373–374.

Harvey, Charles M. 1994. The Reasonableness of Non-Constant Discounting. *Journal of Public Economics* 53(1): 31–51.

Hausman, Jerry. 1979. Individual Discount Rates and the Purchase and Utilization of Energy-Using Durables. *Bell Journal of Economics* 10(1): 33–54.

Keen, Michael. 1990. Welfare Analysis and Intertemporal Substitution. *Journal of Public Economics* 42(3): 47–66.

Kolb, Jeffrey A., and Joel D. Scheraga. 1990. Discounting the Benefits and Costs of Environmental Regulations. *Journal of Policy Analysis and Management* 9(3): 381–390.

Laibson, David. 1997. Golden Eggs and Hyperbolic Discounting. *Quarterly Journal of Economics* 112(2): 443–477.

Lesser, Jonathan A., and Richard O. Zerbe. 1994. Discounting Procedures for Environmental (and Other) Projects: A Comment on Kolb and Scheraga. *Journal of Policy Analysis and Management* 13(1): 140–156.

Lind, Robert C. 1982. *Discounting for Time and Risk in Energy Policy*. Washington, DC: Resources for the Future.

———. 1990. Reassessing the Government's Discount Rate Policy in Light of New Theory and Data in a World Economy with a High Degree of Capital Mobility. *Journal of Environmental Economics and Management* 18(2, part 2): S-8–S-28.

Loewenstein, George, and Richard H. Thaler. 1989. Anomalies: Intertemporal Choice. *Journal of Economic Perspectives* 3(4): 181–193.

Marglin, Stephen A. 1963a. The Social Rate of Discount and the Optimal Rate of Investment. *Quarterly Journal of Economics* 77(1): 95–111.

———. 1963b. The Opportunity Costs of Public Investment. *Quarterly Journal of Economics* 77(2): 274–289.

Moore, Michael J., and W. Kip Viscusi. 1990. Discounting Environmental Health Risks: New Evidence and Policy Implications. *Journal of Environmental Economics and Management* 18(2): S-51–S-62.

Newell, Richard, and William Pizer. 2001. *Discounting the Benefits of Climate Change Mitigation: How Much Do Uncertain Rates Increase Valuation?* A Report for the Pew Center on Global Climate Change. Washington, DC: Resources for the Future.

O'Donoghue, Ted, and Matthew Rabin. 1999. Doing It Now or Later. *American Economic Review* 89(3): 103–124.

Olson, Mancur, and Martin J. Bailey. 1981. Positive Time Preference. *Journal of Political Economy* 89(1): 1–25.

Page, Talbot. 1977. *Conservation and Economic Efficiency: An Approach to Materials Policy.* Baltimore, MD: The Johns Hopkins University Press for Resources for the Future.

Portney, Paul R., and John P. Weyant. 1999. *Discounting and Intergenerational Equity.* Washington, DC: Resources for the Future.

Ramsey, Frank. 1928. A Mathematical Theory of Saving. *Economic Journal* 138(152): 543–559.

Ruderman, Henry, Mark Levine, and James McMahon. 1986. Energy Efficiency Choice in the Purchase of Residential Appliances. In *Energy Efficiency: Perspectives on Individual Behavior,* edited by Willett Kempton and Max Neiman. Washington, DC: American Council for an Energy-Efficient Economy.

Shefrin, Hersh M., and Richard H. Thaler. 1988. The Behavioral Life-Cycle Hypothesis. *Economic Inquiry* 26(4): 609–643.

Strotz, Robert. 1956. Myopia and Inconsistency in Dynamic Utility Maximization. *Review of Economic Studies* 23: 165–180.

Trostel, Philip A., and Grant A. Taylor. 2001. A Theory of Time Preference. *Economic Inquiry* 39(3): 379–395.

U.S. EPA (U.S. Environmental Protection Agency). 2000. *Guidelines for Preparing Economic Analyses.* EPA 240-R-00-003. Washington, DC: U.S. EPA.

U.S. Office of Management and Budget. 1992. *Guidelines and Discount Rates for Benefit–Cost Analysis of Federal Programs.* Circular no. A-94, revised. Washington, DC: U.S. Office of Management and Budget.

Varian, Hal R. 1990. *Intermediate Microeconomics: A Modern Approach.* 2nd ed. New York: Norton.

Weitzman, Martin L. 1998. Why the Far-Distant Future Should Be Discounted at Its Lowest Possible Rate. *Journal of Environmental Economics and Management* 36(3): 201–208.

CHAPTER

8

Values in an Uncertain World

Up to this point, the discussion in this book has been based at least implicitly on the assumption of perfect certainty. For example, individuals have been assumed to know what prices, income, and environmental quality will prevail over the relevant planning horizon. This clearly cannot be literally true. And for some questions, the essence of the matter is uncertainty about some event or condition, such as the probability of an accident or a natural disaster (say, a flood) occurring or the probability of contracting a serious illness following exposure to a chemical. The topic of this chapter is how to define and measure changes in economic welfare when uncertainty is an important characteristic of the economic world in which people are living and making choices.

The uncertainty faced by individuals can take many forms. For example, people could be uncertain about future prices of consumer goods, future income, or the returns on financial assets. However, because there is already an extensive literature on the welfare effects of price and income variability, I will not treat those problems here. See, for example, Rothschild and Stiglitz 1976; Just et al. 1982; and Helms 1985a, 1985b. The uncertainty that is important in this chapter is that faced by individuals who are users or potential users of some environmental service flow or resource. Individuals might be uncertain as to whether a specific environmental service flow or resource will be available for their use in the future; whether they will actually want to use some resource in the future; and whether their exposures to environmental hazards and pollutants will result in illness or death.

These uncertainties raise two kinds of questions for policymakers. The first is, what form of welfare criterion is appropriate under conditions of uncertainty, or more specifically, what modifications to the Hicks–Kaldor potential compensation test are required? The second kind of question is, when public policies change the uncertainties facing people, how are the resulting welfare changes to be measured? These two questions are connected in that the choice of a welfare criterion has important implications for how welfare changes should be measured.

Most of the analysis in this chapter is based on the now-standard model of individual preferences under uncertainty, which is based on the hypothesis that individuals maximize their expected utility. The theory is very attractive as a normative prescription for behavior and choice. However, there is now substantial evidence that individuals' choices frequently violate expected utility theory. For examples of this evidence and discussions of its implications for the theory of preferences under uncertainty, see Grether and Plott 1979; Kahneman and Tversky 1979; Arrow 1982; Machina 1987; Thaler 1987; and Rabin and Thaler 2001. Despite this evidence on the descriptive inadequacies of expected utility theory, its simplicity and elegance still make it attractive for expositional purposes.

In the next section, I develop the standard theory and use it to define welfare measures in several contexts. I then identify and discuss the several alternative welfare indicators that can be derived from this model of preferences. In the second section I discuss alternative aggregate welfare criteria and their implications for the way individuals' welfare changes are measured. In the third section I present some models for the measurement of individuals' willingness to pay for risk-reduction and risk-prevention policies using revealed preference methods. I also examine the implications of some forms of nonexpected utility preferences for welfare measurement. In the fourth section I define the concepts of option value, quasi-option value, and margin of safety and discuss their relationship to the welfare measures developed in this chapter.

Individual Uncertainty and Welfare

The term *individual uncertainty* refers to situations in which an individual is uncertain about which of two or more alternative states of nature will be realized and is not indifferent to which state actually occurs. I assume here that individuals can assign probabilities to these alternative states of nature and that the probabilities are correct in the sense of summing to one and incorporating all available information. Thus I am following the modern practice of making no distinction between risk (where probabilities are known) and uncertainty (where probabilities are unknowable). For a discussion of this often-cited distinction, which is often attributed to Frank Knight, see LeRoy and Singell 1987. I am also abstracting from questions of risk perception, how individuals make assessments of probabilities, and how these assessments are revised in light of new information. For introductions to some of the issues concerning risk perceptions and probability assessments and for references to the relevant literature in psychology and economics, see Arrow 1982; Viscusi 1985; and Machina 1990, especially 169–175.

As a first step in this analysis I provide a formal characterization of how risk is defined and the ways that environmental policy can affect risk. I then turn to the characterization of individual preferences in situations involving risk, and ways of characterizing welfare changes.

Risk and Environmental Policy

Consider an environmental risk, such as a chemical spill or pollution event, or a natural hazard, such as a flood or an earthquake; other examples of such risks include tornadoes and hurricanes, accidents at nuclear reactors, explosions in chemical plants, releases of toxic materials from hazardous waste storage sites, and air pollution episodes associated with unusual meteorological conditions. Risks such as these can be described in terms of two characteristics: the range of possible adverse consequences and the probability distribution across consequences. In this analysis, adverse consequences are measured in units that reflect the consequences for people (for example, the number of buildings damaged and degree of destruction, or the number of days of illness) rather than in terms of measures such as height of flood stage, Richter scale reading, or maximum atmospheric concentration of a pollutant. To keep the exposition simple, I will consider only one possible adverse event and two states of nature: the event occurs with a given set of consequences, or the event does not occur.

Suppose there is some public policy action that has the effect of reducing the probability of the adverse event, reducing the severity of its consequences, or both. Examples of policy measures that reduce the magnitude of an adverse event include regulations requiring earthquake-resistant construction techniques and the building of public shelters for protection against tornadoes and hurricanes. These are often referred to as *risk-reduction* measures. Examples of policy measures that reduce the probability of an adverse event include regulations for nuclear reactor safety, standards for durable containment techniques at hazardous waste storage sites, and the construction of dams to control stream flow to reduce the probability of flooding. These are often referred to as *risk-prevention* measures.

I grant that the distinction between risk reduction and risk prevention may sometimes be arbitrary. For example, the benefits of wearing a seat belt could be modeled in terms of the belt's effect on the severity of injury associated with accidents of unchanged probability, or in terms of its effects on the probabilities of experiencing injuries of various severities per mile driven. The choice of a modeling strategy may depend upon such pragmatic considerations as availability of data.

Many risk-reduction and risk-prevention measures are public goods in that they have the characteristics of nonrivalry and nonexcludability. The public-good character of these measures means that a private market system will fail to provide them in efficient quantities. Thus there is a case for government intervention to improve the efficiency of resource allocation. To determine whether risk-reduction and risk-prevention measures result in improvements in welfare, it is necessary to define and measure the benefits and costs of changes in risks.

Individual Preferences and Expected Utility

Turning to the characterization of individual preferences in situations involving risk, assume that an individual has a well-behaved preference ordering over bundles of goods $\boldsymbol{X}$ and that there is some adverse environmental event A over which that person has no control. The variable A measures the severity of the adverse event. Let $A = A^*$ represent the occurrence of the adverse event and $A = 0$ represent the absence of the adverse event. A takes the value A^* with probability π and 0 with the probability $1 - \pi$. For any given state of nature, this preference ordering can be represented from an ex post perspective by $u = u(\boldsymbol{X}, A)$ with $u(\boldsymbol{X}, A) > u(\boldsymbol{X}, 0^*)$. I assume that $u_A < 0$ and $u_{\boldsymbol{X}} > 0$, where in this chapter subscripts indicate partial derivatives. To avoid unnecessary complexity, income M and price $\boldsymbol{P}$ are assumed to be certain. I assume that $u(\cdot)$ is the same in both states, but that preferences over $\boldsymbol{X}$ are state-dependent. By this I mean that the consumption bundle actually chosen can depend on the state of nature, that is, on whether $A = 0$ or A^*. This means that the marginal utility of $\boldsymbol{X}$ can vary across states of nature. The model can be easily generalized to many states and to make $u(\cdot)$ depend on the state of nature; see, for example, Graham 1981 and Freeman 1984b, 1985, 1989. An alternative formulation is to write two different utility functions, $u^*(\boldsymbol{X})$ and $u^\circ(\boldsymbol{X})$, to indicate the effect of A on utility. These two formulations are essentially equivalent in that in both cases A acts as a shifter of the marginal utility schedules for $\boldsymbol{X}$ and M.

Given utility maximization, there is an ex post indirect utility function $u = v(M, \boldsymbol{P}, A)$ that shows the maximum attainable utility given M, $\boldsymbol{P}$, and A, where M is income and $\boldsymbol{P}$ is the vector of prices. It has the properties

$$
\begin{aligned}
&v_M > 0 \qquad v_{MM} < 0 \\
&v_A < 0 \qquad v_{AA} < 0
\end{aligned}
\tag{8-1}
$$

From here on the $\boldsymbol{P}$ term is suppressed because prices are assumed to be unchanged through the analysis.

I assume that individuals know the magnitude of the adverse event A^* and the probability of its occurrence. I also assume in this section that there are no opportunities for individual protective activities, that is, no ways of spending money to reduce either π or A^*. Ehrlich and Becker (1972) refer to such activities as *self-protection* (for reducing π) and *self-insurance* (for reducing A^*) and analyze individual behavior in the case of purely monetary losses. To the extent that self-protection and self-insurance activities are possible in the case of risks of nonmonetary losses, they may provide a basis for empirical estimation of the values of risk changes. I investigate this possibility in the section on revealed preference methods in this chapter.

Let D represent the monetary value of the damages caused by the event, given that the adverse event has occurred. D is the maximum sum of money the individual would give up to experience $A = 0$ rather than A^*. It is the solution to

$$v(M, A^*) = v(M - D, 0) \tag{8-2}$$

Thus D is a form of compensating surplus (CS) measure of welfare change where the reference level of utility is the utility realized if A^* occurs. It can also be interpreted as the monetary value of the damage cause by A^*.

People often have to make choices before the state of nature is revealed, in other words, ex ante choices. Let us assume that individuals make these choices so as to maximize their expected utility where expected utility is defined as

$$E[u] \equiv [\pi \cdot u(\mathbf{X}, A^*)] + [(1 - \pi) \cdot u(\mathbf{X}, 0)] \tag{8-3}$$

The expected utility expression provides a basis for an alternative measure of the value of avoiding A^*, namely, the willingness to pay (WTP) ex ante. This state-independent payment is sometimes referred to as option price, or OP. Option price is defined as the maximum payment the individual would make to change from the status quo risk to a situation in which A^* would not occur. It is also a form of CS, but one where the reference point is defined in terms of expected utility. OP is the solution to

$$[\pi \cdot v(M, A^*)] + [(1 - \pi) \cdot v(M, 0)] = v(M - OP, 0) \tag{8-4}$$

OP and $\pi \cdot D$ will not in general be equal. This is because they are measuring two different things, or more precisely, measuring the monetary equivalents of two different forms of utility change, the ex post change in $v(\cdot)$ and the ex ante change in $E[v]$. The relationship between these two measures is the basis of the extensive literature on option value. For reviews of this literature, see Bishop 1982 and Smith 1983. I review the current status of this literature in the fourth section of this chapter. Also, these two measures are not the only possible ways of expressing a compensating measure of the welfare value of preventing the adverse event. This can best be seen with the aid of Graham's (1981) WTP locus.

The Willingness-To-Pay Locus

Given the assumption that individuals maximize expected utility, their behavior can be described as the solution to the following problem:

$$\max\text{: } E[u] \equiv [\pi \cdot u(\mathbf{X}, A^*)] + [(1-\pi) \cdot u(\mathbf{X}, 0)] \tag{8-5}$$

subject to the usual budget constraint that expenditure equal income. The solution to this problem is denoted as E^1.

Now consider a policy that would reduce the magnitude of the adverse event from A^* to 0. If the individual were required to pay D to avoid the adverse effects given that the event occurs, and would pay nothing otherwise, the realized utility in each of the two states would be unchanged. Thus this set of payments would not change expected utility. Alternatively, if the individual were to pay OP before the uncertainty were resolved, by definition expected utility would be unchanged. These are only two of an infinite number of possible pairs of payments that would leave the individual indifferent, in expected utility terms, between the status quo and a situation in which the consequences of the adverse event were avoided with certainty. The set of these payments is denoted as

t^* = payment given the state in which the adverse event occurs
t° = payment given the state in which it does not occur

These payments satisfy the following condition:

$$E^1 = [\pi \cdot v(M - t^*, 0)] + [(1-\pi) \cdot v(M - t^\circ, 0)] \tag{8-6}$$

This equation defines Graham's WTP locus, which is shown in Figure 8-1.

Similarly, for a risk-prevention policy the WTP locus is given by the solution to

$$E^1 = E^2 = [\pi' \cdot v(M - t^*, A^*)] + [(1-\pi') \cdot v(M - t^\circ, 0)] \tag{8-7}$$

where π' is the probability of the adverse event with the risk-prevention policy in place. All of the subsequent analysis pertains to risk-reduction policies, but with suitable modification it can also be applied to the benefit–cost analysis of policies that change probabilities.

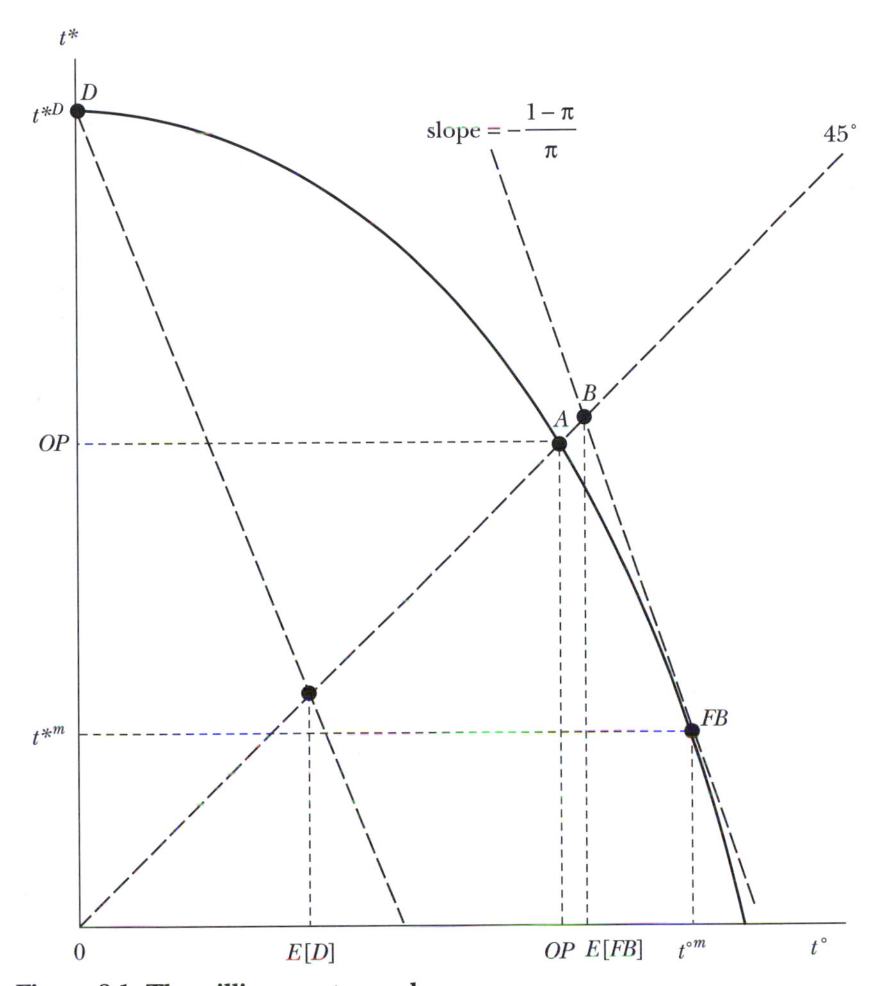

Figure 8-1. The willingness-to-pay locus

By setting the total differential of equation 8-6 equal to zero, we obtain an expression for the slope of the WTP locus:

$$\frac{dt^*}{dt^\circ} = \frac{(1-\pi)\cdot v_{M^\circ}}{\pi \cdot v_{M*}} \tag{8-8}$$

where v_{M*} is the marginal utility of income evaluated at $M - t^*$ given the event has occurred, and v_{M° is evaluated at $M - t^\circ$ given no adverse event. If the individual is risk-averse as assumed above ($v_{MM} < 0$), then the WTP locus will be concave to the origin. This is because increasing t^* (for example) raises v_{M*} relative to v_{M°. Risk neutrality with its constant marginal utility of income would result in a linear WTP locus.

Three points on the WTP locus are of particular interest as possible welfare measures. The first is point D in Figure 8-1. This point involves a payment of zero if the event does not occur and there are no realized damages, and a payment of t^{*D} if the event does occur. This sum represents the maximum payment the individual would make to experience the consequence 0 rather than A^* given that the event occurs. This sum is also equal to the damages (D) from the event were it to occur. There is a line from point D with a slope of $-(1-\pi)/\pi$, which represents the locus of all pairs of monetary values that have the same expected value as t^{*D}. For any t^*, these loci satisfy

$$E[t] = \pi \cdot t^* + (1-\pi) \cdot t^\circ \tag{8-9}$$

Differentiating, this gives

$$\frac{dt^*}{dt^\circ} = -\frac{(1-\pi)}{\pi} \tag{8-10}$$

This line is one of a family of what I will call iso-expected payment lines. The point where the iso-expected payment line through point D intersects the 45° line can be used to find the expected value of damages ($E[D]$). This is the ex post expected CS measure of the welfare change of preventing A^*.

The second point of interest in Figure 8-1 is A. It shows an alternative state-independent payment scheme in which $t^* = t^\circ = OP$. In this example, the state-independent payment is greater than the expected value of damages, but this will not always be the case. The difference between OP and $E[D]$ can be either positive or negative, and in some circumstances it can be quite large; see, for example, Graham 1981 and Freeman 1984b, 1985, 1989.

Finally, there is an iso-expected payment line that is tangent to the WTP locus at point FB in Figure 8-1. This point represents the state-dependent payment scheme, t^{*m}, $t^{\circ m}$, that maximizes the expected value of the individual's payments ($E[FB]$). The tangency of the WTP locus and the iso-expected payment line at point FB implies that the marginal utilities of income are equal in the two states. Because the slopes of the WTP locus and iso-payment lines are equal, making use of equation 8-8 yields

$$-\frac{(1-\pi)}{\pi} = -\frac{(1-\pi) \cdot v_{M^*}}{\pi \cdot v_{M^\circ}} \tag{8-11}$$

and $v_{M^*}/v_{M^\circ} = 1$. This equality is the condition for the efficient distribution of risk and for the optimum purchase of contingent claims at actuarially fair prices.* In this sense, point *FB* represents an optimum contingent payment scheme. In Graham's (1981) terminology, this is the *fair bet point.*

As illustrated in Figure 8-1, these three alternative payment schemes can give quite different summary measures of maximum willingness to pay for the policy. This is because they differ with respect to the implied opportunities for insurance, that is, opportunities to alter the payments in different states of nature to reflect attitudes toward risks. In the example of Figure 8-1, the fair bet point allows the individual to make a larger payment in the state of nature in which no adverse event occurs and in which, therefore, the individual is otherwise better off.

To see the relationship between state-dependent payments of this sort and insurance, let us suppose that actuarially fair insurance is available to the individual. Specifically, suppose that the individual could make a payment I before the uncertainty was resolved to purchase an insurance policy that would pay R if the adverse event occurred. This means that the individual can exchange income across states of nature according to $I/(R-I) = k$, the price of insurance. If the price of insurance is actuarially fair, $k = \pi/(1-\pi)$. In Figure 8-1, the insurance price line has the same slope as the iso-expected payment lines. The condition for the optimal purchase of insurance is, in this case, the equality of the marginal utilities of income in the two states of nature. This is because the individual chooses I (and therefore R) so as to maximize

$$E[u] = \pi \cdot v(M + R - I, A^*) + (1 - \pi) \cdot v(M - I, 0) \tag{8-12}$$

subject to the constraint that $I/(R-I) = k$. Substituting the constraint into the objective function and differentiating gives the first-order condition

$$\frac{\pi}{(1-\pi)} \cdot \frac{v_{M^*}}{v_{M^\circ}} = k \tag{8-13}$$

If the insurance is fair, $k = \pi/(1-\pi)$ and $v_{M^*} = v_{M^\circ}$.

Suppose now that the individual is required to make the payment represented by point D if the adverse event occurs and pays nothing otherwise. Paying D is equivalent to experiencing damages equal to D in this state of nature. Suppose also that the individual can purchase insurance at fair prices. The iso-expected payment line through point D shows that the individual could purchase an insurance contract that would require the individual to pay $E[D]$ if the adverse event did not occur and would reimburse

*A contingent claim is a contract specifying in advance a set of payments or receipts, or both, in which the amounts depend on the state of nature. An insurance policy is a contingent claim.

him or her so that his or her net payment would be $E[D]$ if the adverse event did occur. Because this point lies inside the WTP locus, it results in a higher expected utility than could be realized in the absence of the policy that would reduce A^* to zero. This means that if the individual can also purchase insurance so as to adjust his or her state-dependent payments, he or she would be willing to pay more than D for the policy that reduces A^* to zero. By failing to take into account the individual's opportunities for diversifying risk through insurance, the expected damage measure of welfare change underestimates the true WTP of the individual.

Now suppose that there is some institutional barrier to imposing a payment scheme that varies across states of nature but that the individual can still purchase contingent claims at fair prices. The maximum state-independent payment that could be extracted from the individual ex ante is actually represented by point B in Figure 8-1. Because point B lies outside of the WTP locus, the individual would be left worse off with this payment scheme in the absence of fair insurance. However, with fair insurance, the individual can move down to the right and reach the fair bet point through the optimal purchase of insurance. Thus the availability of fair insurance makes the maximum expected payment of the fair bet point feasible even when varying payments across states of nature are not possible.

The existence of the WTP locus and points such as B (in Figure 8-1) that can be reached by purchasing contingent claims raises an important question. If welfare change is to be measured by the maximum payment that holds expected utility constant, which pair of state-dependent payments is the best welfare measure? We are not yet ready to answer this question because how welfare is to be measured depends in part on the form of the social welfare criterion. The potential Pareto improvement (PPI) form of social welfare criterion requires a comparison of the welfare changes of those who gain and those who lose by the project. When losers face uncertainty about costs, the potential for state-dependent payments and compensation to redistribute risks between gainers and losers must be taken into account, along with the ability of each group to insure itself through the purchase of contingent claims. A full consideration of these issues must be left to the next section. But first we must show how uncertainty about costs can be represented in Graham's state-dependent payment model.

The Required Compensation Locus

The costs of government risk-management activities come ultimately in reductions in the utilities of individuals. This is true whether the activity involves direct government spending or regulations imposing requirements on the private sector. Furthermore, these costs, as correctly measured, may be quite different from the direct expenditures of the government or regulated firms; see, for example, Hazilla and Kopp 1990. The measurement of these costs from an ex ante perspective entails first determining the reductions in expected utility for all those who would bear the costs and then finding the minimum compensation required to restore each person to his or her original level of expected utility in the absence of the government activity.

Any individual who would bear part of the cost of a project has a reference level of expected utility in the absence of the project, say

$$E^2 = \pi \cdot v(M^*) + (1 - \pi) \cdot v(M^\circ) \tag{8-14}$$

The project would impose costs on the individual that could vary across states of nature. These costs could take several forms: a reduction in money income through taxation, the direct disutility of restrictions on nonmarket activities, or changes in product and factor prices.

If these costs vary across states, then compensation must be calculated so as to restore each cost-bearer's expected utility to the status quo level. Let c^* and c° represent all the nonmonetary dimensions of costs and let M^{c^*} and M^{c° represent money income inclusive of project-induced income changes. The problem can then be analyzed graphically, supposing that the probabilities affecting costs and benefits depend on the same random process in nature, so that π and $(1 - \pi)$ represent the relevant probabilities. This assumption will be relaxed later.

Given these assumptions, the project would result in a level of expected utility in the absence of compensation of

$$E^{2\prime} = \pi \cdot v(M^{c^*}, c^*) + (1 - \pi) \cdot v(M^{c^\circ}, c^\circ) \tag{8-15}$$

where M^{c^*} and M^{c° represent net income and c^* and c° represent the direct negative effects of the policy on utility in each state. There is an infinite number of pairs of state-dependent compensation payments, r^* and r°, that can be made to restore expected utility to E^2. These pairs are solutions to

$$E^2 = \pi \cdot v(M^{c^*} + r^*, c^*) + (1 - \pi) \cdot v(M^{c^\circ} + r^\circ, c^\circ) \tag{8-16}$$

The solution to this expression gives the required compensation (RC) locus for an individual who is harmed by the policy. It is shown in Figure 8-2. This is the locus of all possible compensating payments that leave those who bear the costs no worse off in terms of expected utility than if the policy had not been undertaken. The slope of the RC locus is found by setting the total differential of equation 8-16 equal to zero:

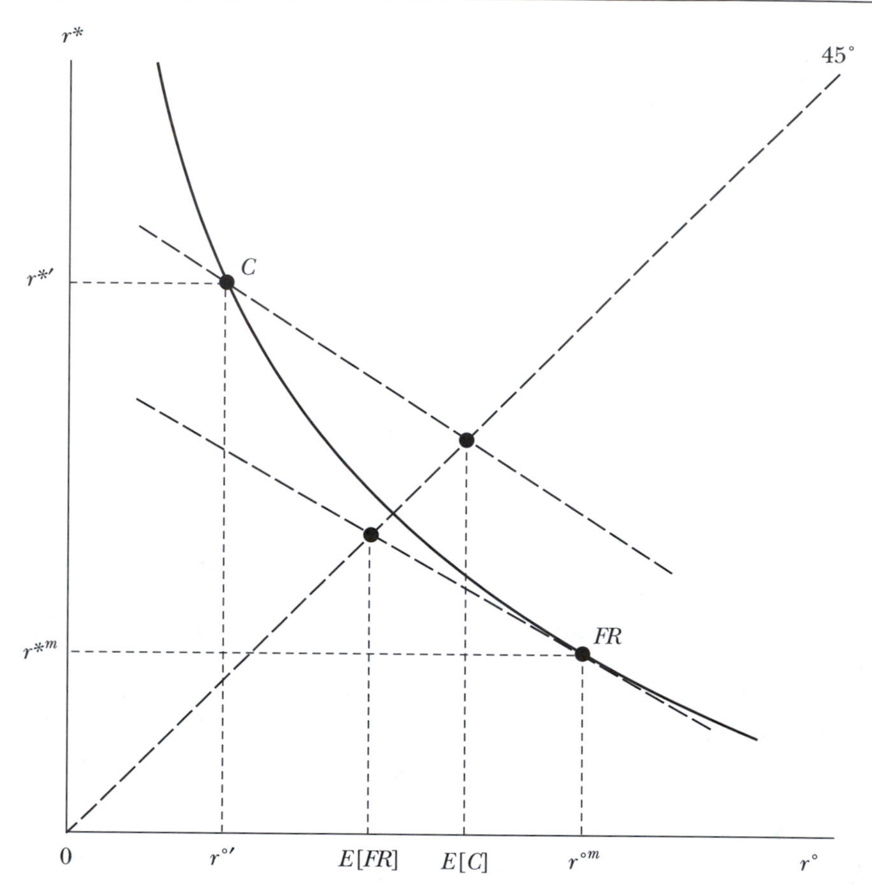

Figure 8-2. The required compensation locus

$$\frac{dr^*}{dr^\circ} = -\frac{(1-\pi)}{\pi} \cdot \frac{v_{M^\circ}}{v_{M^*}} \tag{8-17}$$

Diminishing marginal utility of income ensures that the locus is convex to the origin. The probabilities of the two states also define a family of iso-expected payment lines. They can be used to find the set of compensations that optimally distributes the risk to the cost-bearer across states, that is, that combination of payments that equates the marginal utilities of income in the two states. This is shown as r^{*m}, $r^{\circ m}$, at point *FR* in Figure 8-2. The intersection of the iso-expected payment line through this point with the 45° line gives the expected value of the fair compensation point, *E*[*FR*].

It is also possible to define an expected value of the required compensation from an ex post perspective. Given the occurrence of the event, there is an $r^{*\prime}$ that satisfies

$$v(M^{c*} + r^{*\prime}, c^*) = v(M^*) \tag{8-18}$$

Similarly, $r^{\circ\prime}$ satisfies

$$v(M^{c^\circ} + r^{\circ\prime}, c^\circ) = v(M^\circ) \tag{8-19}$$

These solutions define one point on the RC locus, labeled as C in Figure 8-2. Expected cost is

$$E[C] \equiv \pi \cdot r^{*\prime} + (1 - \pi) \cdot r^{\circ\prime} \tag{8-20}$$

Finally, we can define a state-independent compensation analogous to OP (say, OR). Because it also holds expected utility constant, it is the solution to

$$E^2 = \pi \cdot v(M^{c*} + OR, c^*) + (1 - \pi) \cdot v(M^{c^\circ} + OR, c^\circ) \tag{8-21}$$

But as will become clear in the next section, neither the expected value of cost nor OR will in general be useful as welfare measures for benefit–cost analysis.

Aggregation and the Welfare Criterion*

A major issue in the literature on benefit–cost analysis under uncertainty, at least since Weisbrod's classic article (1964), has been finding the appropriate measure of the welfare change for an individual who benefits from a project. The early option-value literature focused on comparing the expected value of consumer's surplus with option price. Graham (1981) redirected the discussion by showing that there is an infinite number of alternative state-dependent payment vectors and that the choice of a vector depends at least in part on the availability of contingent claims and individuals' opportunities to redistribute risk. Graham added a third candidate to the list of potential welfare measures, the expected value of the fair bet point, and he showed under what conditions the third candidate would be preferable to expected surplus and option price.

The literature since Graham has continued to focus on the question of the availability of contingent claims and whether risks are individual or collective; see, for example, Mendelsohn and Strang 1984; Graham 1984; Cory and Saliba 1987; Colby and Cory 1989; and Smith 1990. For the most part this literature has ignored the cost side of the problem or treated it in a sim-

*This section is adapted with permission from Freeman 1991b.

plified fashion. In particular, most of this literature has neglected explicit treatment of uncertainty in costs, the possibility for changing the nature of risk through state-dependent compensation schemes, and the implications of the latter possibility for the form of the potential Pareto improvement criterion. An exception is the work of Meier and Randall (1991).

Also, for the most part this literature has dealt with only the simplest form of uncertainty: that where only two alternative states of nature are possible. Exceptions include Graham's 1981 and 1992 articles. The latter presents a much more formal, abstract treatment of some of the problems considered here. The simple two-state model and its graphic treatment can be maintained only if both cost and benefit uncertainty result from the same random process and there are only two possible outcomes. If there are n possible benefit outcomes and m possible cost outcomes, the total number of alternative states of nature is $n \cdot m$.

In this section I first discuss alternative forms of the PPI criterion under uncertainty. I then consider a simple model with one gainer and one loser so that the problems of aggregation risk and collective risk as opposed to individual risk are not important. I show that in this simple model none of the three candidates for welfare measurement is a reliable screen for potential Pareto improvements. This is because all three candidates ignore the opportunity for redistribution of risk between the gainer and the loser through state-dependent payments and compensation. I then extend the analysis to a many-person economy. In those cases where the collective nature of risk precludes contingent claims markets, the importance of opportunities for redistributing risk between gainers and losers through state-dependent payment and compensation remains. Neither state-independent payments and compensation nor expected values of fair bet points are reliable indicators of potential Pareto improvements.

Potential Pareto Improvements under Uncertainty

The foundation of the PPI criterion is that a project passes only if there is some way to make redistributive payments so that no one is made worse off by the project and some people are made better off. It is generally agreed that in the context of risk and uncertainty "better off" and "worse off" are defined in terms of expected utility. However, potential compensation tests can be formulated several ways in terms of expected utility. As shown above, there is an infinite number of potential payments and compensations for each individual that will leave expected utility unchanged. However, not all of these payment and compensation vectors are feasible in the aggregate, in the sense that whatever state of nature occurs, the potential payment equals or exceeds the required compensation. In other words, not all of the payment and compensation vectors pass an ex post balanced budget test. Here I develop the PPI criterion in a general form that is consistent with the balanced budget test. In the next subsection I illustrate the implications of this criterion for project selection and for welfare measurement in the simple two-state model. The generalization to many states is straightforward. I then discuss the aggregation of individual welfare measures across many gainers and losers.

A proposed policy can be deemed an improvement in PPI terms only if there is at least one set of state-dependent payments that will finance (in the sense of balancing the budget in all states of nature) a set of state-dependent compensations without making anyone worse off and leaving at least one person better off in expected utility terms. More formally,

> let $i = 1, \ldots, n$ index alternative outcomes of the random process affecting benefits;
>
> let $j = 1, \ldots, m$ index alternative outcomes of the random process affecting costs;
>
> if $\boldsymbol{T}$ and $\boldsymbol{R}$ represent points on the n- and m-dimensional WTP and RC loci respectively, a project passes the PPI criterion if (and only if) there is at least one $\boldsymbol{T}$ and $\boldsymbol{R}$ such that $t_i \geq r_j$ for all i, j with the strict inequality holding for at least one pair.

Graphically, a project represents a PPI only if the WTP locus in the $n \cdot m$ dimensional space intersects the RC locus. Because this expression of the PPI criterion compares t_i and r_j independently of the probabilities associated with alternative outcomes, expected values are irrelevant in screening for PPI projects. An equivalent procedure is to aggregate the positive and negative payments into a net WTP locus. If this locus has a positive segment in the $n \cdot m$ dimensional space, the project passes the PPI test.

This version of the PPI criterion can pass projects that fail more restrictive forms of the criterion. This is because the criterion takes into account the possible benefits of redistributing risks between gainers and losers through state-dependent payments and compensation (Cook and Graham 1977). With this form of the criterion, project evaluation requires consideration of the real effects of both the project and project financing because the latter determines the benefits, if any, of redistributing risks.

Consider a project that would affect only two people. One person would gain, at least in one state of nature. The other would bear the costs. In this simplification, the problem of how to aggregate benefits and costs within the beneficiary and cost-bearer groups is avoided. Assume that it is not possible to buy contingent claims on income across the two states of nature that are relevant. Should the project be built? The answer depends on which version of the PPI compensation test one adopts.

One form of compensation test was offered by Bishop (1986), who called it an ex ante compensation test. For the two-person economy considered here, this test asks whether the option price of the gainer exceeds the required compensation of the loser in every state of nature. Ready (1988) also discussed this criterion; he called it a "weak" form of potential Pareto improvement. The implications of this form of compensation test can be seen in Figure 8-3. Point *A* on the WTP locus shows the option price of the gainer. The ex ante compensation or weak PPI criterion will pass only those projects whose costs lie below and to the left of point *A*. Points *B* and *C* represent projects with uncertain costs. Project *B* would not pass the test because if the adverse event occurred, the person bearing the cost would require compensation that exceeded the amount collected in the form of option price.

Ready (1988) also proposed an alternative, which he called the "strong" potential Pareto criterion. A project passes the strong form of this criterion if there exists some set of state-dependent payments that exceed the state-dependent project costs in every state. The strong form of PPI criterion would pass a project with uncertain costs represented by point *B* in Figure 8-3, but it would reject a project whose costs were represented by point *C*. The strong form of the PPI criterion justifies the use of the expected value of the fair bet point as a benefit measure. If $E[FB]$ exceeds $E[FR]$, then the project passes.

Although both weak and strong forms of the criterion allow for cost uncertainty, they both treat costs as a point in the payment and compensation space. Also, they both ignore opportunities to redistribute risk among cost-bearers or between gainers and losers through state-dependent payments and compensation.

According to the general form of the criterion stated above, a proposed policy is an improvement only if there is at least one set of balanced-budget state-dependent payments that will finance a set of state-dependent compensations without making anyone worse off and leaving at least one person better off. In terms of Figure 8-3, the general form of PPI criterion would pass a project with costs represented by point *C* if the required compensation locus through point *C* intersected the WTP locus. As noted above, this form of the criterion takes account of both the real effects of the project and the redistribution of risks through project financing.

Ready (1993) raised an interesting issue regarding the benefits of redistributing risk. The analysis here counts the benefits of redistributing risk whether or not the compensation is actually paid. Ready argues that they should be counted only if compensation is actually paid because they result from creating a more efficient distribution of risk; this is qualitatively different from project evaluation under certainty, where project finance and compensation only involve the redistribution of wealth. He shows that it would be possible for a project that passes this form of PPI test because of

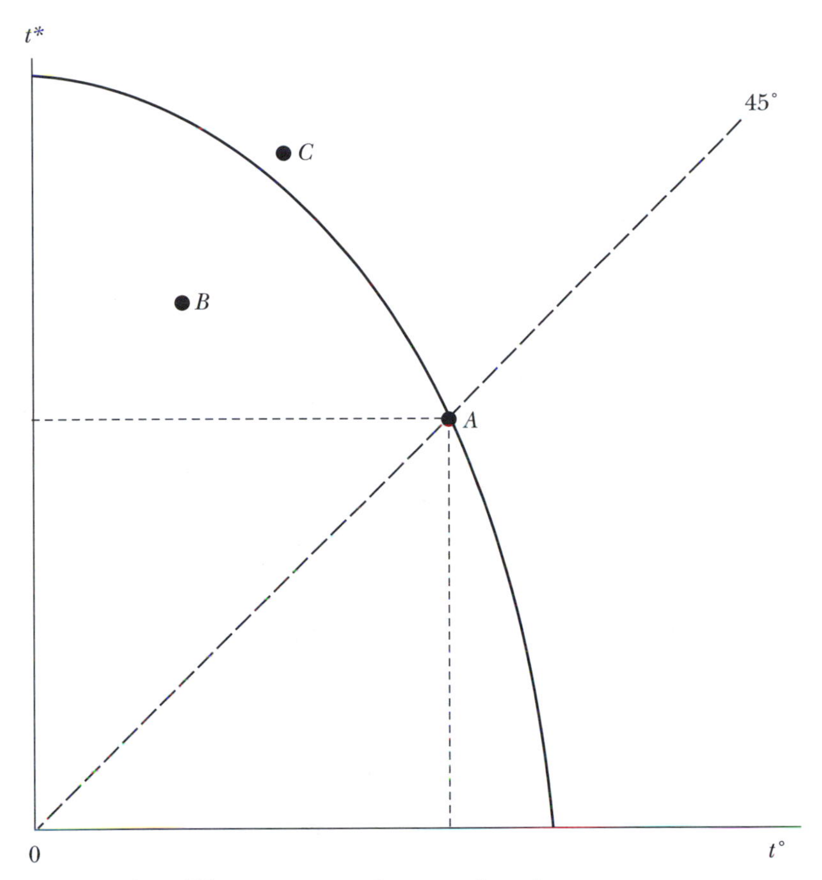

Figure 8-3. The willingness-to-pay locus and project costs

its redistribution of risk benefits to make all parties worse off if the compensation were not in fact paid.

Applying the Criterion in a Two-Person World

Again consider the simplest case of a project that would affect only two people and where it is not possible to buy contingent claims on income across the two states of nature. To provide a specific example, suppose that the project is to divert water from a river through an existing generator to produce hydroelectric power for a beneficiary. The cost would be borne by a farmer who would otherwise use the water for irrigation. In a rainy year, the benefit would be large because of the large flow through the generator. The cost to the farmer would be small because the rainfall would substitute

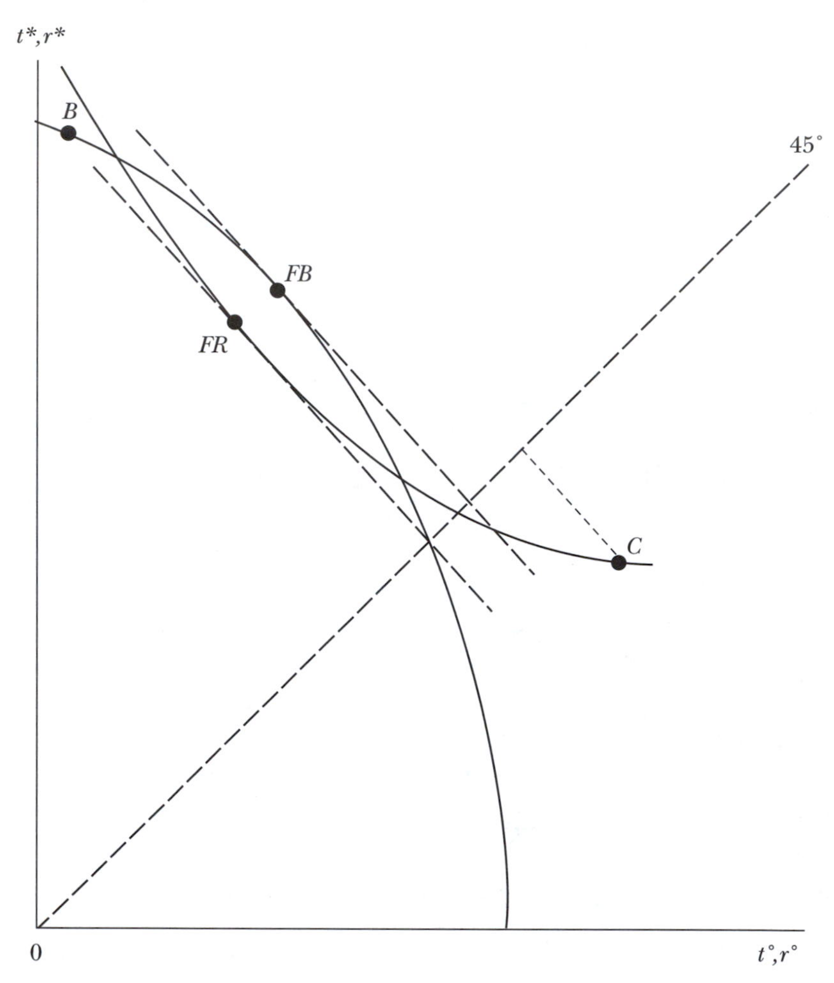

Figure 8-4. A potential Pareto improving project with expected costs exceeding expected benefits

for the irrigation water. Similarly, in a dry year the benefit would be small and the cost to the farmer large.

Should the project be built? Because the benefits and the costs are subject to the same random process, the WTP and RC loci can be combined in a single diagram; Figure 8-4 shows the case. Points *B* and *C* show the state-dependent benefits and costs, respectively. The WTP and RC loci pass through these two points. Another way of stating the PPI criterion is that it asks whether there is a set of state-dependent transfers from the beneficiary to the cost-bearer such that neither party is made worse off and at least one of the parties gains. Any state-dependent transfer is represented by one point in the WTP–RC space. Any such point on or below the WTP locus leaves the beneficiary no worse off; if the point is on or above the RC locus, the cost-bearer is no worse off. Thus any set of transfers that lies in the lens-shaped area between the two loci will successfully finance this project. The project passes the PPI test.

There are several important points to note about this conclusion. First, neither the expected value of cost nor the expected value of benefit is relevant. In the example of Figure 8-4, the expected value of cost substantially exceeds the expected value of benefit. However, the project still passes the PPI test because feasible payment schemes can redistribute the risks associated with the states of nature in ways that can make both parties better off.

The irrelevance of expected benefit and expected cost holds even if benefits and costs are certain, as long as individuals are risk-averse. For example, suppose that points *C* and *B* in Figure 8-4 are both on the 45° line, with *C* above *B*. It is still possible that the aggregate WTP and RC loci through these points could intersect over some range. This demonstrates the importance of considering the manner in which the payment and compensation scheme redistributes risk in project evaluation. For example, suppose that both individuals faced substantial uncertainty about their incomes. The beneficiary would prefer a payment scheme with the larger payment in the state of nature in which income was high. The cost-bearer would prefer to receive more compensation in the low-income state. Project financing provides both individuals with a means of hedging the income uncertainty. The risk of the project also has to be evaluated in the context of both individuals' total risk portfolios.

The second point about the conclusion above is that option prices are not reliable indicators of PPI projects. Again, Figure 8-4 shows that it is possible for the state-independent compensation to exceed the state-independent WTP and yet have the project pass the PPI test because of the ability of state-dependent payments and compensation to redistribute risk in a more favorable manner.

The third point is that neither the expected value of the fair bet point nor the expected value of fair compensation is relevant to project evaluation. Although the potential Pareto improving project in Figure 8-4 has an expected value of the fair bet point exceeding the expected value of compensation, this condition is not sufficient to ensure that the project is a potential Pareto improvement. A case in which the expected value of the fair bet point exceeds the expected value of fair compensation is shown in Figure 8-5. However, because the WTP locus lies below the RC locus everywhere, there is no state-dependent transfer scheme that would make this project potentially Pareto improving. Although on average (that is, in an expected value sense) WTP exceeds RC, there are some states of nature in

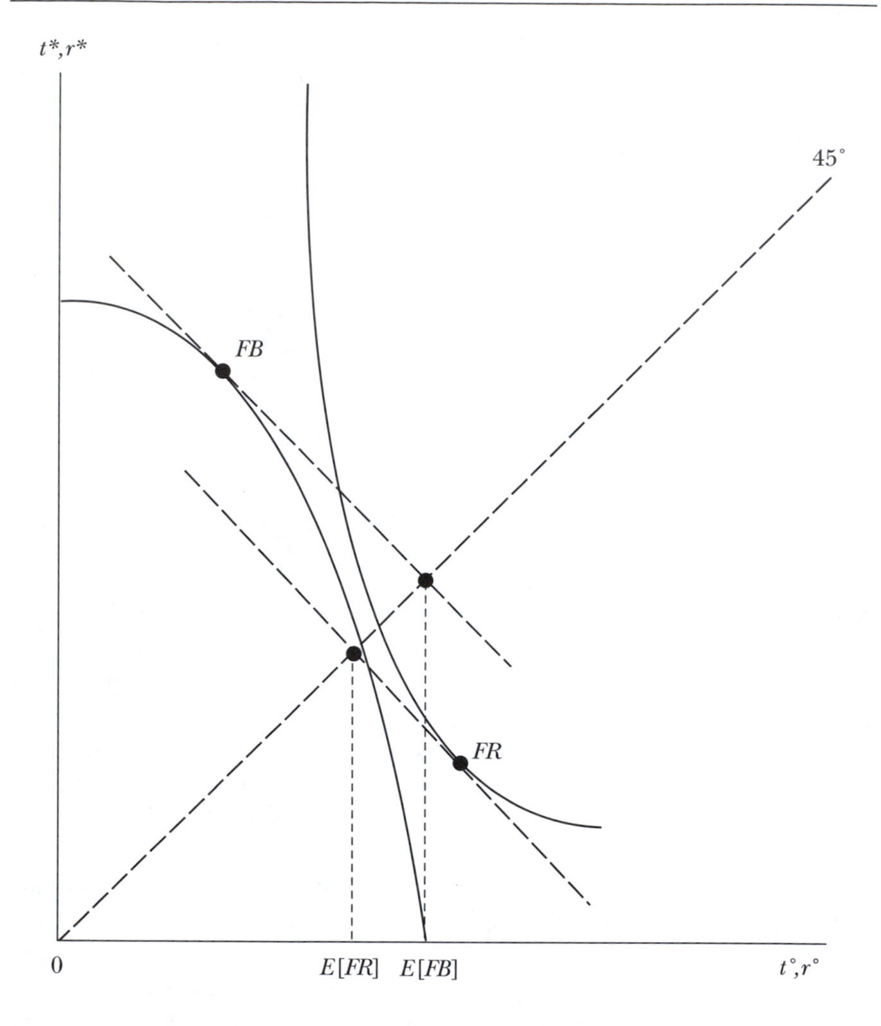

Figure 8-5. A nonpotential Pareto improving project

which the funds for the required compensation could not be collected from the beneficiary.

These results can be summarized in three statements about necessary and sufficient conditions for identifying PPI projects.

1. The expected value of benefits being greater than the expected value of costs is neither necessary nor sufficient for a PPI project.
2. The option price of beneficiaries being greater than the option price of cost-bearers is sufficient but not necessary for a PPI project.
3. The expected value of the fair bet point exceeding the expected value of the fair compensation point is necessary but not sufficient to identify a PPI project.

Up to this point I have not considered the feasibility of collecting or paying out the necessary state-dependent sums. If a social welfare judgement is made not to require payment and compensation, the feasibility of a state-dependent payment and compensation scheme is not an issue. However, if a social welfare judgement is made to require the payment and compensation, then the potential Pareto improvement test must be based on feasible state-dependent payments and compensations. Graham (1981) and Cory and Saliba (1987) provide examples showing that some combinations of state-independent payments and state-dependent payments and rebates can be used to obtain any required state-dependent payment pattern. As Graham puts it, "The trick, of course, is to collect the right payment in every state, but this may well be no more difficult than collecting the same payment in every state" (1984, 1102).

However, if the social welfare judgement requires payment and compensation and it proves to be impossible to devise a state-dependent scheme, then the potential Pareto improvement test must be based on option price measures, as pointed out by Graham (1981, 722). But this is not because option price is a superior welfare indicator in general; rather, it is because an actual compensation welfare criterion is more restrictive than a potential compensation criterion.

Aggregation in the Many-Person Economy

In the many-person economy the principal issue to be addressed is how the WTP and RC loci of individuals should be aggregated for the purposes of social welfare analysis. The answer depends on the nature of the risks faced by gainers and losers, that is, whether risks are collective or individual. By collective risk I mean the case in which, if the event affecting benefits and costs occurs, all the potentially affected individuals experience the event. Individual risk refers to a case in which the risks facing different individuals are independent.

Consider the case of individual risk first. With individual risks, the analysis of aggregate benefits and costs is much simplified. Let one individual who would be a beneficiary have a probability of π^b of experiencing the adverse event. Then if N is the number of potential beneficiaries, $\pi^b \cdot N$ individuals will experience the event with virtual certainty, with sufficiently large N. Similarly, $\pi^c \cdot K$ of the K bearers of the costs will experience the cost state associated with the event with virtual certainty, with sufficiently large K. Although individuals are uncertain as to their own outcomes, aggregate payments and compensation can be calculated with virtual certainty.

Assume for the moment that all people are identical. If those cost-bearers who experience the event are compensated by $r^{*\prime}$ and the rest of the cost-bearers are compensated by $r^{\circ\prime}$, then the aggregate compensation is

$$\pi^c \cdot K \cdot r^{*\prime} + \left(1 - \pi^c\right) \cdot K \cdot r^{\circ\prime} = K \cdot \left[\pi^c \cdot r^{*\prime} + \left(1 - \pi^c\right) \cdot r^{\circ\prime}\right] \tag{8-22}$$

The aggregate compensation is minimized by finding the expected fair compensation for an individual and multiplying by K. The same procedure is used to find the maximum aggregate WTP. If

$$N \cdot \left[\pi^b \cdot t^{*m} + \left(1 - \pi^b\right) \cdot t^{\circ m}\right] > K \cdot \left[\pi^c \cdot r^{*\prime} + \left(1 - \pi^c\right) \cdot r^{\circ\prime}\right] \tag{8-23}$$

then the project passes the PPI test. If individuals are different, all individuals' fair bets and compensations must be found and summed. In general, if each gainer makes the state-dependent payments represented by her fair bet point, the aggregate payment is the sum of the expected values of the fair bet points. Results are similar for the losers. Also, there is a virtual certainty that the gainers can compensate the losers and still be better off.

As long as a social welfare judgement is made not to require actual payment and compensation, this is all that needs to be said on the matter. However, if a social welfare judgement is made to require payment and compensation, then one must address the question of the feasibility of the required state-dependent payment and compensation schemes. If it should turn out that state-dependent payment and compensation is not feasible, then individual option prices are relevant, unless contingent claims could be purchased at actuarially fair prices. In the case of individual risks, it is more likely that such contingent claims will in fact be available.

In the case of collective risk, the full aggregate WTP and RC loci are required for welfare analysis. The procedure for aggregating individuals' WTP loci is explained in Graham 1981, 718–719. For each possible marginal rate of substitution (MRS) between contingent payments in different states as measured by dt^*/dt°, each individual's payment vector is determined, and all payment vectors are summed to obtain

$$T^{*m} = \sum_{i}^{N} t_i^{*m} \quad \text{and} \quad T^{\circ m} = \sum_{i}^{N} t_i^{\circ m} \tag{8-24}$$

where i indexes the N beneficiaries and the superscript m indicates that all individuals' payments are at their fair bet points for a given common MRS. This procedure gives one point on the aggregate WTP locus for each MRS. Each point also represents an efficient distribution of risk within the group of beneficiaries because all individuals have the same MRS for contingent payments. A similar procedure of aggregating individuals' required compensation at different marginal rates of substitution will yield the aggregate RC locus. Each point on the RC locus corresponds to an efficient distribution of cost risk among cost-bearers.

A project passes the potential Pareto improvement test if there exist payment and compensation points, $T^* = R^*$ and $T^\circ = R^\circ$, that lie both on or above the aggregate RC locus and on or below the aggregate WTP locus. The graphic analysis is similar to that in the two-person world shown in Figures 8-4 and 8-5. Aggregate option price is sufficient but not necessary as a screen for PPI projects; the aggregate fair bet points are necessary but not sufficient as indicators of PPI projects.

There are two exceptions to these conclusions. First, if contingent claims could be purchased at actuarially fair prices, the expected values of the fair bet and fair compensation points provide a sufficient test for potential Pareto improvement. This is because individuals' actual payments and compensations can be made at the expected values of the fair bet and fair compensation points. Individuals can then adjust their own positions through contingent claims purchases to reach their individual WTP or required compensation locus at their fair bet point. However, the nature of collective risks is likely to preclude a market for contingent claims at actuarially fair prices.

The second exception to these conclusions occurs if a social welfare judgement is made to require compensation, and if state-dependent payments and compensations are not feasible, then as is the case in the two-person world the welfare criterion must be based on option price measures. If individuals are identical, the appropriate test is based on the aggregate OP^B and OR^C because they are equal to the sums of the individual OPs and ORs. However, if individuals are different, then the sums of the individuals' option price measures will be less than the aggregate measures. This is because the former do not provide for the efficient distribution of risk across individuals within each group. The appropriate welfare test becomes

$$\sum_{i}^{N} OP_i > \sum_{j}^{M} OR_j \qquad (8\text{-}25)$$

In this section I have examined some of the consequences of extending the model of benefit–cost analysis for uncertain projects to reflect the symmetry of benefits and costs as changes in the expected utilities of affected individuals. This analysis implies that there is much work still to be done on the development of measures of benefits and costs to implement the appropriate welfare criteria. Specifically, because currently available methods of benefit and cost estimation under uncertainty generally give only expected cost and benefit and option price measures, we need to develop new methods for estimating the WTP and RC loci.

Revealed Preference Methods for Measuring Values

When individuals have opportunities to adjust to risky positions through transactions in related private goods markets, it may be possible to use one of the revealed preference methods described in Chapter 4 to infer individuals' values for risk changes. In this section I describe how models of averting behavior and hedonic prices can be used to obtain measures of value based on observable behavior. The welfare measures I develop are based on the hypothesis of expected utility maximization. I then show how these models can be extended to take account of other assumptions about individuals' preferences. It must be pointed out that because most observable transactions involve ex ante state-independent payments, the models described here are useful only in estimating option price measures of value.

Expressions for Option Price and Marginal Willingness To Pay

Suppose that public policies are under consideration that will reduce either the magnitude of an adverse event or its probability. The option price for a policy that reduces the magnitude of an adverse event is the set of state-independent payments that results in the same level of expected utility as would have occurred in the original uncertain situation. Denote this payment as OP^A to indicate that it refers to a risk-reduction policy. Making use of the indirect utility function, OP^A for a policy that reduces A from A^* to zero is the solution to

$$\begin{aligned} &\pi \cdot v(M, A^*) + (1-\pi) \cdot v(M, 0) \\ &\quad = \pi \cdot v\left(M - OP^{A^0}, 0\right) + (1-\pi) \cdot v\left(M - OP^{A^0}, 0\right) \\ &\quad = v\left(M - OP^{A^0}, 0\right) \end{aligned} \tag{8-26}$$

Similarly, the value of reducing A from A^* to $A' > 0$ is the solution to

$$\begin{aligned} &\pi \cdot v(M, A^*) + (1-\pi) \cdot v(M, 0) \\ &\quad = \pi \cdot v\left(M - OP^{A'}, A'\right) + (1-\pi) \cdot v\left(M - OP^{A'}, 0\right) \end{aligned} \tag{8-27}$$

For some purposes, it will be more useful to deal with marginal values. The option price for a marginal reduction from A^* is the marginal change in income that holds expected utility constant. This can be found by taking the total differential of the expression for expected utility in its indirect utility function form, setting it equal to zero, letting $dw = 0$, letting $dA = 0$ for $A = 0$, and rearranging terms. We get the individual's marginal WTP, ex ante (w^{A*}), or

$$w^{A*} = \frac{dM}{dA*} = -\frac{\pi \cdot v_{A*}}{\left[\pi \cdot v_{M*} + (1-\pi) \cdot v_{M^\circ}\right]} \tag{8-28}$$

where * and ° indicate the values at which the partial derivatives are evaluated. Recalling that $v^A < 0$, this expression is positive for reductions in $A*$ ($dA* < 0$), indicating a positive willingness to pay to reduce the magnitude of the uncertain event.

Equation 8-28 shows a variant of a standard result from welfare theory. The marginal willingness to pay for a change in $A*$ is equal to the marginal disutility of $A*$ converted to a money measure by using the marginal utility of income. However, because this is an ex ante WTP, the marginal utility terms are the expected values of the relevant marginal utilities, that is, the weighted averages of the marginal utility terms in the two states of nature where the weights are the probabilities of the two states.

The option price for reducing the probability of the event to zero is found by solving the following expression for $OP^{\pi 0}$

$$\begin{aligned} &\pi \cdot v(M, A*) + (1-\pi) \cdot v(M, 0) \\ &\quad = 0 \cdot v\left(M - OP^{\pi^0}, A*\right) + 1 \cdot v\left(M - OP^{\pi^0}, 0\right) \\ &\quad = v\left(M - OP^{\pi^0}, 0\right) \end{aligned} \tag{8-29}$$

Comparison of this expression with equation 8-26 shows that the option price for reducing the probability of an event to zero is equal to the option price for reducing its severity or magnitude to zero. Thus in this case there is no difference between the value of risk reduction and the value of risk prevention because in both cases the policy being valued eliminates the risk entirely.

The option price for a reduction in the probability of the event from π to π' is found by solving the following expression for $OP^{\pi'}$:

$$\begin{aligned} &\pi \cdot v(M, A*) + (1-\pi) \cdot v(M, 0) \\ &\quad = \pi' \cdot v\left(M - OP^{\pi'}, A*\right) + (1 \cdot \pi') \cdot v\left(M - OP^{\pi'}, 0\right) \end{aligned} \tag{8-30}$$

The marginal value for a change in π can be derived by taking the total differential of equation 8-30, setting it equal to zero, and holding $dA*$ at zero. The result is a standard result in the analysis of the value of risk prevention (Jones-Lee 1974; Cook and Graham 1977; Machina 1983; Smith and Desvousges 1988).

$$w^{\pi} = \frac{dM}{d\pi} = \frac{\left[v(M,0) - v(M,A^*)\right]}{\left[\pi \cdot v_{M^*} + (1-\pi) \cdot v_{M^\circ}\right]} \tag{8-31}$$

where w^{π} is the willingness to pay ex ante for a change in the probability of A^*. This expression is positive, indicating a positive willingness to pay for reductions in π (where $d\pi < 0$).

A comparison of equations 8-28 and 8-31 is instructive. Whereas equation 8-28 is based on the marginal disutility, as given by the term v_{A^*}, equation 8-31 is based on the nonmarginal difference in utility between the two states of nature, converted to monetary units by a weighted average of the marginal utilities of income in the two states of nature.

Averting Behavior

Observations of an individual's averting behavior might provide a basis for inferring the values of changes in risk. Suppose that, ex ante, an individual can select a level of private spending R, that will reduce the magnitude of A^* given that the event occurs according to the relationship $A^* = A(R, G)$, where G is the level of government protective spending. An example would be purchasing and wearing a seat belt to reduce the severity of injury given the occurrence of an accident. Assume that this function has the following properties:

$$A(0,0) = A^* \quad \text{and} \quad A_R^* < 0 \quad \text{and} \quad A_G^* < 0$$

The individual chooses R, given G, to maximize expected utility:

$$E[u] = \pi \cdot v[M - R, A(R, G)] + (1 - \pi) \cdot v(M - R, 0) \tag{8-32}$$

The first-order conditions include

$$\frac{1}{A_R^*} = \frac{\pi \cdot v_A}{\pi \cdot v_{M^*} + (1-\pi) \cdot v_{M^\circ}} \tag{8-33}$$

where v_{M^*} is the marginal utility of income evaluated at the level of A associated with the given level of G. The term $1/A_R^*$ is the reciprocal of the marginal productivity of expenditure on risk reduction, or equivalently, the marginal private cost of reducing A^*. The right-hand side is the marginal value of reducing A^* that was derived above.

The relevant value for policymakers is the value to the individual of an increase in public spending on risk reduction, dG. To find the option price measure, we take the total differential of equation 8-32, set it equal to zero, and substitute the first-order condition for the choice of private protective spending into this expression. After some simplification, we have

$$\frac{dM}{dG} = -\frac{A_G^*}{A_R^*} = \frac{\partial R}{\partial G} \tag{8-34}$$

This means that the individual's marginal willingness to pay for a small increase in government spending is the ratio of the marginal productivities of private spending and public spending in reducing A^* or the marginal rate of technical substitution between R and G in reducing A^*. This measure can be calculated if the technical relationship $A(R, G)$ is known. The relationship is observable in principle. The welfare change is also given by the marginal rate of substitution between private and public spending, holding expected utility constant. This is not the same thing as the observed change in private spending. For example, if G increases, the individual will reduce R but will also attain a higher level of expected utility. As discussed in Chapter 4, if R enters the utility function directly, then the welfare measure will include unobservable marginal utility terms, and the measure derived here will be an underestimate (or overestimate) if R provides positive (or negative) utility.

Similar results can be obtained for the case where individual ex ante spending has the effect of reducing the probability of the adverse event. Now let the production function relating private and public expenditures to the probability of the adverse event be

$$\pi = \pi(R, G) \tag{8-35}$$

where $\pi(0, 0) = \pi^*$ and $\pi_R < 0$, $\pi_G < 0$. The individual chooses R to maximize expected utility given by

$$E[u] = \pi(R, G) \cdot v(M - R, A^*) + [1 - \pi(R, G)] \cdot v(M - R, 0) \tag{8-36}$$

The first-order conditions include

$$\frac{1}{\pi_R} = -\frac{v(M - R, 0) - v(M - R, A^*)}{\pi(R, G) \cdot v_{M^*} + [1 - \pi(R, G)] \cdot v_{M^\circ}} \tag{8-37}$$

Again, the left-hand side is the reciprocal of the marginal productivity of private expenditure on reducing the probability, or, equivalently, the marginal private cost of reducing π. The right-hand side is the marginal value of reducing π as given by equation 8-31.

The marginal value of an increase in public spending to reduce the probability of the event is found by totally differentiating equation 8-36, setting the result equal to zero, and substituting the first-order condition where appropriate. The result is the following:

$$\frac{dM}{dG} = -\frac{\pi_G}{\pi_R} = \frac{\partial R}{\partial G} \tag{8-38}$$

This result is similar to the case of public spending to reduce A^*. The individual's marginal willingness to pay for public spending is equal to the ratio of the marginal productivities of private and public spending to reduce π, or to the marginal rate of technical substitution between R and G, holding A^* constant. Again, this result is analogous to those derived in the existing literature on protective spending in the absence of uncertainty.

Unfortunately, these results do not carry over to the case where the averting activity jointly produces reductions in π and A^*. Repeating the steps described above but making both A^* and π functions of R and G leads to the following expression (Shogren and Crocker 1991):

$$\frac{dM}{dG} = -\frac{\pi \cdot v_{A^*} \cdot A_G + \pi_G \cdot \left(v^* - v^\circ\right)}{\pi \cdot v_{A^*} \cdot A_R + \pi_R \cdot \left(v^* - v^\circ\right)} \tag{8-39}$$

The unobservable utility terms do not cancel out of this expression, so marginal WTP cannot be inferred from information on the averting technology. As shown in Chapter 4, the inability to use the averting behavior model in this case is due to the jointness of the implicit production technology. It is not a consequence of introducing risk into the analysis.

Hedonic Prices

If either the probability or the magnitude of a risk (or both) is a characteristic of heterogeneous goods such as housing, hedonic price estimation can be used to obtain the relevant ex ante marginal values for risk changes. Suppose that the magnitude of the adverse event varies across the space used for residential housing. For example, the dose of a toxic chemical from an accidental release would depend on the distance from the source of the release. If people are aware of this spatial variation, then they should be willing to pay more for houses in those areas with lower-magnitude risks. Competition for these more attractive houses would result in a systematic inverse relationship between the price of housing, P_h and A_i^*, where i indexes the location of the house. For simplicity, suppose that the magnitude of the event is the only relevant characteristic of housing. Then the price of a house at location i can be found from the hedonic price function $P_h(A_i^*)$. Given income, the probability of the event, and the magnitude of the event, the individual chooses a location to maximize expected utility:

$$E[u] = \pi \cdot v\left[M - P_h\left(A_i^*\right), A_i^*\right] + (1-\pi) \cdot v\left[M - P_h\left(A_i^*\right), 0\right] \tag{8-40}$$

The first-order condition is

$$-\frac{\partial P_h}{\partial A_i^*} = \frac{\pi \cdot v_{A_i^*}}{\left[\pi \cdot v_{M^*} + (1-\pi) \cdot v_{M^\circ}\right]} \tag{8-41}$$

Because the right-hand side of this condition is the ex ante marginal value of a reduction in A_i^* (see equation 8-28 again), this condition says that expected utility maximization calls for setting the marginal value of risk reduction equal to its marginal implicit price, the slope of the hedonic price function. Thus if individuals and the housing market are in equilibrium, the estimated marginal implicit price of risk reduction for each individual reveals each individual's marginal ex ante valuation for risk reduction. However, because a house is a long-lived asset and P_h is an asset price, equation 8-41 yields a compensating wealth measure of the lifetime welfare change associated with a permanent change in π.

If the relevant housing characteristic that varies across space is the probability of the adverse event, and housing prices reflect differences in π_i, the results are similar. The expression for expected utility is

$$E[u] = \pi_i \cdot v(M - P_h(\pi_i), A^*) + (1 - \pi_i) \cdot v(M - P_h(\pi_i), 0) \tag{8-42}$$

and the first-order condition is

$$\frac{\partial P_h}{\partial \pi_i} = -\frac{\left\{v\left[M - P_h\left(\pi_i\right), 0\right] - v\left[M - P_h\left(\pi_i\right), A^*\right]\right\}}{\left[\pi_i \cdot v_{M^*} + \left(1 - \pi_i\right) \cdot v_{M^\circ}\right]} \tag{8-43}$$

Again, the right-hand side is the ex ante marginal value of the probability change in equation 8-31. Thus the observed implicit price of probability reduction also reveals the individual's marginal ex ante value of risk reduction (Smith 1985). If both the probability and the magnitude of the event vary independently across space, housing prices will be a function of both characteristics. Both equations 8-41 and 8-43 must be satisfied in equilibrium. Hedonic price functions that do not include both characteristics as explanatory variables will be misspecified.

Welfare Change with Nonexpected Utility Preferences*

The models described above for revealed preference benefit measurement are based on expected utility as a representation of individuals' preferences under uncertainty. As discussed above, there is quite a bit of empirical evidence against expected utility maximization as a description of behavior. An important question, therefore, is whether models of value of the sort described here can be modified for use with nonexpected utility preferences.

Revising the models in this way is straightforward and involves no additional complications, at least in certain circumstances. This follows from the key features of revealed preference methods for estimating individuals' values from data on behavior. Generally speaking, the models involve first deriving the expression for welfare change, finding the first-order conditions for optimization, and substituting them into the expression for welfare change. Given the assumptions of the models described here, the substitution allows for canceling out any observable utility terms. For many nonexpected utility representations of preferences, the same result occurs, so that the derived observable welfare measures are independent of the particular form of preferences. This is a straightforward consequence of the envelope theorem. In what follows I first show this for the general case and then for two specific forms of nonexpected utility preferences.

Let I be some general index of preferences where the preferences depend on income, prices (implicitly), the probabilities of different states of nature, and the magnitudes of the adverse event in different states. Thus,

$$I = f(M, A, \pi) \tag{8-44}$$

I assume that this function is convex and twice differentiable. This expression could be nonlinear in the probabilities or incorporate regret and rejoice terms or other deviations from the standard expected utility function, or both. Expected utility preferences also fit this general formulation.

Consider the averting behavior model where $A^* = A(R, G)$. The first-order condition for the optimum R is

*This section is adapted with permission from Freeman 1991a.

$$\frac{\partial I}{\partial R} = -f_{M*} + f_{A*} \cdot A_R^* = 0 \qquad (8\text{-}45)$$

or

$$f_M = f_{A*} \cdot A_R^* \qquad (8\text{-}46)$$

To find the marginal welfare measure for a policy that reduces A^*, totally differentiate equation 8-44, rearrange terms, and substitute in the first-order condition to obtain

$$dI = f_{M*} \cdot dM + \left(f_{A*} \cdot A_R^* - f_M\right) dR + f_{A*} \cdot A_G^* \cdot dG = 0 \qquad (8\text{-}47)$$

$$\frac{dM}{dG} = -\frac{f_{A*} \cdot A_G^*}{f_{M*}} = -\frac{A_G^*}{A_R^*} \qquad (8\text{-}48)$$

Thus the marginal willingness to pay for publicly supplied risk reduction is equal to the marginal rate of technical substitution between public and private risk reduction.

Suppose that in the hedonic model probabilities vary across space. Then the general index of preferences would be

$$I = f[M - P_h(\pi_i), A, \pi_i] \qquad (8\text{-}49)$$

The first-order condition for the selection of the risk characteristic of housing is

$$\frac{\partial I}{\partial \pi_i} = -f_M \cdot \frac{\partial P_h}{\partial \pi_i} + f_{\pi_i} = 0 \qquad (8\text{-}50)$$

or

$$f_{\pi_i} = f_M \cdot \frac{\partial P_h}{\partial \pi_i} \qquad (8\text{-}51)$$

Totally differentiating equation 8-49 to obtain the welfare measure for the change in π and substituting the first-order condition gives

$$dI = \left(f_M \cdot dM\right) + \left(f_{\pi_i} - f_M \cdot \frac{\partial P_h}{\partial \pi_i}\right) d\pi_i + \left(f_{\pi_i} \cdot dG\right) = 0 \qquad (8\text{-}52)$$

$$\frac{dM}{dG} = -\frac{f\pi_i}{fM} = \frac{\partial P_h}{\partial \pi_i} \tag{8-53}$$

The marginal willingness to pay for publicly supplied risk prevention is equal to the observable marginal implicit price of the risk characteristic of housing.

To illustrate this general result, consider the prospect theory model of Kahneman and Tversky (1979). In the two-state model based on the indirect utility function, the index of preferences takes the following form:

$$I = g(\pi) \cdot v(M, A^*) + g(1-\pi) \cdot v(M, 0) \tag{8-54}$$

where

$$g(0) = 0,\ g(1) = 1 \tag{8-55}$$

and

$$g(\pi) + g(1-\pi) < 1 \text{ for } 0 < \pi < 1 \tag{8-56}$$

Assume that the magnitude of the adverse event depends both on the level of expenditure on a private averting activity R and public expenditure G. The value of reducing A^* is

$$\frac{dM^*}{dA^*} = \frac{g(\pi)\cdot v_{A^*}}{g(\pi)\cdot v_{M^*} + g(1-\pi)\cdot v_{M^\circ}} \tag{8-57}$$

which is not directly observable because of the utility and probability weighting terms, but it can be inferred. Given the level of G, the individual's optimal level of the private averting activity is given by

$$\frac{\partial I}{\partial R} = -\left[g(\pi)\cdot v_{M^*}\right] + \left[g(\pi)\cdot v_{A^*}\cdot A_R^*\right] - \left[g(1-\pi)\cdot v_{M^\circ}\right] = 0 \tag{8-58}$$

Thus,

$$\frac{1}{A_R^*} = \frac{g(\pi)}{g(\pi)\cdot v_{M^*} + g(1-\pi)\cdot v_{M^\circ}}\cdot v_{A^*} \tag{8-59}$$

The marginal value to the individual of a change in G is found by totally differentiating equation 8-54, setting it equal to zero, and solving for

$$\frac{dM}{dG} = \frac{dR}{dG} - \frac{g(\pi)}{[g(\pi)\cdot v_{M*}] + [g(1-\pi)\cdot v_{M^\circ}]}\cdot v_{A*}\cdot A_R^*\cdot\frac{dR}{dG} - \frac{g(\pi)}{[g(\pi)\cdot v_{M*}] + [g(1-\pi)\cdot v_{M^\circ}]}\cdot v_{A*}\cdot A_G \tag{8-60}$$

After substituting in equation 8-59, this becomes

$$\frac{dM}{dG} = -\frac{A_G^*}{A_R^*} \tag{8-61}$$

Similarly, if π can be reduced by private and public expenditure, the value of a reduction in π is

$$\frac{dM}{d\pi} = \frac{[g(1-\pi)\cdot v(M-R,0)] - [g(\pi)\cdot v(M-R,A^*)]}{[g(\pi)\cdot v_{M*}] + [g(1-\pi)\cdot v_{M^\circ}]} \tag{8-62}$$

The first-order condition for private averting expenditure is

$$-\frac{1}{\pi_R} = \frac{v(M-R,0) - v(M-R,A^*)}{[g(\pi)\cdot v_{M*}] + [g(1-\pi)\cdot v_{M^\circ}]} \tag{8-63}$$

After substitution, the value of the public risk prevention expenditure is

$$\frac{dM}{dG} = -\frac{\pi_G}{\pi_R} \tag{8-64}$$

Similar results can be derived for other forms of preferences, for example the regret theory of Loomes and Sugden (1982).

Extending these results to the case of weak complementarity is straightforward. Suppose that the consumption of some market good x_i increases the probability of occurrence of some adverse event for the purchaser. Suppose further that there is some public policy action represented by G that can reduce the risk associated with consuming x_i for all consumers. So

$$\pi = \pi(x_i, G) \tag{8-65}$$

with

$$\partial\pi/\partial x_i > 0$$

and

$$\partial \pi / \partial G < 0$$

This policy will increase the general preference index for all consumers of x_i and will cause the demand curve for x_i to shift out. Formally, the general preference index can be written as

$$I = f(M, A, \pi, p_i) \tag{8-66}$$

where p_i is the price of the complementary good. The general form of the expenditure function is

$$e = e(p_i, A, \pi, I^*) \tag{8-67}$$

Weak complementarity requires that there be a choke price for x_i and that at the choke price

$$\frac{\partial e}{\partial \pi} = 0 \tag{8-68}$$

As is shown in Chapter 4, when the conditions for weak complementarity are satisfied, the monetary equivalent of the increase in well-being is the area between the compensated demand curves for the good before and after the public policy change. Similarly, if the market good affects the severity of the adverse event, weak complementarity requires that $\partial e / \partial A^*$ or $\partial I / \partial A^*$ be zero when the good is not purchased. Because these results are independent of any particular specification of the preference function (other than the conditions of weak complementarity), they will hold for expected utility and nonexpected utility preferences as well as in the case of certainty.

I have shown that the three broad classes of models that have been developed to measure the benefits of environmental change from revealed preferences under certainty can be easily generalized to apply to valuing changes in risk. This generalization of the models does not require that individual preferences take the expected utility form. The principal requirement is that individuals be maximizing some objective function. Then by the envelope theorem, welfare measures that contain unobservable preference terms can be reduced to functions of observable relationships by substitution of the first-order conditions for preference maximization. Thus if the conditions for using these models are satisfied, there is no particular need to be concerned with how people make their choices under uncertainty.

Additional Topics in the Welfare Economics of Uncertainty

In this section, I briefly discuss four other topics that are related to the welfare economics of uncertainty and that bear on environmental policymaking. The first topic is the relationship between option price and the expected damage associated with some risk. The latter has a long history of use in applied benefit–cost analysis. The question is whether it is a useful proxy for the preferred ex ante welfare measure. The second topic is option value and whether option value can be measured and added to other types of values when welfare measures are computed. The third topic is quasi-option value, its source, and its validity as a welfare measure. The fourth topic is the margin of safety provided by a particular environmental regulation and whether the welfare economics of uncertainty can provide any guidance as to the "adequacy" of a specific margin of safety.

Comparing Option Price and Expected Damage Measures

There is a long history of using ex post expected damage measures in benefit–cost analysis. Perhaps the earliest example of this approach to measuring the value of risk changes is the U.S. Army Corps of Engineers' method for estimating the benefits of flood control projects. For many years, the Corps calculated the reduction in the costs of replacement, repair, and cleaning up after a flood for each possible flood stage, multiplied the monetary damages by the probability associated with that level of flooding, and summed across all possible flood stages (Eckstein 1961). We would now add to the repair and cleaning costs some monetary measure of the loss of utility associated with the flooding events.

Expected damage measures have the virtue of being relatively easy to calculate from experience with risky events. For example, for most river basins in the United States there is a large body of data on hydrology, land use, and the distribution of structures that can be used to calculate expected flood damages under alternative proposed flood control projects. However, even if these losses (including lost utility) were accurately measured, we have shown that they will not in general correctly measure welfare change from the ex ante perspective.

Because at least in some circumstances expected damage measures may be relatively easy to obtain, an important question is whether they can be taken as useful approximations of the other forms of ex ante values when policymakers would prefer to have the latter. Are the differences between the two types of measure likely to be small or large? Is the expected damage measure likely to be larger or smaller than the option price measure? What factors determine the sign of the difference and its magnitude?

The derivation of an expected damage measure for risk reduction is straightforward. The first task is to establish a measure of the value of avoiding an event given that A^* will occur with certainty. The willingness to pay to avoid the consequences of the event given its occurrence is a compensating surplus (CS) measure of damage, D. It is found as the solution to

$$v(M, A^*) = v(M - CS, 0) \tag{8-69}$$

The expected damage measure of the value of avoiding the event that has probability of π of occurring is the mathematical expectation of the WTP given by equation 8-69, that is, $\pi \cdot D$. The marginal value of reducing A^* given that the event has occurred can be found by totally differentiating the left-hand side of equation 8-69, setting it equal to zero, and solving for

$$\frac{dM}{dA^*} = -\frac{v_{A^*}}{v_{M^*}} \tag{8-70}$$

Let this be denoted by d^A to indicate that it is a marginal value and that it refers to changes in A. Equation 8-70 is a standard result showing that ex post the marginal value is the marginal utility of A^* converted to a money measure by using the marginal utility of income. Taking the expectation yields the ex post value of a marginal reduction in A^*:

$$E\left[d^A\right] = -\pi \frac{v_{A^*}}{v_{M^*}} \tag{8-71}$$

The percentage difference between the option price and expected damage measures is defined according to the following expression, using equations 8-28 and 8-71:

$$\% \text{ difference} \equiv \frac{w^{A^*} - E[d^A]}{w^{A^*}} = (1-\pi) - \left[(1-\pi)\frac{v_{M^\circ}}{v_{M^*}}\right] \tag{8-72}$$

This expression shows that if the marginal utility of income is independent of the event, so that $v_{M^\circ} = v_{M^*}$, the difference in the measures is zero. This would be the case, for example, if the individual were risk-neutral or could purchase actuarially fair insurance.

If the adverse event reduces the marginal utility of income so that v_{M^*} is less than v_{M°, then the expected damage measure will exceed the option price measure, and the difference will be negative. On the other hand, if the adverse event increases the marginal utility of income, the option price measure will be larger than the expected damage measure. In both cases the magnitude of the difference will be large when the difference in the marginal utilities of income is large and when the probability of the adverse event is low. To summarize, the sign of the difference between the option price and expected damage measures depends on how the marginal utility of income varies with A. The magnitude of the difference depends on how much it varies and on the probability of the event.

In Freeman 1989, I computed these differences for several explicit forms of von Neuman–Morgenstern cardinal utility functions with different degrees of risk aversion. Examples included

$$v = (M - A^*)^b \text{ with } 0 < b < 1 \tag{8-73}$$

$$v = -e^{-b(M-A^*)} \text{ with } b > 0 \tag{8-74}$$

$$v = -e^{-bM(1-A^*)} \tag{8-75}$$

$$v = \ln[(1 - A^*)M] \tag{8-76}$$

For equations 8-73 and 8-74, the price flexibility of income for A^* is zero, meaning that the marginal value of a change in A is independent of M. Risks of this sort are essentially equivalent to financial risks. An arbitrary choice of units for measuring A^* can be made such that A^* is equal to the ex post D. For some of these functional forms, the marginal utility of income is higher given the adverse event. One way of interpreting this last characteristic is that income and the absence of A^* are substitutes in consumption such that if A^* does not occur, the marginal utility of the consumption that M allows is diminished. This might be the case if the event increased the marginal utilities of ameliorating activities such as cleaning up and repairing damages after a flood. Option price values for reducing A^* will be greater than expected damages, and more so for larger losses occurring with lower probabilities.

The findings for utility functions with this characteristic can be summarized by the following statements:

1. Because the differences are positive as predicted by equation 8-72, empirical valuation measures based on changes in expected damages will understate the option prices.
2. The difference is an increasing function of CS because a larger CS means a larger difference between v_{M° and v_{M^*}.
3. The differences are for the most part trivially small for small losses (equivalent to up to 1% of income). Only for high degrees of risk aversion does the difference approach 10% for low-probability losses.
4. For relatively large losses (equivalent to, say, 10% of income), the differences range between 5% and 20%, depending on the probability and

for moderate degrees of risk aversion. For relatively high degrees of risk aversion, the differences range above 60%.

5. For the catastrophic loss (equivalent to 50% of income), the differences are large for all but the least risk-averse form of utility function and most likely event. For the more risk-averse forms of utility function, this means that option prices for risk reduction can exceed expected damages by factors of 2.5 to more than 100.

This last conclusion might be a possible explanation for a phenomenon that some observers have noted, that is, that the public seems unwilling to accept some very low-probability, high-consequence risks that experts have judged to be acceptable. If the experts' judgements are based on ex post valuation measures and if people are relatively risk-averse and their preferences take this form, the experts could be vastly underestimating the true potential welfare costs that these risks would impose on people.

For other forms of utility functions, $v_{M^{\circ}}$ is greater than v_{M^*}. Examples include:

$$v = (1 - A^*)^c \cdot M^b \quad \text{with } 0 < b,\ c < 1 \tag{8-77}$$

and

$$v = (1 - A^*)\ln M \tag{8-78}$$

One way of interpreting this characteristic is that income and the absence of A^* are complements in consumption such that if the event does not occur, the marginal utility of the consumption that M allows is high. But if the event does occur, the capacity for consumption to generate utility is diminished. For example, the event "broken leg" decreases the marginal utility of expenditures for hiking and skiing trips. For these functions, expected damages exceed option prices.

Calculations with these functions show that differences are negative as predicted and are less than 10% except for low probabilities of very large losses (50% of income). For most functional forms, the magnitude of the difference is larger for the most risk-averse utility functions. However, for one functional form the difference is largest when the degree of risk aversion is zero. And for one functional form, even the sign of the difference depends on the degree of risk aversion.

As these results demonstrate, option price and expected damage measures of the value of risk reduction are likely to be different. Expected damage measures are unreliable proxies for option price measures. The sign of the difference depends on specific features of the functions chosen to represent preferences. The degree of risk aversion is not a reliable predictor of the size or even the sign of the differences. Values for risk reduction based on expected damages are not likely to be useful and could be seriously misleading as guides for risk management decisions.

I now turn to the value of risk prevention. The expected damage measure of the value of reducing the probability of a loss is the reduction in the expected value of the compensating surplus associated with the loss. As explained above, the expected loss, $E[D]$, is $\pi \cdot D$. By differentiation, we have

$$dE[D] = d(\pi \cdot CS) = CS \cdot d\pi \qquad (8\text{-}79)$$

To examine the relationship between the option price and expected damage measures of the value of a probability change, we need to compare the right-hand sides of equations 8-31 and 8-79, but they are not easily comparable. The option price measure starts with the difference between two utility levels and converts it to a money measure by using the weighted average of the marginal utilities of income in the two states. The expected damage measure is already in monetary units; specifically, it is the difference in money expenditure necessary to achieve the same level of utility under two different sets of conditions.

Again, in Freeman 1989, I assumed specific functional forms and parameters and calculated differences. The results are difficult to summarize except to say that differences can be positive or negative, large or small, and can be large even in the absence of risk aversion.

In estimating the values of reducing the probability of adverse events, expected damage measures are unreliable indicators of the desired option prices. Expected damage measures can be either underestimates or overestimates, and the errors involved in using them as proxies for option prices can be large. However, they cannot be predicted without detailed knowledge of the specific characteristics of individuals' preferences.

Is Option Value a Value?

The concept of option value was introduced by Weisbrod in a much-cited paper almost 40 years ago (Weisbrod 1964). This concept has had an interesting history, both in terms of its development as a theoretical construct and because of its relationship to and role in policy discussions concerning environmental resources. Weisbrod argued that an individual who was unsure of whether he or she would visit a site such as a national park would be willing to pay a sum over and above his or her expected consumer's surplus to guarantee that the site would be available should he or she wish to visit it. Weisbrod called this extra sum the option value of the site. Option value was seen to arise when an individual was uncertain about whether he or she would demand a good in some future period and was faced with uncertainty about the availability of that good. If option price (*OP*) is defined as the maximum sum the individual would be willing to pay to preserve the option to visit the site before his or her own demand uncertainty is resolved, then the excess of option price over expected consumer's surplus can be called option value (*OV*). It was thought that option value should be measured, if it were possible, and added to expected consumer's surplus to obtain the full measure of the value of providing an environmental service.

Weisbrod apparently viewed the existence of positive option value as intuitively obvious. Indeed, there is no formal mathematical or logical proof of the existence of option value in Weisbrod's paper. However, as the subsequent literature has shown, the concept has several subtleties and complications, as well as traps for the unwary investigator. For example, Cicchetti and I thought we had proved that option value was positive for risk-averse individuals (Cicchetti and Freeman 1971), but our proof was wrong (see Schmalensee 1972; Anderson 1981; and Bishop 1982). Schmalensee showed that even for a risk-averse individual, option value could be greater than, equal to, or less than zero depending upon the particular circumstances. This is essentially a matter of the relationship between the expected damage (or surplus) measure and the option price measure discussed in a preceding section of this chapter. As Bohm (1972) pointed out in a comment on Schmalensee, what matters is the relationship between the marginal utilities of income in the different states of nature.

The matter seemed to rest there until Hartman and Plummer (1987) and I (Freeman 1984b) provided specific characterizations of the nature of demand uncertainty and examined the implications of different types of demand uncertainty for the relationship between the marginal utilities of income for different states of nature. For example, Hartman and Plummer showed that if an individual was uncertain about future income and the demand for the good in question was a positive function of income, option value is unambiguously negative for risk-averse individuals. I showed that in this case risk-lovers would have positive option values, and that one plausible form of state-dependent preferences ensured positive option values for risk-averse individuals.

At about the same time Bishop (1982) suggested that useful insights might be obtained by considering the simpler case in which an individual is certain of demand but faces uncertain supply of the good. He showed that option value is greater than zero for a project that eliminates the uncertainty of supply. Thus option value appeared to be resurrected as a benefit associated with a guaranteed future supply of goods, such as national parks. However, I was able to show that Bishop's conclusion did not hold for all possible forms of reduction in uncertainty of supply. What I called supply-side option value could also be either positive or negative (Freeman 1985). See also Wilman 1987.

Providing a rationale for preservation of wilderness and scenic beauties seems to have been the motivation for the original investigation of option value. Milton Friedman (1962) in *Capitalism and Freedom* had argued that because there were no externalities associated with uses of national parks, all of the relevant economic values could be captured by the owners of parks through admission fees. Thus the allocation of land to national parks should be subjected to a market test. Weisbrod's paper (1964) was clearly at least in part a response to this argument. Krutilla (1967) included option value in his list of reasons why markets might fail to achieve allocative efficiency for unique environmental resources. Option value was presented as an economic value over and above the expected use values, and it could not be captured through admission fees. The preservation of the option to visit the park was a form of public good that might justify the preservation of a natural area even when expected use values were less than opportunity costs.

Schmalensee's (1972) theoretical proof that the sign of option value is ambiguous seems to have reduced interest in option value as a policy-relevant concept, at least temporarily. However, in 1983 the U.S. Environmental Protection Agency (EPA) listed option value as one form of intrinsic benefit that could be included in the benefit–cost analyses of proposed regulations required under the terms of Executive Order 12291 (U.S. EPA 1983). Measurement of option value was also part of the work done in several EPA-sponsored research projects during the 1980s and early 1990s. In light of the state of the theory, this continuing effort to measure option value and to use it in policymaking is an interesting comment on the power of the idea.

The idea of option value may have been most useful through stimulating the more rigorous analysis of the theory of welfare measurement under uncertainty. Because of the theoretical contributions of authors such as Schmalensee (1972), Graham (1981), Bishop (1982), and Smith (1987a, 1987b), we can now see that what has been called an option value is really just the algebraic difference between the expected values of two different points on a WTP locus. Specifically, it is the algebraic difference between the expected value of the consumer's surplus and the state-independent WTP (option price). Because these two points represent alternative ways of measuring the same welfare change, the difference between their expected values cannot be a separate component of value. Furthermore, option value cannot be measured separately; it can only be calculated if we have enough information on preferences to calculate both option price and expected surplus. Finally, as I have shown in this chapter, neither of these points on the WTP locus has any particular claim as a superior welfare measure. Perhaps in recognition of this, option value is not mentioned in EPA's most recent set of guidelines for economic assessment (U.S. EPA 2000).

Planners' Uncertainty and Quasi-Option Value

When policymakers are uncertain about the magnitude of the benefits or the costs of alternative courses of action, or both, decision rules and procedures should be modified to reflect this uncertainty. One possible modification entails altering the time sequence of choices so as to take advantage of information that might become available in the future. The term quasi-option value was coined by Arrow and Fisher (1974) to describe the welfare gain associated with delaying a decision when there is uncertainty about the payoffs of alternative choices and when at least one of the choices involves an irreversible commitment of resources. They showed that quasi-option value is not dependent on risk aversion. It can be present even when decisionmakers make choices on the basis of expected values of monetary benefits and costs.

Most of the literature on the role of this concept in environmental decisionmaking has concluded that consideration of quasi-option value would lead to relatively less irreversible development and relatively more preservation of natural environments. However, the conclusion that there is a quasi-option value benefit to preserving a natural area or to delaying its development springs from a specific feature of the models used by Arrow and Fisher (1974), Conrad (1980), Miller and Lad (1984), Fisher and Hanemann (1987), and others. In these models quasi-option value stems from the value of the information gained by delaying an irreversible decision to develop a natural area. However, it is not difficult to imagine situations where the relevant information to guide future decisions can be gained only by undertaking now a little development. For example, suppose there is uncertainty about the magnitude of a mineral deposit underlying a wilderness area. Perhaps the only way that the uncertainty about the magnitude of the benefits of development relative to preservation can be resolved is through exploratory drilling, that is, through a little bit of development. In such cases there can be positive quasi-option value to development, or, what is the same thing, a negative quasi-option value to preservation (Miller and Lad 1984; Freeman 1984a).

Whether quasi-option value exists or whether it is positive or negative for preservation depends on the nature of the uncertainty, the opportunities for gaining information, and the structure of the problem. Quasi-option value is not a component of the values individuals attach to resource changes. Even if individuals' utility functions were known, quasi-option value could not be estimated separately and added into a benefit–cost calculation. Quasi-option value is a benefit of adopting better decisionmaking procedures. Its magnitude can only be revealed by comparing two strategies where one of the strategies involves optimal sequential decisionmaking to take advantage of information obtained by delaying irreversible resource commitments. The decisionmaker who knows how to use an optimal sequential decisionmaking strategy has no reason to calculate quasi-option value. The calculation would be redundant because the best decision is already known.

The Margin of Safety

Section 109 of the Clean Air Act of 1970 calls for setting primary national ambient air-quality standards to protect human health, "allowing an *adequate margin of safety*" (emphasis added). The question addressed in this section is whether the phrase "adequate margin of safety" can be given a meaningful interpretation in the context of the conventional theory of welfare measurement under uncertainty. A basic assumption that underlies the process for setting air-quality standards as established by Congress is that there are threshold levels of pollution below which there would be no observable adverse effects on human health. The existence of a threshold means that the risk to health associated with air pollution is zero even at some positive pollution levels. Under the act, protection of human health was to be achieved by setting the standard for air quality at or below the threshold for each pollutant. The margin of safety was intended to take into account uncertainties in determining the exact position of the threshold.

In considering the question of meaningful interpretation of "adequate margin of safety," the framework developed by Lichtenberg and Zilberman (1988) is useful. Suppose that some standard of performance has been established as a policy goal; in the case of the Clean Air Act, the standard is zero excess adverse health effects caused by air pollution. Furthermore, suppose that uncertainty concerning the parameters of the function relating ambient air quality to incidence of disease or mortality makes it impossible to determine with certainty the minimum levels of air pollution that are necessary to achieve this target. For any given air-quality standard, there is some non-zero probability that the target of zero excess adverse health effects will not actually be achieved. The probability of failure to meet the target can be reduced by setting a more stringent air-quality standard.

The margin of safety can be interpreted as one minus the probability of failing to achieve the target. The question of what is an adequate margin of safety can be restated as, what is the maximum acceptable probability of failure to meet the target? From an economic perspective, the answer to this question is found by comparing the additional benefits to people of reducing the probability of failure with the costs. The costs are what it takes to meet the higher air-quality standard that is consistent with the lower probability of failure. The benefits are the willingness to pay to reduce this probability. The optimum is where the marginal benefits just equal the additional costs. To implement this economic approach to determining the adequacy of the margin of safety, we need a way to measure people's willingness to pay for reduced probability of failure.

A key feature of the approach described here is that for any pollution level the individual's risk of mortality or illness is not known with certainty. Rather, there is some probability distribution of risks, with greater uncertainty about risks being reflected in a higher variance of that probability distribution. In contrast, the models of individual choice described in this chapter are based on the assumption that the risk is known with certainty. Thus in the case of certainty there is no need for policymakers to consider a margin of safety. To give economic content to the term "adequate margin of safety" in the context of placing values on reductions in risk would require building more sophisticated models of individual choice. One approach is to assume that people have preferences over probability distributions of risk. Given distributions with the same mean, people would prefer distributions with smaller variances. Given distributions with the same variance, people would prefer distributions with lower means for risk. An improvement in air quality that reduced the probability of failing to achieve the target of no excess adverse health effects would shift each individual's probability distribution of risk, thereby conveying benefits to each person. None of the empirical models for estimating values of risk reduction take into account this added level of complexity.

Summary

In this chapter I have explored the various ways that welfare changes can be defined and measured for changes in risks and how these measures relate to the potential Pareto improvement criterion when it is used in benefit–cost analyses of policies dealing with risk. The methods presented here are based on an important maintained assumption, which is that individuals can assign probabilities to alternative outcomes that are consistent with probability theory and contain all relevant information. However, there is a substantial body of evidence that people find it hard to think in probabilistic terms and often have serious misperceptions about the magnitude of important environmental risks. For references to some of this evidence and discussions of its implications for economic analysis, see Machina 1990 and Arrow 1982. Other important references include Slovic et al. 1979, 1980, 1982. Also, as noted above, people often make choices that are not consistent with expected utility theory.

The observation that our models of behavior may not be good descriptions of how people actually behave toward risk raises some interesting and difficult questions concerning the evaluation of public policies dealing with risks. Two of these questions are, whose probabilities should be used in policy analysis, those of the experts or those of the people facing the risks? and, whose preferences should count, those of the rational expert, or those of the people facing the risks? Portney (1992) posed these questions in the starkest manner with his hypothetical example of Happyville's contaminated water supply, in which the experts are utterly convinced that the contaminating chemical is benign but the citizens believe that it is responsible for the above-average incidence of cancer in their town. Each citizen is willing to pay $1,000 for what the experts say will be unnecessary and costly treatment.

Arrow (1983) and Machina (1990) have both considered these questions in their more general form. They agree that the probabilities are part of the description of the policy problem and the effect of the policy, and that it is important that this description be accurate. Thus they recommend reliance on the experts' probability assessments. Machina also recommends that individuals' preferences should be recognized even if they are not consistent with expected utility theory.

This is good advice, but many problems come in a form in which it may not be possible to follow that advice. Consider Portney's Happyville example. On the information given, we cannot determine whether the people have misperceived the risk of cancer (errors in probabilities) or whether they place a high value on preventing a small but catastrophic risk (preferences). All we know is that each citizen is willing to pay $1,000 for treatment of the water supply. If that will cover the cost, it would seem to be welfare-enhancing to build the treatment plant.

When revealed preference measurement methods are used to estimate the values people place on changes in risks, these questions of probabilities and preferences may be less important. Suppose for the sake of argument that individuals assign probabilities to adverse outcomes that are consistently higher than the probabilities assessed by risk experts. Which probabilities should be used, those of the experts or those of the people? Here I think that the answer is pretty clear. As long as the objective of the research is to learn about people's preferences and WTP, we have to know what they think they are getting in return for their expenditures on risk reduction and risk prevention. This means relying on their own subjective probabilities. In any case, when revealed preference methods such as averting behavior and hedonic price analysis are used to infer values of risk changes, it is not generally possible for the researcher to determine whether a high marginal value results from the subject's overestimate of the risk change or from a high valuation of a correctly perceived change. All that is observed is the risk–dollar trade-off.

References

Anderson, Robert J. 1981. A Note on Option Value and the Expected Value of Consumer's Surplus. *Journal of Environmental Economics and Management* 8(2): 187–191.

Arrow, Kenneth J. 1982. Risk Perception in Psychology and Economics. *Economic Inquiry* 20(1): 1–9.

———. 1983. Behavior under Uncertainty and Its Implications for Policy. In *Foundations of Utility and Risk Theory with Applications*, edited by B. T Stigum and F. Wenstop. Dordrecht, Germany: Reidel.

Arrow, Kenneth J., and Anthony C. Fisher. 1974. Environmental Preservation, Uncertainty, and Irreversibility. *Quarterly Journal of Economics* 88(1): 312–319.

Bishop, Richard C. 1982. Option Value: An Exposition and Extension. *Land Economics* 58(1): 1–15.

———. 1986. Resource Evaluation under Uncertainty: Theoretical Principles for Empirical Research. In *Advances in Applied Microeconomics*. Vol. 4, edited by V. Kerry Smith. Greenwich, CT: JAI Press.

Bohm, Peter. 1972. Option Demand and Consumer's Surplus: Comment. *American Economic Review* 65(3): 233–236.

Cicchetti, Charles J., and A. Myrick Freeman III. 1971. Option Demand and Consumer's Surplus: Further Comment. *Quarterly Journal of Economics* 85(3): 528–539.

Colby, Bonnie E., and Dennis C. Cory. 1989. Valuing Amenity Resources under Uncertainty: Does the Existence of Fair Contingent Claims Markets Matter? *Journal of Environmental Economics and Management* 16(1): 149–155.

Conrad, Jon M. 1980. Quasi-Option Value and the Expected Value of Information. *Quarterly Journal of Economics* 94(2): 813–820.

Cook, Philip J., and Daniel A. Graham. 1977. The Demand for Insurance and Protection: The Case of Irreplaceable Commodities. *Quarterly Journal of Economics* 91(1): 143–156.

Cory, Dennis C., and Bonnie Colby Saliba. 1987. Requiem for Option Value. *Land Economics* 63(1): 1–10.

Eckstein, Otto. 1961. *Water Resources Development: The Economics of Project Evaluation.* Cambridge, MA: Harvard University Press.

Ehrlich, Isaac, and Gary S. Becker. 1972. Market Insurance, Self Insurance and Self Protection. *Journal of Political Economy* 80(4): 623–648.

Fisher, Anthony C., and W. Michael Hanemann. 1987. Quasi-Option Value: Some Misconceptions Dispelled. *Journal of Environmental Economics and Management* 14(2): 183–190.

Freeman, A. Myrick, III. 1984a. The Quasi-Option Value of Irreversible Development. *Journal of Environmental Economics and Management* 11(3): 292–295.

———. 1984b. The Sign and Size of Option Value. *Land Economics* 60(1): 1–13.

———. 1985. Supply Uncertainty, Option Price, and Option Value. *Land Economics* 61(2): 176–181.

———. 1989. Ex Ante and Ex Post Values for Changes in Risks. *Risk Analysis* 9(3): 309–317.

———. 1991a. Indirect Methods for Valuing Changes in Environmental Risks with Non-expected Utility Preferences. *Journal of Risk and Uncertainty* 4(2): 153–165.

———. 1991b. Welfare Measurement and the Benefit–Cost Analysis of Projects Affecting Risks. *Southern Economic Journal* 58(1): 65–76.

Friedman, Milton. 1962. *Capitalism and Freedom.* Chicago, IL: University of Chicago Press.

Graham, Daniel A. 1981. Cost–Benefit Analysis under Uncertainty. *American Economic Review* 71(4): 715–725.

———. 1984. Cost–Benefit Analysis under Uncertainty: Reply. *American Economic Review* 74(5): 1100–1102.

———. 1992. Public Expenditure under Uncertainty: The Net Benefit Criteria. *American Economic Review* 82(4): 882–846.

Grether, David M., and Charles R. Plott. 1979. The Economic Theory of Choice and the Preference Reversal Phenomenon. *American Economic Review* 69(4): 623–638.

Hartman, Richard, and Mark Plummer. 1987. Option Value under Income and Price Uncertainty. *Journal of Environmental Economics and Management* 14(3): 212–225.

Hazilla, Michael, and Raymond J. Kopp. 1990. The Social Cost of Environmental Quality Regulations: A General Equilibrium Analysis. *Journal of Political Economy* 98(4): 853–873.

Helms, L. Jay. 1985a. Errors in the Numerical Assessment of the Benefits of Price Stabilization. *American Journal of Agricultural Economics* 69(1): 93–100.

———. 1985b. Expected Consumer's Surplus and the Welfare Effects of Price Stabilization. *International Economic Review* 26(3): 603–617.

Jones-Lee, Michael W. 1974. The Value of Changes in the Probability of Death or Injury. *Journal of Political Economy* 99(4): 835–849.

Just, Richard E., Darrell L. Hueth, and Andrew Schmitz. 1982. *Applied Welfare Economics and Public Policy.* Englewood Cliffs, NJ: Prentice-Hall, Inc.

Kahneman, Daniel, and Amos Tversky. 1979. Prospect Theory: An Analysis of Decisions under Risk. *Econometrica* 47(1): 263–291.

Krutilla, John V. 1967. Conservation Reconsidered. *American Economic Review* 57(4): 777–786.

LeRoy, Stephen F., and Larry D. Singell, Jr. 1987. Knight on Risk and Uncertainty. *Journal of Political Economy* 95(2): 394–406.

Lichtenberg, Erik, and David Zilberman. 1988. Efficient Regulation of Environmental Health Risks. *Quarterly Journal of Economics* 103(2): 167–178.

Loomes, Graham, and Robert Sugden. 1982. Regret Theory: An Alternative Theory of Rationale Choice under Uncertainty. *Economic Journal* 92(368): 805–824.

Machina, Mark J. 1983. Generalized Expected Utility Analysis and the Nature of Observed Violations of the Independence Axiom. In *New Foundations of Utility and Risk Theory with Applications,* edited by B. T. Stigum and F. Wenstop. Dordrecht, Germany: Reidel.

———. 1987. Choice under Uncertainty: Problems Solved and Unsolved. *Journal of Economic Perspectives* 1(1): 121–154.

———. 1990. Choice under Uncertainty: Problems Solved and Unsolved. In *Valuing Health Risks, Costs, and Benefits for Environmental Decision Making*, edited by P. Brett Hammond and Rob Coppock. Washington, DC: National Academy Press.

Meier, Charles E., and Alan Randall. 1991. Use Value under Uncertainty. *Land Economics* 67(4): 379–389.

Mendelsohn, Robert, and William J. Strang. 1984. Cost–Benefit Analysis under Uncertainty: Comment. *American Economic Review* 74(5): 1096–1099.

Miller, Jon R., and Frank Lad. 1984. Flexibility, Learning and Irreversibility in Environmental Decisions: A Bayesian Approach. *Journal of Environmental Economics and Management* 11(2): 161–172.

Portney, Paul R. 1992. Trouble in Happyville. *Journal of Policy Analysis and Management* 11(1): 131–132.

Rabin, Matthew, and Richard H. Thaler. 2001. Anomalies: Risk Aversion. *Journal of Economic Perspectives* 15(1): 219–232.

Ready, Richard C. 1988. "Cost–Benefit Analysis under Temporal Uncertainty." Ph.D. dissertation, University of Wisconsin, Madison.

———. 1993. The Choice of a Welfare Measure under Uncertainty. *American Journal of Agricultural Economics* 75(54): 896–904.

Rothschild, M., and J.E. Stiglitz. 1976. Equilibrium in Competitive Insurance Markets. *Quarterly Journal of Economics* 90(4): 629–650.

Schmalensee, Richard. 1972. Option Demand and Consumer's Surplus: Valuing Price Changes under Uncertainty. *American Economic Review* 62(5): 813–824.

Shogren, Jason F., and Thomas D. Crocker. 1991. Risk, Self-Protection, and Ex Ante Economic Value. *Journal of Environmental Economics and Management* 20(1): 1–15.

Slovic, Paul, Baruch Fischhoff, and Sarah Lichtenstein. 1979. Rating the Risks. *Environment* 21: 14–20, 36–39.

———. 1980. Facts versus Fears: understanding Perceived Risk. In *Societal Risk Assessment: How Safe Is Safe Enough*? edited by W.A. Albers. New York: Plenum.

———. 1982. Response Mode, Framing, and Information-Processing Effects in Risk Assessment. In *Question Framing and Response Consistency*, edited by Robin Hogarth. San Francisco, CA: Jossey-Bass.

Smith, V. Kerry. 1983. Option Value: A Conceptual Overview. *Southern Economic Journal* 49(3): 654–668.

———. 1985. Supply Uncertainty, Option Price, and Indirect Benefit Estimation. *Land Economics* 61(3): 303–307.

———. 1987a. Non-Use Values in Benefit–Cost Analysis. *Southern Economic Journal* 54(1): 19–26.

———. 1987b. Uncertainty, Benefit–Cost Analysis, and the Treatment of Option Value. *Journal of Environmental Economics and Management* 14(3): 283–292.

———. 1990. Valuing Amenity Resources under Uncertainty: A Skeptical View of Recent Resolutions. *Journal of Environmental Economics and Management* 19(2): 193–202.

Smith, V. Kerry, and William H. Desvousges. 1988. The Valuation of Environmental Risks and Hazardous Waste Policy. *Land Economics* 64(3): 211–219.

Thaler, Richard G. 1987. The Psychology of Choice and the Assumptions of Economics. In *Laboratory Experimentation in Economics: Six Points of View,* edited by Alvin E. Roth. Cambridge, U.K.: Cambridge University Press.

U.S. EPA (U.S. Environmental Protection Agency). 1983. *Guidelines for Performing Regulatory Impact Analyses.* Washington, DC: Office of Policy Analysis.

———. 2000. *Guidelines for Preparing Economic Analyses.* Washington, DC: U.S. EPA.

Viscusi, W. Kip. 1985. Are Individuals Bayesian Decision Makers? *American Economic Review* 75(2): 381–385.

Weisbrod, Burton A. 1964. Collective Consumption Services of Individual Consumption Goods. *Quarterly Journal of Economics* 77(3): 71– 77.

Wilman, Elizabeth A. 1987. Note on Supply-Side Option Value. *Land Economics* 63(3): 284–289.

CHAPTER

9

Environmental Quality as a Factor Input

In addition to providing services directly to individuals as consumers, environmental and resource systems can affect the costs and output levels in the production sector of an economy. Costs and output levels can be affected, for example, by changes in the flow of minerals and petroleum from the ground, the negative effects of air pollution on the flow of food and fiber from agriculture, and the effect of pollution on the costs of manufactured goods through requirements for more frequent cleaning, repair, and replacement of materials. The effects of these changes will be transmitted to individuals through the price system in changes in the costs and prices of final goods and services and changes in factor prices and incomes. All these examples involve a common economic mechanism. Improvements in the resource base or environmental quality lower costs and prices and increase the quantities of marketed goods, leading to increases in consumers' and perhaps producers' surpluses. Similarly, increases in pollution can cause economic harm to producers and consumers by decreasing their surpluses.

Most of the early studies of the effects of air and water pollution on producers were based on the damage function approach (Freeman 1982, Chapters 5 and 9). This approach involves (1) estimating a dose–damage function that relates some measure of pollution to a physical measure of damage; (2) applying this function to estimates of the inventory of materials exposed or at risk; and (3) multiplying the result by some unit value.

For example, in the case of damages to materials and structures, the physical damage might be corrosion, soiling, or loss of paint thickness. This measure would have to be translated into an estimate of the increase in the frequency of some repair or replacement activity. The unit cost of this activity would provide the basis for estimating monetary damage. The problem with this approach is that paint thickness and corrosion rates are not economically relevant effects. Rather, we require some understanding of the responses of producers in terms of increased frequency of repair, degree of degradation of performance, and so forth, all of which translate into increases in the cost of production. Similarly, in the case of agriculture the damage function approach focuses on reduction in harvestable yield and multiplies this by a market price. However, this approach ignores adaptive behavior on the part of farmers and effects on consumers resulting from possible changes in market price and in production cost of a crop.

Properly specified economic models of the effects of pollution on producers use cost functions or production functions to link the physical effects of changes in environmental quality to changes in market prices and quantities, and ultimately to changes in consumers' and producers' surpluses. Either directly in the case of the production function approach, or implicitly in the case of the cost function approach, these models incorporate the whole range of possible producers' responses to changes in pollution levels (through, for example, material substitution, increased protection activities, and changes in maintenance and repair schedules).

The basic theory for estimation of productivity benefits is outlined in Chapter 4. In the first section of this chapter, I lay out the theory in more detail for the case of single-product firms in a competitive industry. In the second section I develop the analysis for the more likely case of multiproduct firms. I then examine (in the third section) how the welfare effects on factor owners and consumers are passed through vertically linked markets for inputs and intermediate products when there are several intermediate stages of production. In the fourth section I consider the effects that market distortions such as monopoly power and regulated prices have on welfare measures. In the fifth section I show how the methods presented in this chapter can be used to value changes in the productivity of natural resource systems such as commercially exploited forests and fisheries. For another treatment of many of these topics, with a greater emphasis on applications, see Bockstael and McConnell (forthcoming).

Basic Theory*

This section derives welfare measures for changes in some parameter, q, that enters directly into the production functions of single-product firms. This parameter can be interpreted as a measure of environmental quality or the quality of a resource input into production. Alternatively, it could be

*This section is adapted with permission from Freeman and Harrington 1990.

a measure of innovation or technological change. Whatever the interpretation, an increase in this variable is assumed to increase the output attainable with any given set of inputs. Following on the earlier contributions of Anderson (1976), Schmalensee (1976), and Just and Hueth (1979), Just and others (1982) provided a rigorous analysis of how to measure changes in welfare due to price distortions in factor and product markets. These models provide a basis for analyzing the effects of productivity-induced changes in product and factor prices. As we will see, in the case of single-product firms the results that they derive for price distortions carry over in a straightforward manner to the case of parametric shifts of cost and supply functions caused by environmental changes.

In this section I derive two alternative measures of welfare change for both marginal and nonmarginal changes in q for the case of single-product firms. One measure is the value of the marginal product of q derived from the production function. The other measure is based on the aggregate cost function for the industry and can be interpreted as the area between the old and new supply curves. The analysis in this section is essentially short-run, focusing on changes in quasi-rents to firms and on consumers' surpluses. In the long run, quasi-rents are competed away, except for those accruing to specialized factors owned by the firms. These rents can be viewed as increases in the prices of such factors.

Consider a competitive industry with n firms producing a single good, x. The ith firm is assumed to have a production function $x_i = x_i(V_i, k_i, q)$, $(i = 1, \ldots, n)$, where V_i is a vector of variable factor inputs to firm i, v_{ij} $(j = 1, \ldots, m)$, k_i is a fixed factor, and q measures environmental or resource quality. For example, q could be a measure of air or water quality or natural soil fertility. Alternatively, q could be some measure of technical efficiency such as the rate at which an electrical generating station converts thermal energy to electricity. In this case, the model developed here would measure the benefits of investments in increasing technical efficiency. To simplify the notation, the k_i terms will be suppressed in all that follows. Both V_i and q have positive marginal products, but q is given exogenously to the industry.

Assume that the industry faces perfectly elastic supplies for all factor inputs at prices f_j, or in vector notation, F. If factor supplies are less than perfectly elastic, the changes in rents to factor owners as factor prices change must also be included in the social welfare measure. Also, as Just and others (1982, Chapter 9 and Appendix D) show in their analysis of multimarket changes, general equilibrium supply and demand curves must be used. This is discussed below.

Let aggregate industry output be denoted by $y = \Sigma_i x_i$. The industry faces an inverse demand function for its product, $p = p(y)$, where p is the market price and income and all other prices are assumed constant. Define an aggregate production function

$$y = y(v_{11}, \ldots, v_{nm}, q) \tag{9-1}$$

to be the sum of the n firm production functions.

The Production Function Approach

Now assume that demand functions are compensated so that consumer welfare changes can be measured by the appropriate areas. Then the social welfare W associated with producing y is the area under the demand curve for y, less the cost of the inputs:

$$W\left(v_{11}, \ldots, v_{nm}, q\right) = \int_0^y p(u)du - \sum_i \boldsymbol{F} \cdot \boldsymbol{V}_i \tag{9-2}$$

The first-order conditions for choosing the v_{ij} to maximize social welfare are

$$\frac{\partial W}{\partial v_{ij}} = \frac{\partial y}{\partial v_{ij}} \cdot p\left(y\right) - f_j = 0 \quad \text{for all } i, j \tag{9-3}$$

Because each firm is a price taker, this welfare optimum is also the competitive equilibrium. These first-order conditions define input demand functions $v_{ij}^*(f_j, q)$, and, in turn, an output function $y^*(q) = y[v_{ij}^*(f_j, q)]$ and a social welfare function $W(q) = W[v_{ij}^*(f_j, q)]$. The asterisks indicate optimally chosen quantities. In what follows, the f_j will not appear as function arguments except where factor prices are assumed to be variable.

Using the envelope theorem, we have

$$\frac{\partial W}{\partial q} = p(y^*) \cdot \frac{\partial y\left[v^*(q), q\right]}{\partial q} \tag{9-4}$$

The net welfare gain is, in effect, the value of the marginal product of q in the production function. Note that $\partial y / \partial q$ is not the observed increase in y^*. Rather, it is the increase in y that would occur holding all other inputs constant. Knowledge of the production function is required to implement this measure.

The Cost Function Approach

The variable cost functions for firms can be added to obtain an aggregate variable cost function for the industry:

$$C = C[y^*(q), q] \tag{9-5}$$

Assuming profit maximization with price equal to marginal cost, this also defines a market supply curve. The total contribution to social welfare made by the production and consumption of this good is the sum of compensating surpluses (*CS*) and quasi-rents to firms:

$$W = \int_0^y p(u)\,du - C\left(y^*, q\right) \tag{9-6}$$

This is maximized when output is set where price equals marginal cost,

$$p(y) = \frac{\partial C(y^*, q)}{\partial y} \tag{9-7}$$

and, as before, the optimal output is a function of the parameter q.

Again applying the envelope theorem, we have the following expression for welfare change associated with marginal changes in q:

$$\frac{\partial W}{\partial q} = -\frac{\partial C(y^*, q)}{\partial q} \tag{9-8}$$

This marginal value can be calculated if the cost function is known. The duality of the cost function and the production function ensure that the two measures of marginal welfare change given by equations 9-4 and 9-8 are equivalent.

The change in total welfare is positive if increasing q reduces costs. However, this fact does not necessarily imply that both producers' surplus and *CS* must increase. *CS* is given by the area under the compensated demand curve less the actual expenditure, or

$$CS\left(q\right) = \int_0^{y^*} p\left(u\right)du - \left[p\left(y^*\right) \cdot y^*\left(q\right)\right] \tag{9-9}$$

The marginal effect on *CS* of a change in q is the derivative of this expression, or

$$\frac{\partial CS}{\partial q} = -\frac{\partial p}{\partial y^*} \cdot \frac{\partial y^*}{\partial q} \cdot y^* \tag{9-10}$$

Of the three terms in equation 9-10, $\partial p/\partial y^* \leq 0$ by the law of demand and $y^* > 0$. Thus *CS* will decrease if and only if equilibrium output y^* decreases with an increase in q. This possibility cannot be ruled out.* It is possible for the marginal cost curve to increase over the relevant range, leading to an increase in price, even though total cost is reduced. Therefore, it is possible for consumers to be made worse off by an increase in q.

The marginal change in producers' quasi-rent, R, is the marginal welfare change given by equation 9-8 less the change in CS given by equation 9-10, that is,

$$\frac{\partial R}{\partial q} = -\frac{\partial C}{\partial q} + \left(\frac{\partial p}{\partial y^*} \cdot \frac{\partial y^*}{\partial q} \cdot y^*\right) \tag{9-11}$$

The first term is positive, and the second term is usually negative. Thus quasi-rent could decrease only if the second term were larger in absolute value. The second term can also be expressed in terms of the price elasticity of demand:

$$\varepsilon = \frac{\partial y^*}{\partial p} \cdot \frac{p}{y^*} < 0 \tag{9-12}$$

When this expression is substituted into equation 9-11, we see that the magnitude of the second term varies inversely with the elasticity of demand. For sufficiently inelastic demands, the fall in price brought about by the increase in quantity could actually harm producers by reducing their quasi-rents.

Nonmarginal Changes in q

For nonmarginal changes in q, say a change from q^0 to q^1, the aggregate benefit can be found by integrating either equation 9-4 or equation 9-8.

*To determine the sign of $\partial y^*/\partial q$, differentiate the first-order condition 9-7:

$$\frac{\partial p}{\partial y^*} \cdot \frac{\partial y^*}{\partial q} = \left(\frac{\partial^2 C}{\partial y^{*2}} \cdot \frac{\partial y}{\partial q}\right) + \frac{\partial^2 C}{\partial y^* \partial q}$$

and rearrange to obtain

$$\frac{\partial y^*}{\partial q} = \frac{(\partial^2 C / \partial y^* \partial q)}{[(\partial p / \partial y^*) - (\partial^2 C / \partial y^{*2})]}$$

Because the denominator of this expression is always negative, $y^*(q)$ is decreasing whenever $\partial^2 C/\partial y^* \partial q$ is positive.

$$\begin{aligned} W_q &= \int_{q^0}^{q^1} p(y^*) \cdot \left\{ \frac{\partial y\left[V^*(q), q\right]}{\partial q} \right\} dq \\ &= -\int_{q^0}^{q^1} \left\{ \frac{\partial C\left[y^*(q), q\right]}{\partial q} \right\} dq \end{aligned} \tag{9-13}$$

As equation 9-13 indicates, both of these measures require integration along a path. To implement them, not only do we need to know either the production function or the cost function, but we also must know how equilibrium output changes with q or how the levels of inputs change with q, or both.

However, W_q may be calculated directly from the demand curve and the cost function if the initial and final output levels y^{*0} and y^{*1} are known. This calculation is the change in the area bounded by the demand curve and cost curve:

$$W_q = \int_0^{y^1} p(y)dy - C(y^1, q^1) - \int_0^{y^0} p(y)dy - C(y^0, q^0) \tag{9-14}$$

This is shown in Figure 9-1, where $p_y(y)$ is the inverse demand curve and $MC_y(q^0)$ and $MC_y(q^1)$ indicate the industry marginal cost curves as a function of y^* and q. Graphically, equation 9-14 is equivalent to

$$\begin{aligned} W_q &= (a+b+c+d+e)-(c+e)-(a+b+c)+(b+c) \\ &= b+d \end{aligned} \tag{9-15}$$

The welfare change can be represented diagrammatically by the area between the old and new cost curves bounded by the demand curve. The division of the gain between producers and consumers is also shown graphically in Figure 9-2. Consumers gain the areas $u + v + w$. Producers lose area w but gain $s + t$. The net social gain is $s + t + u + v$. If the cost and the demand functions are known, these areas can be calculated by taking the appropriate integrals.

Conclusions

In this section, two alternative measures have been identified for the welfare change associated with a change in a nonmarket input such as q. One is based on shifts in the cost or supply curves (the cost function approach); the other is based on the direct effect of the change in q on the production function (the production function approach). Because of duality, they are

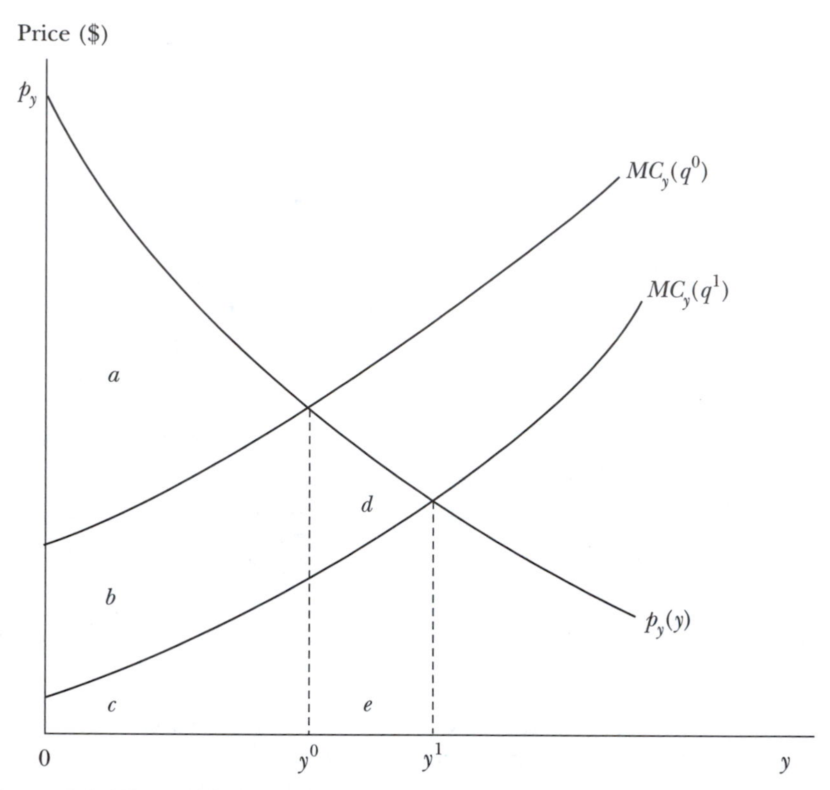

Figure 9-1. The welfare change for single-product firms

both based on essentially the same information and must give the same results. As a practical matter, the choice of a measure depends on the availability of data.

In some cases it may be possible to estimate the production function from data on physical inputs and outputs and environmental quality. For example, the availability of experimentally derived crop loss functions for ozone (see Heck and coauthors 1983) makes it feasible to use the production function approach to measure the benefits to agriculture of reducing ozone air pollution. Alternatively, one could estimate the cost function for an industry (Mathtech 1982) or the rent or profit function (Mjelde et al. 1984).

The production function and cost function methods are based on observations of the optimizing behavior of producers. An alternative to observing this behavior is to simulate it through formal optimization models. One example of this approach is the model used by Adams, Hamilton, and McCarl (1984) and Adams and McCarl (1985) to estimate the effects on the agricultural sector of controlling ozone air pollution. They com-

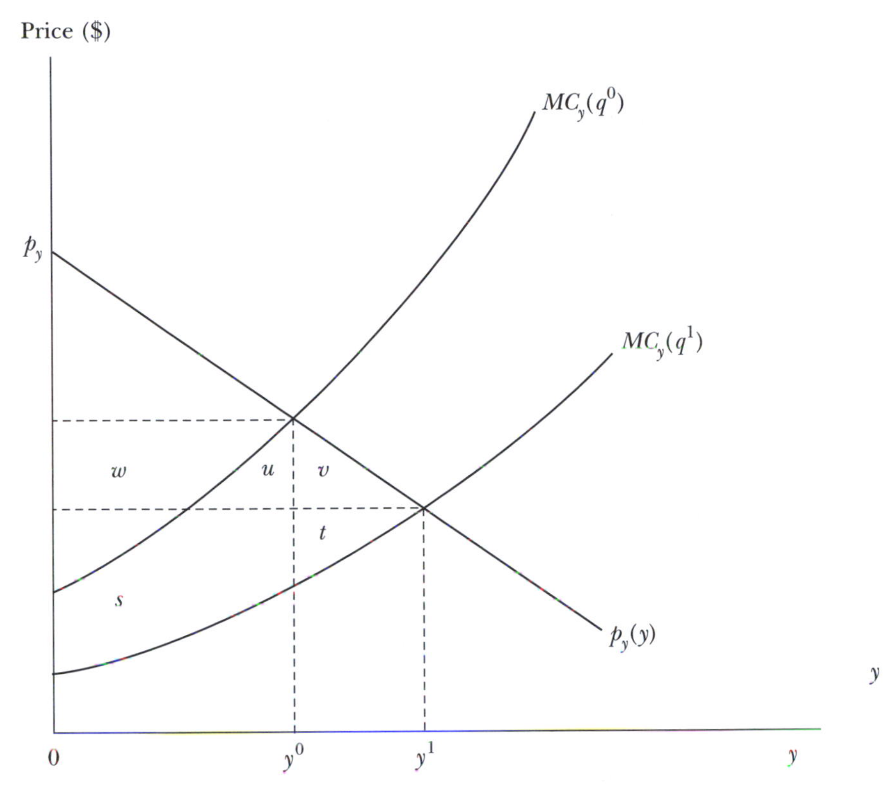

Figure 9-2. The division of the welfare gain between consumers and producers

bined a mathematical programming model of representative farms in the five Corn Belt states with an experimentally derived damage function to simulate producers' responses to changes in ozone levels. In another study of the economics of controlling ozone air pollution and agriculture, Kopp and coauthors (1985) constructed supply functions from region-specific farm budget and farm practice data. This model lacks the formal optimization characteristics of the Adams and McCarl model, but it allows for more detailed treatment of differences in farm economics across regions.

Multiproduct Firms

In the case of multiproduct firms with joint production technologies, things are not so simple. For marginal changes, the production function and cost function approaches can still be applied, with appropriate modification. However, for nonmarginal changes, it is not correct simply to add up areas between observed shifts of marginal cost or supply curves. When production is characterized by jointness, the model used for measurement must account for the interconnectedness of the marginal cost or supply functions for the various outputs. The gain in welfare for an increase in q is measured by the sum of the changes in compensating surpluses in the markets for all the affected products plus the aggregate change in quasi-rents to the affected firms. However, in measuring the changes in quasi-rents, it is necessary to account for how changes in the output of one good affect the marginal cost curves of the other goods being jointly produced.

For notational simplicity, I assume here only two products. The other notation is unchanged. The production function for the ith firm is given by

$$x_i(x_{1i}, x_{2i}, v_{i1}, \ldots, v_{im}, q) \geq 0 \tag{9-16}$$

That is, $x_i \geq 0$ is feasible production and $x_i = 0$ is efficient production. I also assume $\partial x_i/\partial q > 0$, which ensures that increasing q will increase the welfare associated with the production of x_{1i} and x_{2i}.

Both the aggregate production function and the cost function approaches still provide a straightforward basis for measuring the marginal welfare change. As before, assume that each firm is a price taker in all its product and factor markets and that the industry faces infinitely elastic factor supply curves. In a fashion similar to the analysis leading to equations 9-4 and 9-8, we can define the aggregate production function as the sum of the n firm production functions:

$$y(y_1, y_2, V, q) = 0 \tag{9-17}$$

where $y_1 = \Sigma^i x_{1i}$, $y_2 = \Sigma^i x_{2i}$, and $V = (v_{11}, \ldots, v_{nm})$ is the vector of all inputs for all firms. As before, the social welfare function is

$$W(V,q) = \int_0^{y_1} p_1(u_1)\,du_1 + \int_0^{y_2} p_2(u_2)\,du_2 - \Sigma \boldsymbol{F} \cdot \boldsymbol{V} \tag{9-18}$$

subject to $y(y_1, y_2, V, q) = 0$.

The first-order conditions for a maximum of social welfare are

$$W_1 = p_1(y_1) - \lambda \partial y/\partial y_1 = 0 \tag{9-19}$$

$$W_2 = p_2(y_2) - \lambda \partial y/\partial y_2 = 0 \tag{9-20}$$

$$W_{vj} = -f_j - \lambda \partial y/\partial v_j = 0 \tag{9-21}$$

where λ is the Lagrangian multiplier on the production constraint. These conditions define optimal $y_1^*(q)$, $y_2^*(q)$, and $V^*(q)$ in the usual way. By the envelope theorem,

$$\partial W/\partial q = -\lambda \cdot \partial y/\partial q \tag{9-22}$$

If we now differentiate the production function with respect to q, we have

$$\left(\frac{\partial y}{\partial y_1} \cdot \frac{\partial y_1^*}{\partial q}\right) + \left(\frac{\partial y}{\partial y_2} \cdot \frac{\partial y_2^*}{\partial q}\right) + \frac{\partial y}{\partial q} + \left(\sum_j \frac{\partial y}{\partial v_j} \cdot \frac{\partial v_j^*}{\partial q}\right) = 0 \tag{9-23}$$

Combining this with equation 9-22 yields

$$\begin{aligned} \frac{\partial W}{\partial q} &= -\lambda \cdot \frac{\partial y}{\partial q} \\ &= \left(\lambda \cdot \frac{\partial y}{\partial y_1} \cdot \frac{\partial y_1^*}{\partial q}\right) + \left(\lambda \cdot \frac{\partial y}{\partial y_2} \cdot \frac{\partial y_2^*}{\partial q}\right) + \frac{\partial y}{\partial q} + \left(\sum_j \lambda \cdot \frac{\partial y}{\partial v_j} \cdot \frac{\partial v_j^*}{\partial q}\right) = 0 \end{aligned} \tag{9-24}$$

Substituting terms from the first-order conditions yields

$$\frac{\partial W}{\partial q} = \left[p_1\left(y_1^*\right) \cdot \frac{\partial y_1^*}{\partial q}\right] + \left[p_2\left(y_2^*\right) \cdot \frac{\partial y_2^*}{\partial q}\right] - \sum_j f_j \frac{\partial v_j^*}{\partial q} \tag{9-25}$$

Alternatively, in terms of the aggregate cost function, $C(y_1^*, y_2^*, q)$, the benefits are

$$\frac{\partial W}{\partial q} = -\frac{\partial C\left(y_1^*, y_2^*, q\right)}{\partial q} \tag{9-26}$$

In the case of nonmarginal changes in q, it is tempting to say that we need only to add up the areas between the new and old supply curves for each of the products of the multiproduct firms, but that would be wrong. Because of the interdependencies of the marginal cost curves under joint production, adding up these areas will not give a correct measure of the change in quasi-rents. The benefit of a nonmarginal change is the increase in the social values of the outputs, net of any changes in the joint cost of production. Again, for a change from q^0 to q^1 we have

$$W_q = \int_{y_1^0}^{y_1^1} p_1\left(y_1^*\right) dy_1 + \int_{y_2^0}^{y_2^1} p_2\left(y_2^*\right) dy_2 + \left[C\left(y_1^0, y_2^0, q^0\right) - C\left(y_1^1, y_2^1, q^1\right)\right] \tag{9-27}$$

The change in joint costs, ΔC, can be decomposed into three steps:

$$\Delta C = C\left(y_1^0, y_2^0, q^0\right) - C\left(y_1^0, y_2^0, q^1\right) + C\left(y_1^0, y_2^0, q^1\right) - C\left(y_1^1, y_2^0, q^1\right) + C\left(y_1^1, y_2^0, q^1\right) - C\left(y_1^1, y_2^1, q^1\right) \tag{9-28}$$

Substituting this into equation 9-27 gives

$$W_q = \int_{y_1^0}^{y_1^1} p_1\left(y_1^*\right) dy_1 + \int_{y_2^0}^{y_2^1} p_2\left(y_2^*\right) dy_2 - \int_{q^0}^{q^1} \left[\frac{\partial C\left(y_1^0, y_2^0, q\right)}{\partial q}\right] dq - \int_{y_1^0}^{y_1^1} \left[\frac{\partial C\left(y_1, y_2^0, q^1\right)}{\partial q}\right] dy_1 - \int_{y_2^0}^{y_2^1} \left[\frac{\partial C\left(y_1^1, y_2, q^1\right)}{\partial q}\right] dy_2 \tag{9-29}$$

In each of the markets, one of the two marginal cost curves used in defining the welfare measure presented here is not actually observed. Implementation of equation 9-29 requires full knowledge of the joint cost function. Basing measures on comparisons of observed marginal cost curves before and after the change will result in error, the sign and magnitude of which will depend on the specific characteristics of the joint technology.

A graphic interpretation of equation 9-29 may prove helpful. First consider the case where y_1 and y_2 are substitutes in production in the sense that $\partial^2 C/\partial y_1 \partial y_2 = \partial^2 C/\partial y_2 \partial y_1 > 0$. This case is shown in Figure 9-3. The solid cost curves represent the pre- and postchange observed marginal cost curves. The dashed cost curve in panel A corresponds to the fourth integral in equation 9-29, where q has changed holding y_2 at y_2^0.

The geometric areas in panels A and B corresponding to the five integrals in equation 9-29 are given by the following:

$$\begin{aligned} W_q &= (a+b+c)+(f+g)+(d+e)-c-g \\ &= a+b+d+e+f \end{aligned} \tag{9-30}$$

The welfare change is measured by the area between the two cost curves for y_1 holding y_2 at y_2^0 plus the observed welfare triangle f in the market for y_2 (panel B). An empirical measure based on areas between observed cost curves would yield an underestimate of the true welfare change.

Figure 9-4 shows the case where y_1 and y_2 are complements in production, that is, $\partial^2 C/\partial y_1 \partial y_2 = \partial^2 C/\partial y_2 \partial y_1 < 0$. The geometric equivalent to equation 9-29 is

$$\begin{aligned} W_q &= (a+b+c)+(f+g)+d-(b+c+e)-g \\ &= a+d-e+f \end{aligned} \tag{9-31}$$

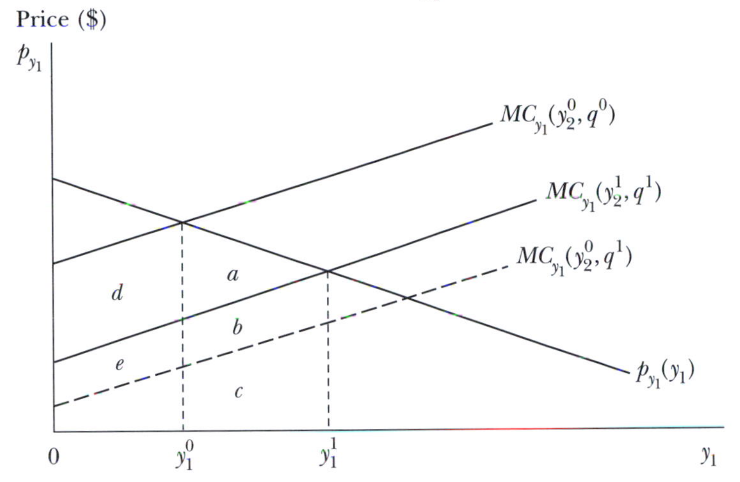

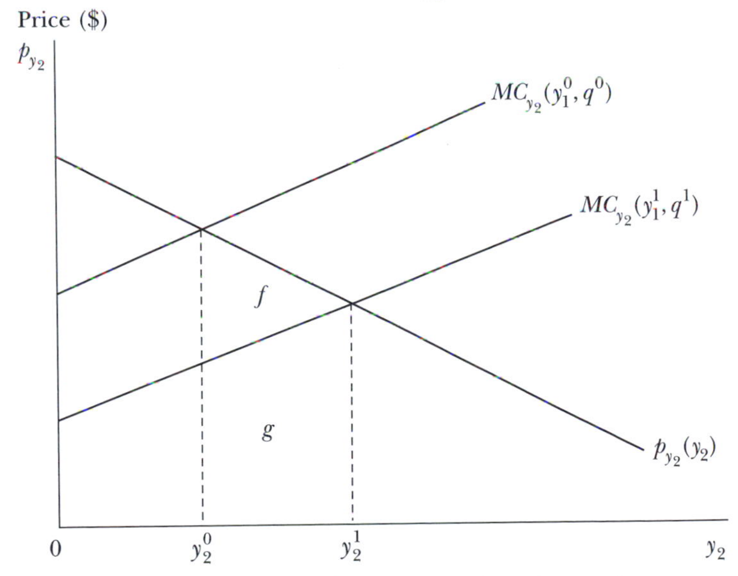

Figure 9-3. The welfare measure for multiproduct firms when outputs are substitutes in production

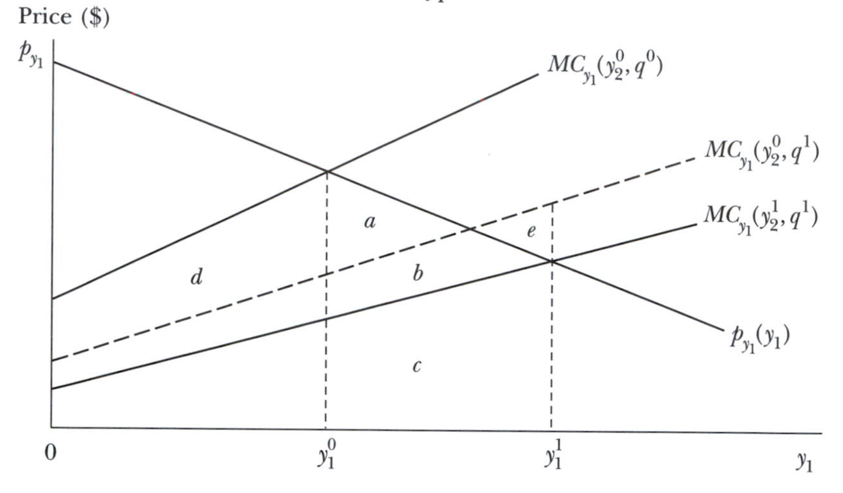

Panel B: Demand and Cost Curves for y_2

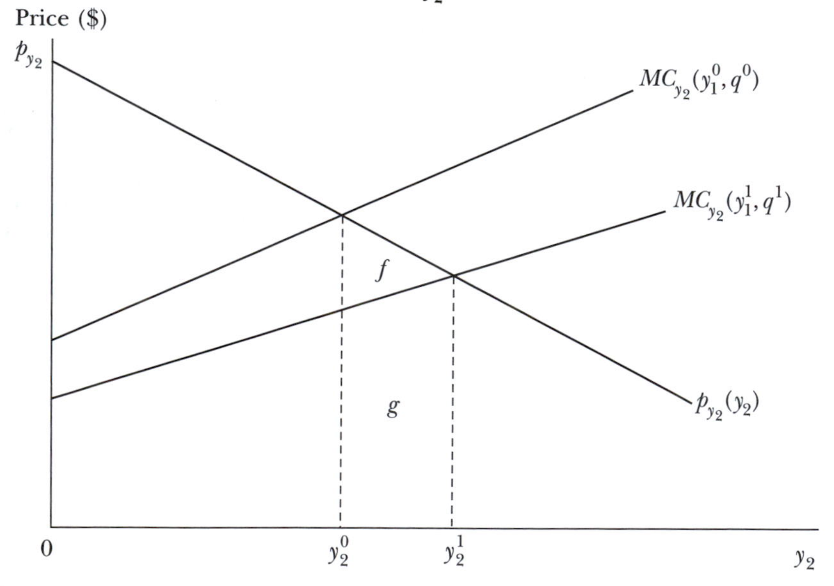

Figure 9-4. The welfare measure for multiproduct firms when outputs are complements in production

Again, measuring the first term requires knowledge of the unobserved cost curve for y_1 after the change in q but holding y_2 at y_2^0.

It is of interest to know the magnitude of the error that would result from failure to recognize the existence of joint costs. To get an approximation of the error, form a second-order Taylor polynomial expanded around (y_1^0, y_2^0, q^0). The approximate change in cost is

$$\Delta C = C\left(y_1^1, y_2^1, q^1\right) - C\left(y_1^0, y_2^0, q^0\right) = \Delta C^1 + \Delta C^2 + \Delta C^J \tag{9-32}$$

where

$$\begin{aligned}\Delta C^1 = &\left[\left(y_1^1 - y_1^0\right)\cdot\frac{\partial C}{\partial y_1}\right] + \left[\frac{1}{2}\left(y_1^1 - y_1^0\right)^2\cdot\frac{\partial^2 C}{\partial {y_1}^2}\right] \\ &+ \left[\left(y_1^1 - y_1^0\right)\left(q^1 - q^0\right)\frac{\partial^2 C}{\partial y_1 \partial q}\right]\end{aligned} \tag{9-33}$$

$$\begin{aligned}\Delta C^2 = &\left[\left(y_2^1 - y_2^0\right)\cdot\frac{\partial C}{\partial y_2}\right] + \left[\frac{1}{2}\left(y_2^1 - y_2^0\right)^2\cdot\frac{\partial^2 C}{\partial y_2}\right] \\ &+ \left[\left(y_2^1 - y_2^0\right)\left(q^1 - q^0\right)\frac{\partial^2 C}{\partial y_2 \partial q}\right]\end{aligned} \tag{9-34}$$

$$\Delta C^J = \left(y_1^1 - y_1^0\right)\left(y_2^1 - y_2^0\right)\frac{\partial^2 C}{\partial y_1 \partial y_2} \tag{9-35}$$

The third term, ΔC^J, is the contribution to the change in cost by the jointness in y_1 and y_2.

There is another way of looking at the difficulties involved here. Recall that one component of the welfare change is the change in quasi-rents to firms. With joint production, in general it is not possible to measure the quasi-rent of a multiproduct firm from data from just one market. However, Just and others (1982) showed that in the case of price changes there are circumstances in which it is possible to measure the change in quasi-rents in either a single factor market or a single product market. Their analysis can be extended to the case of parametric shifts in production technology. There must be either a necessary input or a necessary output. A necessary output is one for which there is some positive minimum price at which the firm will choose to stop producing not only that output but all other products as well; in other words, production will shut down completely. All of the quasi-rent to a firm can be attributed to the necessary output and can be measured by integrating above the supply curve for that output from the shutdown price to the current market price. Alternatively, the quasi-rent can be measured by the area under the demand curve for any necessary factor input. An input is deemed necessary if there is some price for that input at which its derived demand falls to zero and if all of the firm's outputs fall to zero when that input is set at zero.

The measurement of changes in quasi-rents can be shown graphically for the case of two necessary outputs, y_1 and y_2, and one necessary input, v_1. Suppose that, as shown in Figure 9-5, an increase in q shifts the supply functions for both outputs and the demand function for the necessary input to the right. The compensating surplus component of the welfare change is the areas $w_1 + v_1 + w_2 + v_2$ in panels B and C. The increase in the quasi-rents to all firms can be measured alternatively by $u_1 + t_1 - w_1$ in panel B, or $u_2 + t_2 - w_2$ in panel C, or b in panel A.

To see this, first suppose that y_1 is the necessary output. At q^0 if p_1 is at p_1^0, other things being equal, firms shut down and quasi-rents are zero. If p_1 increases to p_1^1, firms produce both goods in positive quantities, but all the quasi-rents are attributable to the increase in p_1 and are measured by the area $w_1 + z_1$. Comparing quasi-rents at q^1 and q^0 gives $\Delta R = R^1 - R^0 = u_1 + t_1 + z_1 - w_1 - z_1 = u_1 + t_1 - w_1$. For the necessary input v_1, if its price is f_1^1, firms shut down and quasi-rents are zero. If f_1 falls, firms produce both goods. All of the quasi-rents are attributable to the increased use of v_1 and are measured by the area a in panel A.

The three alternative measures of the total welfare gain are the following:

y_1 is necessary:
$$\begin{aligned} W_{1q} &= w_1 + v_1 + t_1 + w_2 + v_2 + u_1 - w_1 \\ &= u_1 + v_1 + t_1 + w_2 + v_2 \end{aligned} \tag{9-36}$$

y_2 is necessary:
$$\begin{aligned} W_{2q} &= w_1 + v_1 + w_2 + v_2 + u_2 + t_2 - w_2 \\ &= w_1 + v_1 + u_2 + v_2 + t_2 \end{aligned} \tag{9-37}$$

v_1 is necessary:
$$W_{vq} = w_1 + v_1 + w_2 + v_2 + b \tag{9-38}$$

Where empirical measures of producers' benefits have been reported in the literature, production has typically been modeled either implicitly or explicitly as being nonjoint, even in those cases where multiproduct firms are typical. For example, Adams, Hamilton, and McCarl (1984), Adams and McCarl (1985), Kopp and coauthors (1985), and Adams, Crocker, and Katz (1984) all provided estimates of the benefits to agriculture of controlling ozone air pollution, using variations of the cost function approach. The first two studies used supply functions derived from models of farm behavior rather than from econometric estimates. Independence of the marginal cost or supply curves by crop was a characteristic of these models.

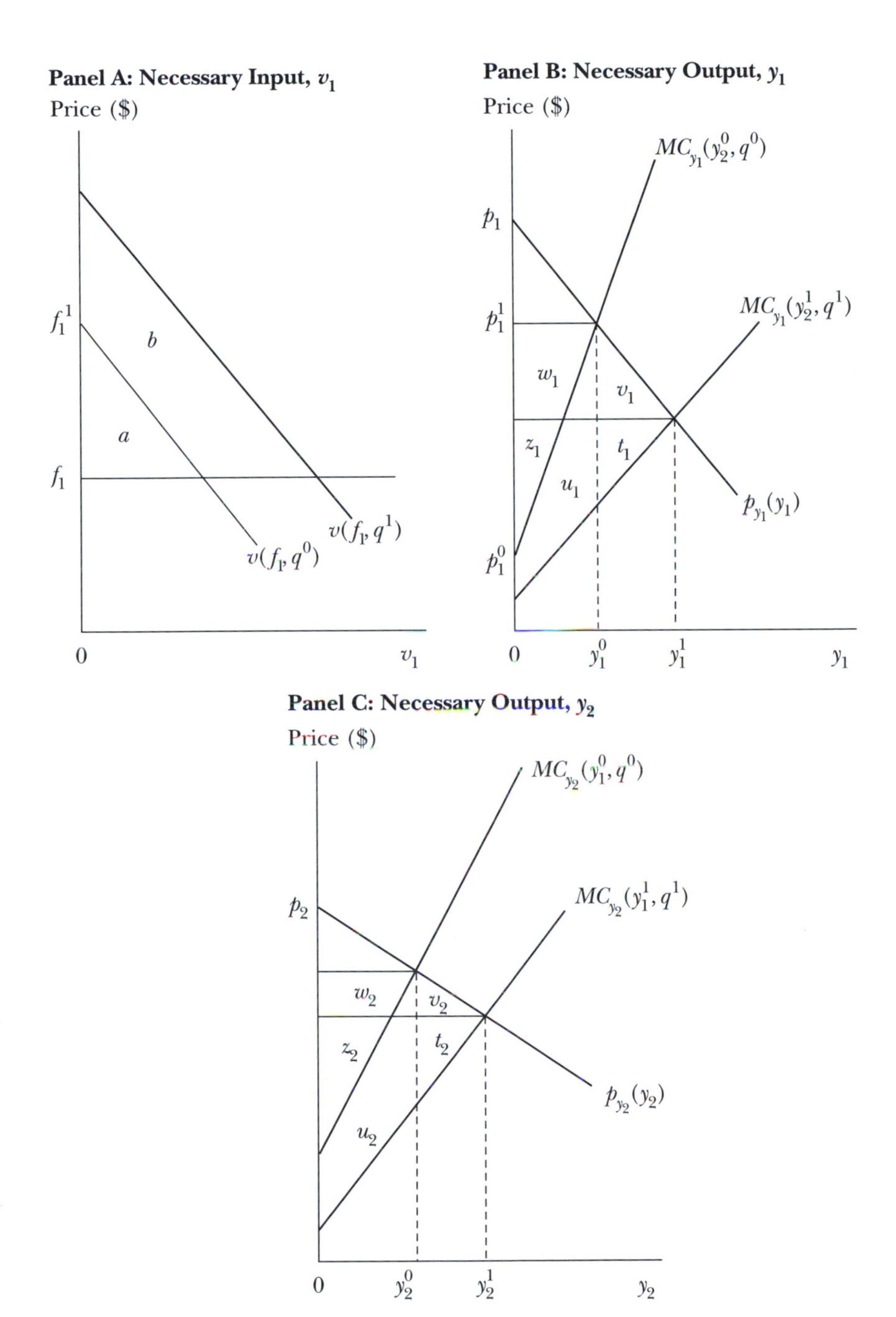

Figure 9-5. The welfare gain for multiproduct firms with necessary outputs

Adams, Crocker, and Katz (1984) used econometric estimates of supply functions for four crops, but the estimating equations imposed nonjointness on the production technology even though many farms produced at least two of the four crops modeled (corn and soybeans). Other researchers (Mathtech 1982; Mjelde et al.1984; Garcia et al. 1986) have aggregated across products to obtain industry profit or cost functions. However, consistent aggregation of this sort is possible only if marginal cost functions are independent, a condition that is not satisfied when there is joint production. The analysis presented here suggests that in the future it will be important to confront the question of joint production directly. If the industries being studied are characterized by joint production, the empirical models of production must reflect this, and the measures of welfare change must account for this in the way just described.

Vertically Linked Markets

The next case to consider is that of vertically linked markets in which the output of one set of firms is purchased as an input by another set of firms. Assume that q affects costs and prices in one industry that is part of a set of vertically linked industries. The model used to derive welfare measures in this case must take account of the fact that every price change affects both buyers and sellers, but in different directions; that is, for a price increase, the buyer loses while the seller gains, and vice versa for price decreases. For simplicity, assume that all firms are single-product firms. Some firms are producers that purchase primary factor inputs and sell an intermediate product y_w. Other firms buy the intermediate product and factor inputs and sell y_r at retail to consumers. Assume that all factor supply curves but one are infinitely elastic; the exception is the supply curve for labor used in the wholesale industry, v_w. Assume also that the compensated supply curve for v_w is upward sloping and that y_w and y_r are produced and sold in perfectly competitive markets. Figure 9-6 shows the case.

Assume now that q affects the producers of y_w. Suppose that at the initial level of q the supply function for y_w is MC_w^0 and the demand curve is D_w^0, as shown in panel B of Figure 9-6. The first effect of an increase in q is to shift MC_w down to MC_w^1. This results in a similar downward shift of MC_r and a decrease in the retail price, as shown in panel C of Figure 9-6. In the wholesale market, the demand curve for y_w is derived based on the assumption that the retail price is constant. When the retail price falls, the derived demand for y_w shifts to the left. The wholesale producers respond to the changes in demand along a path that reflects the adjustments in price in the retail market. The path of this adjustment can be described by what Just and others (1982) call a general equilibrium demand curve. This

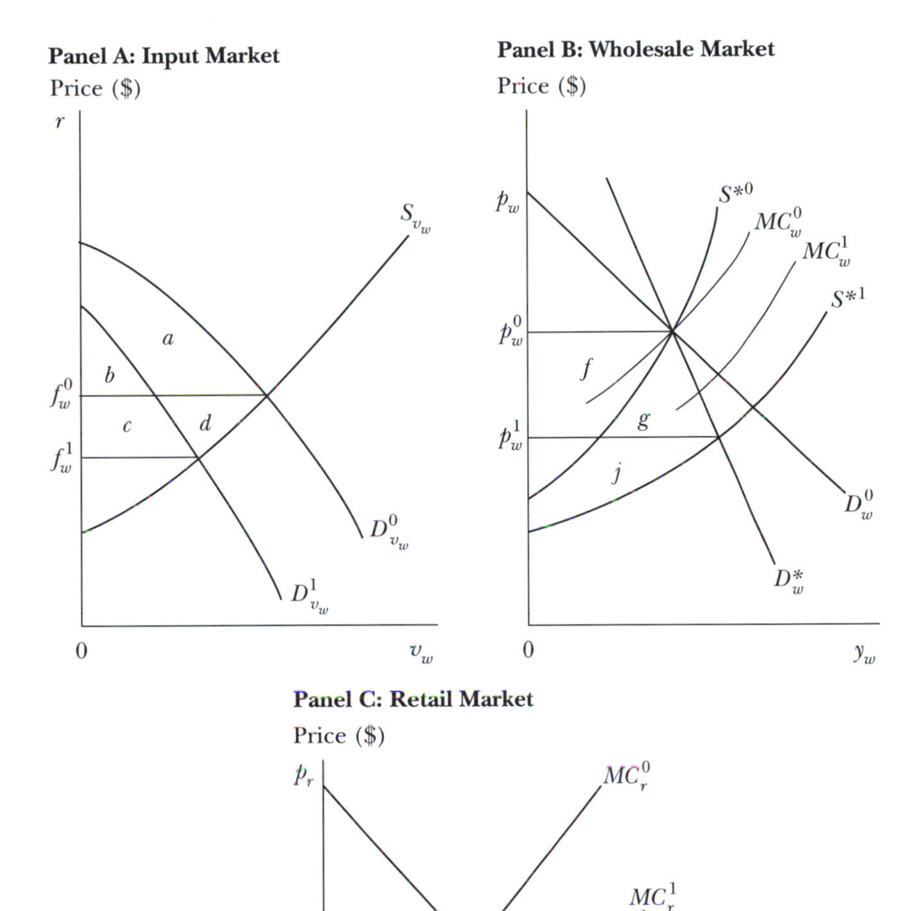

Figure 9-6. Measuring welfare gains in vertically linked markets

demand curve shows the maximum willingness to pay by the y_r industry, given its costs and the retail price at which it must sell its product, and assuming that all other input prices are constant. This is D_w^* in panel B of Figure 9-6.

At the same time, the demand curve for v_w in panel A of Figure 9-6 is shifting, for two reasons. First, the decrease in p_w pushes the demand curve for v_w to the left. Second, the change in q also affects the productivity of and demand for v_w. The direction of this effect depends on the specific characteristics of the production function and on the elasticity of demand for y_w. As the price of v_w changes, this shifts MC_w in the same direction. There are also two general equilibrium supply functions, S^{*0} and S^{*1}, in panel B, one for each level of q. They describe the supply adjustments to changing factor prices. They show the minimum supply price for y_w for each level of q, taking account of changes in factor prices as the derived demand for v_w shifts. The net result is an equilibrium in this market at the intersection of D_w^* and S^{*1}, as shown in Figure 9-6 in panel B.

Three groups of economic agents are affected by the change in q, so there are three components to the measure of welfare gain. The first group, the consumers of y_r, receive an increase in consumers' surplus, shown as $w + v$ in Figure 9-6 in panel C. The second group, the owners of v_w, experience a change in their rents because of the change in price. They may either gain or lose depending upon how the demand for v_w changes; in Figure 9-6, panel A, they lose the area $c + d$. The third group consists of the firms in the wholesale and retail industries that experience changes in their quasi-rents. The change in quasi-rents in the retail industry can be measured directly in the retail market by the area $u - w$ in panel C. As discussed above, if v_w is a necessary input, then the change in quasi-rents to the wholesale industry can be measured in the factor market by the area $c - a$ in panel A. Considering all of the changes, and netting out transfers between buyers and sellers resulting from price changes, we have

$$\begin{aligned} W_q &= \Delta PS_v + \Delta R_w + \Delta R_r + \Delta CS_r \\ &= -(c+d)+(c-a)+(u-w)+(w+v) \\ &= -(d+a)+(u+v) \end{aligned} \tag{9-39}$$

Following Just and others (1982), it can be shown that the welfare gain can also be measured in the directly affected wholesale market, provided that the measurements are based on the general equilibrium supply and demand curves. These curves reflect the adjustments of prices in linked markets that transfer gains and losses among buyers and sellers. Specifically,

$$W_q = \Delta CS_w^* + \Delta R_w^* \tag{9-40}$$

where CS_w^* (= $\Delta R_r + \Delta CS_r$) and R_w^* (= $\Delta R_w + \Delta PS_v$) are the areas behind the general equilibrium demand and supply curves, respectively. So from equation 9-40,

$$W_q = f + g + j - f = g + j \tag{9-41}$$

Thus if general equilibrium demand and supply functions for the directly affected market can be estimated, it may be more convenient to measure welfare gains using this expression than to attempt to measure separately the changes in quasi-rents and surpluses in all the remaining vertically linked markets.

Consider the case of multiproduct firms that are part of a chain of vertically linked markets. As I have just shown, the welfare gain can be measured in the directly affected market. However, the welfare measure must be modified to take account of the multiproduct characteristic of the producing industry. Specifically, if there is any jointness in production, it is not correct simply to add the areas $g + j$ across all products. Rather, the correct measure is the sum of the consumers' surplus changes as measured by areas like $f + g$, and the change in producing firms' quasi-rents as measured by $j - f$, in the market for one essential output.

Market Distortions

The welfare measures derived in the preceding sections come from models in which all markets are perfectly competitive. In this section I briefly examine two kinds of market distortions. The first is distortion caused by monopoly power in the output market; the second is the distortion caused by price support and commodity production control programs such as those affecting U.S. agricultural markets.

Monopoly

The correct welfare measure in the case of monopoly can be derived graphically in a straightforward manner. Measurement is another question. Figure 9-7 shows the demand, marginal revenue, and marginal cost curves for a monopolist. If an increase in q shifts the marginal cost curve outward, the welfare measure is the sum of the increases in consumers' surplus and monopoly quasi-rents:

$$W_q = \Delta CS + \Delta R \tag{9-42}$$

where

$$\Delta CS = a + b + c \tag{9-43}$$

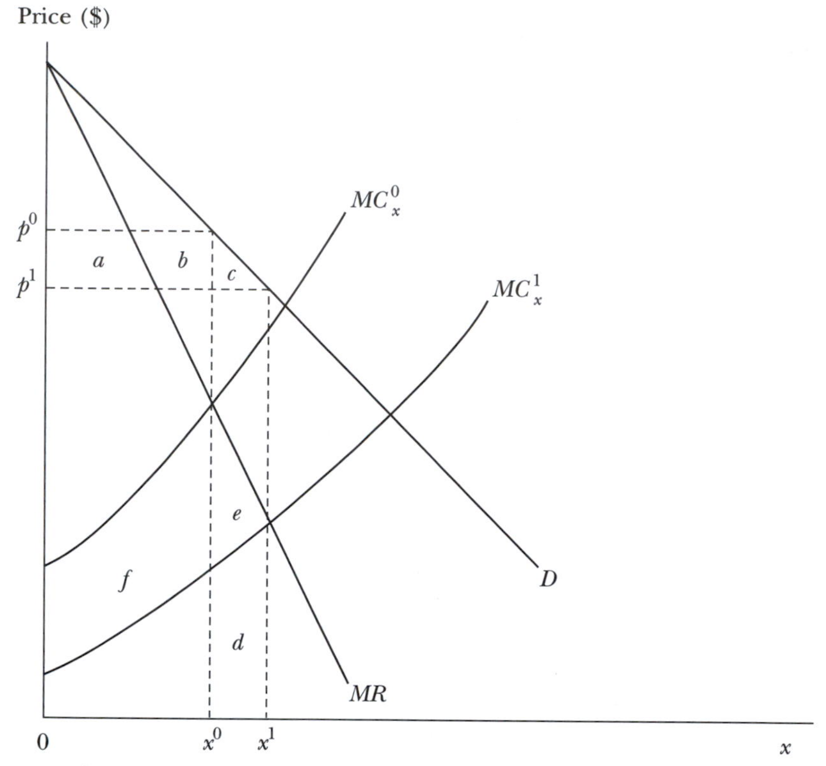

Figure 9-7. The welfare change for a monopoly firm

$$\Delta R = \Delta TR - \Delta C = d + e - (d - f) = e + f \tag{9-44}$$

Thus,

$$W_q = a + b + c + e + f \tag{9-45}$$

As in the case of perfect competition, consumers always gain because of the price reduction. However, in this case the monopolist also always gains from an increase in q. The monopolist's profit is

$$I = p(x) \cdot x - C(x, q) \tag{9-46}$$

$$\frac{\partial I}{\partial q} = \left(p + x\frac{dp}{dx} - \frac{\partial C}{\partial x} \right) \frac{dx}{dq} - \frac{\partial C}{\partial q} \tag{9-47}$$

The first-order condition for profit maximization requires that the term in parentheses be equal to zero. So

$$\frac{\partial I}{\partial q} = -\frac{\partial C}{\partial q} > 0 \tag{9-48}$$

The problem posed for measurement is that the marginal cost curve of the monopolist cannot be observed from market data on optimally chosen prices and quantities. One possible approach to deriving marginal cost functions is to construct models of the firm based on engineering or technological data (see, for example, Russell and Vaughan 1976).

Price Supports: The Case of Agriculture

In the United States, the federal government intervenes in the markets for several major crops to maintain a price above the market clearing price and to support the incomes of farmers. This intervention and distortion of the markets for crops complicates the analysis of the benefits of controlling air pollution–induced damages to agriculture, the benefits of technological change in agriculture, and the costs of regulations to protect other environmental values (for example, the costs of controls on pesticide applications and the costs of changes in farming practices to control erosion and nonpoint-source pollution). Here I will use a fairly simple example based on studies of farm policies of the late 1980s to illustrate the issues that must be dealt with to arrive at a defensible measure of welfare change in the case of market distortions in agriculture.

As described in McGartland (1987), one of the major mechanisms through which the government intervened in agricultural markets at that time was the "target price/deficiency payment," through which farmers receive a subsidy equal to the difference between a target price and the actual market clearing price for each unit of the crop they produce. The target price, which is the effective average revenue for farmers, determines the quantity produced according to the supply function for the crop.

Assume that the government has established p_t as the target price as shown in Figure 9-8. The supply curve for the crop is initially S^0; farmers therefore produce x^0. When this quantity is placed on the market, the equilibrium price is p_m^0.* The government makes up the difference between the

*Models for welfare measurement are sometimes constructed by assuming a particular functional form and elasticity for the demand function, and selecting parameter values to "force" the demand curve through some assumed price and quantity equilibrium such as point F in Figure 9-8. For examples, see Adams, Hamilton, and McCarl 1984 and Kopp et al. 1985. Lichtenberg and Zilberman (1986) and McGartland (1987) point out some of the problems that arise if the calculations and calibration process do not accurately reflect the form of the government intervention in the market.

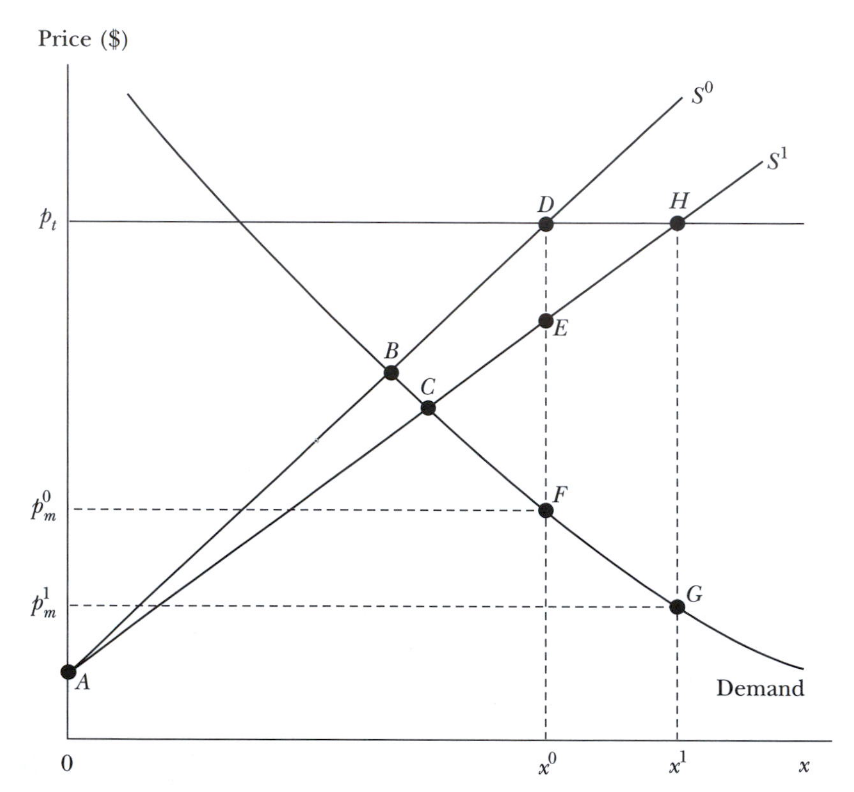

Figure 9-8. Welfare measures in price-supported markets: the case of agriculture

target price and the market price, resulting in an aggregate deficiency payment to farmers equal to the area $p_m^0FDp_t$. Because the efficient level of output would be at the intersection of the demand curve and the supply curve (point *B*), the excess production results in a dead-weight loss equal to the area *BDF*. Note that the deficiency payment is a transfer from taxpayers to producers and that as such it does not enter into the welfare calculation.

Now suppose that an air pollution control program increases productivity for this crop, resulting in the new supply curve S^1. Producers respond by increasing output to x^1, given the unchanged target price p_t. What is the social value of this extra output? The most conservative answer is that the social value is zero, and that, therefore, the "benefit" (negative) of the pollution control is the extra cost of producing the additional output, measured by the area x^0EHx^1. This "conservative" answer would make sense in some circumstances, for example, if the additional output were simply destroyed to keep it off the market. However, under the target price/deficiency payment system, the increment to output is placed on the market and drives the market clearing price down to p_m^1. The value to consumers of the extra output is the area x^0FGx^1. However, the value to consumers is less than the cost of this increment to production. Although the additional output costs more than it is worth to consumers (the difference being the area *FEHG*), the policy also reduces the cost of producing the original output level, x^0. This is clearly a benefit and is measured by the area *ADE*. The net welfare gain is the cost reduction (*ADE*) minus the excess cost of the increase in output (*FEHG*). This amount can be either positive or negative.

The proper welfare measure can also be seen in another way. If the market were in equilibrium, the net gain in producers' and consumers' surpluses would be the area *ABC*. But given the policy, there is also an increase in the dead-weight loss from *BDF* to *CHG*. From this perspective, the net welfare change is the efficiency gain (*ABC*) less the increase in dead-weight loss (*CHG* minus *BDF*). As can be seen from Figure 9-8, these two ways of describing the welfare change are geometrically equivalent.

The welfare measure provided here is based on two key assumptions. The first is that farmers actually respond to the reduction in pollution as predicted by the new supply curve. The second is that there are no changes in government policy in response to the increase in agricultural productivity. Concerning the first assumption, Kopp and Krupnick (1987) argued that farmers might not fully perceive the increase in productivity that lies behind the downward shift in the supply curve, especially given the natural variability of yields due to vagaries of weather, rainfall, and so forth. Because the supply curve is a long-run relationship, it incorporates the predicted increases in the inputs of labor, capital, and land on the part of farmers. However, if farmers do not perceive the increase in productivity, they will not make these increases in factor inputs. Kopp and Krupnick (1987) presented calculations of social welfare gains for a 10% reduction in ambient ozone for four crops under alternative assumptions about policy responses. In all cases, the benefits of the cost reduction outweighed the increases in the dead-weight loss. They show that benefit estimates can vary by up to 30%, depending upon the assumptions about policy responses.

In conclusion, the form that an agricultural price support policy takes can have a substantial effect on the benefits actually realized from any change in agricultural productivity. Measures of welfare change should be based on realistic assumptions concerning the relevant features of agricultural policy. Welfare measures should be interpreted as conditional upon the existence of the agricultural policy assumed. Given present agricultural policy, the benefits of pollution control are partly dissipated by the costs of the additional excess production of farm crops. Conversely, the costs associated with environmental regulations that raise the cost of agricultural production tend to be partly offset by the reduction in the dead-weight losses because of excess production.

Valuing Changes in the Productivity of Natural Resource Systems

The productivity of commercially exploited natural biological systems can depend on such things as the flow of nutrients into the system, the population of a predator species, or the level of a pollutant. Any of these factors might be subject to human manipulation. Therefore, it would be useful to have an economic framework for evaluating the welfare consequences of policies to change these things. Because the outputs of such resource systems are traded in markets, the framework presented in this chapter is applicable. The welfare consequences come in changes in producers' and consumers' surpluses. However, some special features of natural resource systems need to be taken into account.

In this section, I provide a brief description of how an environmental quality variable can be introduced into the standard economic models of two types of natural resource systems: the commercial forest and the commercial fishery. In the case of the forest, an investment in environmental quality may alter the optimum time of harvest, so the intertemporal features of the forest optimization model must be examined. Moreover, an environmental quality variable might affect the economic value of a nonmarket output such as recreation. Thus the interaction between the nonmarket and market outputs must be analyzed in a multiple-use framework.

In the case of the fishery, the literature has two alternative bio-economic models, the Schaefer–Gordon growth model and the Beverton–Holt stock-recruitment model. Environmental parameters can be easily introduced into both models. However, the economic implications of a change in environmental quality are more transparent in the case of the Schaefer–Gordon model, as we will see. In addition, the economic value of an environmental quality change will depend upon the institutions for ownership or management of the fishery. I will show the results of a comparison of the economic value of an environmental change under the alternatives of optimum management and open-access exploitation.

Commercial Forests and the Role of Time

The models described in the preceding sections of this chapter were timeless in the sense that environmental changes and changes in prices and quantities were contemporaneous. Intertemporal effects did not need to be modeled explicitly. Where the linkages between environmental changes and market changes are not contemporaneous, time must be built into the model explicitly. An example is the case of the effects of changes in air quality on commercial forest productivity.

One of the things that makes the case of the effects of environmental change on commercial forests interesting from an economic perspective is that a change affecting a stand of young trees today will not have an effect on marketed outputs for perhaps 40 years, when the trees are harvested and sold. If there were an active market in forest land, any increase in the growth rate of young trees now would have an immediate effect on the market price of land where there are standing, growing trees. But even so, any consumers' surplus benefit from lower prices of forest products would not be realized until the increased harvests actually took place. In what follows, I ignore the product price effects on consumers and focus only on rents accruing to forests and owners.

Suppose that the trees on a plot of land grow in net value over time according to

$$G_t = G(t, q) \tag{9-49}$$

where G_t is the stumpage value, that is, the market price of the harvested volume at age t, net of harvest costs and transportation to the mill. This general formulation allows for growth in harvestable volume as well as for changes in the price per unit of volume because of changes in quality. I also assume that replanting and management costs are zero.

Assuming no economically relevant alternative uses for a unit of land, the landowner's economic problem is to choose a sequence of harvesting dates to maximize the present value (V) of the stream of net receipts at each harvest, that is,

$$\begin{aligned} \max_t : V &= \sum_{h=1}^{\infty} e^{-h \cdot r \cdot t} G(t,q) \\ &= \frac{G(t,q)e^{-r \cdot t}}{1 - e^{-r \cdot t}} \end{aligned} \tag{9-50}$$

where t is the age of the stand at the time of harvest, r is the interest rate, and h indexes the generation of the stand. For simplicity, I assume that there are no costs for planting, thinning, and other management activities during the rotation. For a more complete treatment of the forestry optimization problem, see Samuelson 1976; Hyde 1980; and Bowes and Krutilla 1989.

The stand should be harvested at the age that satisfies

$$\frac{\partial V}{\partial t} = \frac{\dfrac{\partial G}{\partial t} - \left[r \cdot G(\cdot)\right] - \dfrac{r \cdot G(\cdot)e^{-r \cdot t}}{1 - e^{-r \cdot t}}}{1 - e^{-r \cdot t}} = 0 \tag{9-51}$$

or

$$\frac{\partial G}{\partial t} = \left[r \cdot G(\cdot)\right] + \frac{r \cdot G(\cdot)e^{-r \cdot t}}{1 - e^{-r \cdot t}} \tag{9-52}$$

The stand should be allowed to grow as long as the marginal gain in value through growth ($\partial G/\partial t$) exceeds the interest forgone by not realizing G through harvest ($r \cdot G$) plus the opportunity cost of postponing the stream of returns from future rotations. Harvest should occur when the marginal gain from waiting just equals the marginal opportunity cost.

Two questions of interest are the effects of changes in q on V and on the optimum rotation length. From equation 9-50,

$$\frac{\partial V}{\partial q} = \frac{\partial G}{\partial q} \cdot \frac{e^{-r \cdot t}}{1 - e^{-r \cdot t}} \tag{9-53}$$

The increase in q increases G at the time of the next harvest; the second term in this expression gives the present value of the stream of these increases over the infinite future. So if the growth function were known, including the effect of q, then calculating the effect of changes in the steady-state level of q on the stream of rents would be straightforward. However, where q varies over time, because, for example, of trends in pollution associated with economic growth, such calculations would require more detailed knowledge of the growth function than is currently available. Answering the second question appears to be more difficult. If t^* is the solution to equation 9-52, it appears to be impossible to determine unambiguously the sign of $\partial t^*/\partial q$.

Environmental Quality and Multiple-Use Management of Forests

Suppose that in addition to the periodic harvest of marketable products, the forest unit provides a flow of nonmarket services such as recreation that depends on the age of the forest and on q. Models for measuring the value of recreational resources are discussed in Chapter 13. See also Chapter 5 for a discussion of the possibility of nonuse values arising from natural resources.

Let the value of the nonmarket service be given by $R(t, q)$. Over one harvest cycle, the present value of the nonmarket service flow is

$$V_n = \int_0^{t^*} R(t, q)\,dt \tag{9-54}$$

The objective function becomes

$$\max_t : V = \sum_{h=1}^{\infty} e^{-h \cdot r \cdot t} \left[G(t, q) + \int_0^t R(t, q) dt \right] \tag{9-55}$$

As Hartman (1976) first showed, including the value from the standing forest in the objective function will either increase the optimum age at harvest (t^*) or make harvesting entirely uneconomical. This is because the standing forest adds a second term to the left-hand side of equation 9-52, the marginal benefit of delaying the harvest. The optimum harvest (if it exists) occurs when

$$\frac{\partial G}{\partial t} + R(\cdot) = r \cdot G(\cdot) + \frac{r \cdot G(\cdot) e^{-r \cdot t}}{1 - e^{-r \cdot t}} \tag{9-56}$$

An increase in q can be valuable both because of its increase in the present value of the flow of harvestable product and because of its effect on the value of the flow of services from the standing forest. For further discussion of the Hartman result and multiple-use forest management in general, see Bowes and Krutilla 1989; Swallow and Wear 1993; and Swallow et al. 1997.

One benefit of modeling exercises of this sort is to provide guidance for the natural science research required to support future economic analyses. As these models suggest, it is important to know how changes in q affect the whole time pattern of the growth of trees and the flow of nonmarket services over the life of the forest.

Environmental Quality and Commercial Fisheries

The economic analysis of fisheries rests on a foundation of a biological model of the growth and mortality of a species. Two alternative biological models have dominated the literature on the economics of fisheries. The first is the so-called Schaefer–Gordon model, which makes the growth rate of the aggregate biomass of the species at any time a function of the current level of biomass. The alternative Beverton–Holt model explicitly describes both the number of fish in each age cohort and their weight or size. The Beverton–Holt model may be more realistic from a biological perspective, but it is also more complex from an economic perspective.

The Schaefer–Gordon Model. In this model the relationship between the growth of the aggregate stock, g, and the aggregate size of the stock, z, takes the following form:

$$g = b \cdot z \left(1 - \frac{z}{k}\right) = b \cdot z - \frac{b}{k} \cdot z^2 \tag{9-57}$$

where b is the intrinsic growth rate and k is the carrying capacity of the environment. As this equation shows, when the stock has grown to the carrying capacity ($z = k$), the growth rate is zero. This quadratic growth function results in a logistic time path for the stock in the absence of harvest, with z approaching the carrying capacity asymptotically.

Environmental parameters have been incorporated into the Schaefer–Gordon growth model in several empirical studies of fisheries. Either the intrinsic growth rate or the carrying capacity (or both) can be made a function of an environmental quality parameter, q. For example, in his study of the North American lobster fishery, Bell (1972) included sea water temperature as one variable helping to explain the annual harvest. Lynne and others (1981) examined the effects of changes in the acreage of marine wetlands on annual harvests of blue crabs in the Florida Gulf Coast fishery. For other examples, see Bell 1989; Swallow 1994; and Barbier and Strand 1998.

The Schaefer–Gordon model can be incorporated into an economic model to determine the optimum levels of harvest and stock. This model can also be used to trace the effects of changes in environmental parameters on harvest, stock, and economic welfare. If an environmental quality parameter can be increased by public policy at some cost, optimum levels of investment in environmental quality can also be determined.

The standard approach is to specify a production function that makes the fishery industry's annual harvest, h, a function of economic inputs and the stock of fish to be caught, z. For simplicity, the economic inputs are aggregated into a measure of effort, e. A unit of effort can be called a "boat" and can be interpreted as an optimal combination of labor and capital. Assume that each unit of effort has a cost of p_e. For any chosen level of h, the production function can be solved for the required level of effort, given the size of the stock. Because the size of the stock at any place and time is the net result of past growth (which depends in part on q) and past harvest, the cost function can be written as

$$C = p_e \cdot e = C\left(h, z, q\right) \tag{9-58}$$

Managing the fishery for economic objectives is a dynamic problem because current harvests and costs depend on past harvest decisions and how they have affected the size of the stock. The economic objective is to maximize the present value of the net economic return from the fishery over time, subject to the biological constraint imposed by the growth function. If $p_h(h)$ is the inverse demand function for fish, the objective function is

$$\max_{h_t} : \int_0^{\infty} \left[\int_0^{h} p_h\left(h_t\right) dh - C\left(h_t, z_t, q\right) \right] e^{-r \cdot t} \cdot dt \tag{9-59}$$

subject to

$$\frac{dz}{dt} = g(z_t,\ q) - h_t \tag{9-60}$$

For any given initial conditions, this problem can be solved for the sequence of harvests that maximizes the objective function. The economic dimensions of the problem are easiest to see if we focus on the conditions for the long-run, steady-state equilibrium, where growth and harvest are equal and the stock is of constant size. The first-order conditions for this optimum are

$$p_h = \frac{\partial C}{\partial h} + \rho \tag{9-61}$$

and

$$r = \frac{\partial g}{\partial z} - \left(\frac{1}{\rho} \cdot \frac{\partial C}{\partial z}\right) \tag{9-62}$$

where r is the interest rate and ρ is the shadow value of the stock, reflecting its effect on the rate of growth and the cost of harvest. See, for example, Fisher 1981 or Clark 1976, 1985. Equation 9-61 says that the price of fish must equal the full marginal cost of their harvest, where this includes the opportunity cost of decreasing the stock, ρ. Equation 9-62 defines the optimum intertemporal trade-off. The interest rate is the opportunity cost of forgoing $1 worth of harvest now. It must equal the benefit of forgoing harvest, which has two components. The first component is the contribution the additional stock makes to future growth; the second reflects the contribution that the extra stock makes to lowering harvest costs.

Valuation of Changes in q in the Schaefer–Gordon Model. An increase in q shifts the cost function in equation 9-59 and changes the growth function of equation 9-60. This change leads to new solution values for h and z in every period. Assuming optimal management and the satisfaction of equations 9-61 and 9-62, the welfare value in each period is the increase in the net value term in brackets in equation 9-59, that is, the area between the marginal cost curves bounded by the demand curve. The stream of changes in that value must be discounted to its present value to obtain the total welfare value of the change in q.

Taking wetland acreage as a measure of q, Ellis and Fisher (1987) drew on the earlier work by Lynne and others (1981) to calculate the net welfare value of increases in wetlands in the blue crab fishery of the Gulf Coast of Florida. They abstracted from the dynamic, intertemporal dimension of the fishery management problem by focusing only on a single-period optimum and by assuming that the cost of harvest was independent of this stock. This is equivalent to asserting that $\rho = 0$, so that an optimum is defined by the price of harvest equal to its marginal cost.

However, the economic value of a change in environmental quality depends not only on the economic and biological parameters of the model, but also on the institutional arrangements for ownership or public management of the resource. The preceding discussion (and the Ellis–Fisher analysis) are based implicitly on the assumption of either private ownership with perfect competition or public regulation to achieve the economic optimum. However, most fishery resources are characterized by absence of private ownership and more or less open access to the resource. Where there is public regulation, it is seldom designed to achieve an economically efficient outcome. As a consequence, the net economic value of the resource will be lower; the economic value of changes in q will also be affected by the ownership and management arrangements.

I will now present a simple static model to show the effect of open access on the welfare value of a change in the productivity of the fishery resource such as might result from an improvement in water quality. (See also McConnell and Strand 1989.) Under open access without regulation, competition drives rent to zero (Gordon 1954; Scott 1955). Let MC_x^0 in Figure 9-9 be the marginal cost of harvesting fish at the original level of q. If the fishery is privately owned, or if it is optimally managed, the output and price will be given by the intersection of the marginal cost of harvest function and the demand curve at p_m^0 and x_m^0. An improvement in q would shift the marginal cost of harvest function outward to the right, resulting in a higher quantity and a lower equilibrium price. The new marginal cost curve is not shown in Figure 9-9. The welfare measure would be the same as that presented above.

If there is open access to the fishery, the condition for equilibrium is that each fisher would earn zero profits. This requires that price be equal to the average cost of harvesting fish. Compared to the efficient price and output under private ownership, more fishers enter in pursuit of profit. The increase in fish caught decreases the stock, raising costs or decreasing the price of fish, or both. Entry continues until these forces eliminate the incentive for entry. At the initial level of q, the average cost of harvesting fish is AC_x^0, as shown in Figure 9-9. The open-access equilibrium is at p_c^0 and x_c^0.

An improvement in q also shifts the average cost curve outward, resulting in a lower price and a higher quantity. However, because price equals average cost both before and after the change in environmental quality, there is no change in producers' surplus. The benefit consists entirely of

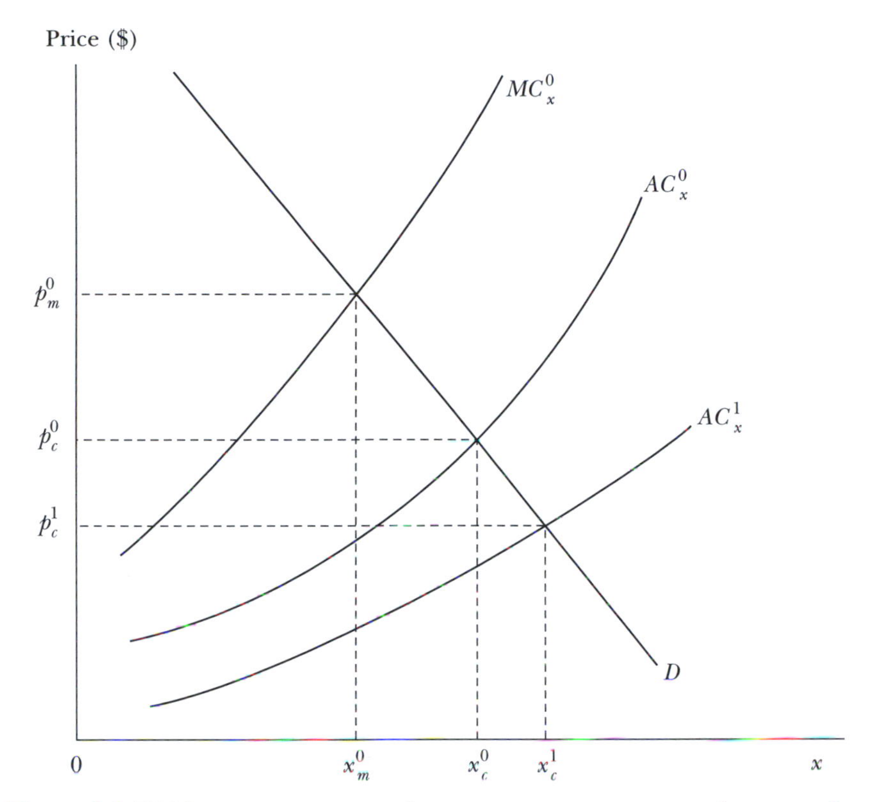

Figure 9-9. Welfare measurement for open-access resources: the case of a fishery

the increase in consumers' surplus associated with the price decrease. The more elastic is the demand curve for fish, the smaller is the welfare gain associated with the environmental improvement. In Freeman (1991) I presented some illustrative calculations of the magnitude of these effects based on the Ellis–Fisher data. In the limit, if this fishery is small relative to the market and the demand for fish is perfectly elastic, there is no welfare gain under open access. The physical improvement in productivity brought about by the higher water quality is entirely dissipated by the uneconomic competition of fishers for the potential increase in rents. However, with inelastic demand the welfare gain from an increase in q is slightly higher under open access than under optimal regulation.

The Beverton–Holt Model. The Schaefer–Gordon model has been criticized as too aggregated and not biologically realistic for many species. Also, one of its key predictions is not supported by observation for some species. The prediction is that if effort increases, eventually the stock must decrease, leading to a decline in growth and harvest and eventually to the biological and economic collapse of the fishery. Yet for several fisheries, plots of effort and harvest over time show increasing effort associated with almost no change in harvest. Townsend (1986) showed this to be the case for the North American lobster fishery originally analyzed by Bell (1972). McClelland (1991) also showed this to be the case for several of the Florida Gulf Coast fisheries analyzed by Bell (1989). This observed pattern of harvest and effort is entirely consistent with the Beverton–Holt biological model of the fishery.

The Beverton–Holt model combines a model of the growth of individual members of a species and the number of individuals in a single expression. The first element of the model is the recruitment of k individuals of the same age into a cohort at time $t = 0$. In the basic model, k is considered to be exogenous, but clearly it could be a function of environmental factors. Consider, for example, the oyster. If recruitment is defined as the attachment of larva to solid surfaces where shell development begins, the limiting factor in this stage of development may be the quantity of suitable surface available for attachment, not the number of eggs released and fertilized in the spawning stage.

The second element of the model is an expression for the population of the cohort at any point in time. Over the lifetime of the species, the total population changes because of natural mortality, m, and the harvest of fish, f. The latter is sometimes referred to as fishing mortality. Both of these forms of mortality can vary over time. Thus the population of the cohort at time t is given by

$$n_t = k \cdot e^{-(f_t + m_t)^t} \tag{9-63}$$

The third element of the model is an expression giving the weight of the typical member of the cohort at any point in time. Each fish grows over time, so its weight is given by

$$w_t = w_\infty \left[1 - e^{-\alpha(t)}\right]^\beta \tag{9-64}$$

The total biomass of the cohort at time t is the product of equations 9-63 and 9-64.

The last element of the model shows the effect of fishing effort on the cohort. The total weight of fish harvested at any point in time is

$$h_t = w_t \cdot n_t \cdot f_t \tag{9-65}$$

where f_t depends at least in part on the level of effort.

Cost and revenue functions can be specified, and the optimum time path of the exploitation of a cohort can be derived (see, for example, Clark 1976, 1985). However, because the number of recruits in one cohort is largely independent of the fishing mortality experienced by earlier cohorts, the intertemporal properties of the Beverton–Holt model are not nearly as interesting as those of the Schaefer–Gordon model.

The number of places at which environmental quality parameters can enter into the biological model interests us. For example, environmental variables can affect either the number of fish available for harvest at a point in time or their individual weights, or both. Both recruitment and natural mortality may depend upon environmental quality. The individual growth rate, α, and the upper limit on size are also likely to depend on environmental quality. An interesting and potentially fruitful area for future research is the role of environmental variables in the Beverton–Holt model and the development of expressions for the economic value of these changes.

Summary

When an environmental quality variable affects the production costs of firms, the welfare value of the change is measured by changes in the surpluses of producers and consumers. Measuring these values requires a model of the market for the output. In the simplest case of single-product firms in a competitive industry, developing such a model and deriving estimates is straightforward, given knowledge of how q affects the production functions, cost functions, or both. In this chapter I have shown how the simple model can be extended to deal with several kinds of complications, including the cases of multiproduct firms, vertically linked markets, monopoly power, and markets with price floors, such as those for some agricultural products. I have also shown how the models for optimum management of biological resources such as forests and fisheries can be adapted for the purposes of welfare measurement and how environmental quality parameters can be incorporated into these models.

References

Adams, Richard M., Thomas D. Crocker, and Richard W. Katz. 1984. Assessing the Adequacy of Natural Science Information: A Bayesian Approach. *Review of Economics and Statistics* 66(4): 568–575.

Adams, Richard M., Scott A. Hamilton, and Bruce McCarl. 1984. *The Economic Effects of Ozone on Agriculture.* Corvallis, OR: U.S. Environmental Protection Agency.

Adams, Richard M., and Bruce A. McCarl. 1985. Assessing the Benefits of Alternative Ozone Standards on Agriculture: The Role of Response Information. *Journal of Environmental Economics and Management* 12(3): 264–276.

Anderson, James E. 1976. The Social Cost of Input Distortions: A Comment and a Generalization *American Economic Review* 66(1): 235–238.

Barbier, Edward B., and Ivar [E.] Strand. 1998. Valuing Mangrove–Fishery Linkages: A Case Study of Campeche, Mexico. *Environmental and Resource Economics* 12(2): 151–166.

Bell, Frederick W. 1972. Technological Externalities and Common Property Resources: An Empirical Study of the U.S. Northern Lobster Fishery. *Journal of Political Economy* 80(1): 148–158.

———. 1989. *Application of Wetland Evaluation Theory to Florida Fisheries.* Tallahassee, FL: The Florida Sea Grant College.

Bockstael, Nancy E., and Kenneth E. McConnell. Forthcoming. Valuing the Environment as a Factor of Production. In *Handbook of Environmental Economics,* edited by Karl-Göran Mäler and Jeffery R. Vincent. Amsterdam: North-Holland.

Bowes, Michael D. and John V. Krutilla. 1989. *Multiple-Use Management: The Economics of Public Forestlands.* Washington, DC: Resources for the Future.

Clark, Colin W. 1976. *Mathematical Bioeconomics: The Optimal Management of Renewable Resources.* New York: John Wiley.

———. 1985. *Bioeconomic Modelling and Fisheries Management.* New York: John Wiley.

Ellis, Gregory M., and Anthony C. Fisher. 1987. Valuing the Environment as Input. *Journal of Environmental Management* 25(2): 149–156.

Fisher, Anthony C. 1981. *Resource and Environmental Economics.* New York: Cambridge University Press.

Freeman, A. Myrick, III. 1982. *Air and Water Pollution Control: A Benefit–Cost Assessment.* New York: John Wiley.

———. 1991. Valuing Environmental Resources under Alternative Management Regimes. *Ecological Economics* 3(3): 247–256.

Freeman, A. Myrick, III, and Winston Harrington. 1990. Measuring Welfare Values of Productivity Changes. *Southern Economics Journal* 56(4): 892–904.

Garcia, Phillip, Bruce L. Dixon, James W. Mjelde, and Richard M. Adams. 1986. Measuring the Benefits of Environmental Change Using a Duality Approach: The Case of Ozone and Illinois Cash Grain Farms. *Journal of Environmental Economics and Management* 13(1): 69–80.

Gordon, H.S. 1954. The Economic Theory of a Common Property Resource: Fisheries. *Journal of Political Economy* 62(2): 124–142.

Hartman, Richard C. 1976. The Harvesting Decision when a Standing Forest Has Value. *Economic Inquiry* 14(1): 52–58.

Heck, Walter W., Richard M. Adams, and others. 1983. A Reassessment of Crop Loss from Ozone. *Environmental Science and Technology* 17: 572A–581A.

Hyde, William F. 1980. *Timber Supply, Land Allocation and Economic Efficiency.* Washington, DC: Resources for the Future.

Just, Richard E., and Darrell L. Hueth. 1979. Multimarket Welfare Measurement. *American Economic Review* 69(5): 947–954.

Just, Richard E., Darrell L. Hueth, and Andrew Schmitz. 1982. *Applied Welfare Economics and Public Policy.* Englewood Cliffs, NJ: Prentice-Hall.

Kopp, Raymond J., and Alan J. Krupnick. 1987. Agricultural Policy and the Benefits of Ozone Control. *American Journal of Agricultural Economics* 69(5): 956–962.

Kopp, Raymond J., William J. Vaughan, Michael Hazilla, and Richard T. Carson. 1985. Implications of Environmental Policy for U.S. Agriculture: The Case of Ambient Ozone Standards. *Journal of Environmental Management* 20(4): 321–331.

Lichtenberg, Erik, and David Zilberman. 1986. The Welfare Economics of Price Supports in U.S. Agriculture. *American Economic Review* 76(5): 1135–1141.

Lynne, Gary D., Patricia Conroy, and Frederick J. Prochaska. 1981. Economic Valuation of Marsh Areas for Marine Production Processes. *Journal of Environmental Economics and Management* 8(2): 175–186.

Mathtech. 1982. *Benefits Analysis of Alternative Secondary National Ambient Air Quality Standards for Sulfur Dioxide and Total Suspended Particulates, Final Analysis.* Princeton, NJ: Mathtech, Inc.

McClelland, John D. 1991. On Selecting Models for Fishery Management. Honors paper for the Department of Economics, Bowdoin College, Brunswick, ME.

McConnell, Kenneth E., and Ivar E. Strand. 1989. Benefits for Commercial Fisheries when Demand and Supply Depend on Water Quality. *Journal of Environmental Economics and Management* 17(3): 284–292.

McGartland, Albert M. 1987. Implications of Ambient Ozone Standards for U.S. Agriculture: A Comment and Some Further Evidence. *Journal of Environmental Management* 24(2): 139–146.

Mjelde, James W., Richard M. Adams, Bruce L. Dixon, and Phillip Garcia. 1984. Using Farmers' Actions To Measure Crop Loss Due to Air Pollution. *Journal of the Air Pollution Control Association* 34(4): 360–364.

Russell, Clifford S., and William J. Vaughan. 1976. *Steel Production: Processes, Products, and Residuals.* Baltimore, MD: The Johns Hopkins University Press for Resources for the Future.

Samuelson, Paul A. 1976. Economics of Forestry in an Evolving Society. *Economic Inquiry* 14(4): 466–492.

Schmalensee, Richard. 1976. Another Look at the Social Valuation of Input Price Changes. *American Economic Review* 66(1): 239–243.

Scott, A.D. 1955. The Fishery: The Objectives of Sole Ownership. *Journal of Political Economy* 63(1): 116–124.

Swallow, Stephen K. 1994. Renewable and Nonrenewable Resource Theory Applied to Coastal Agriculture, Forest, Wetland, and Fishery Linkages. *Marine Resource Economics* 9: 291–310.

Swallow, Stephen K., Piyali Talukdar, and David N. Wear. 1997. Spatial and Temporal Specialization in Forest Ecosystem Management under Sole Ownership. *American Journal of Agricultural Economics* 79(2): 311–326.

Swallow, Stephen K., and David N. Wear. 1993. Spatial Interactions in Multiple-Use Forestry and Substitution and Wealth Effects for the Single Stand. *Journal of Environmental Economics and Management* 25(2): 103–120.

Townsend, Ralph E. 1986. A Critique of Models of the American Lobster Fishery. *Journal of Environmental Economics and Management* 13(3): 277–291.

CHAPTER

10

Valuing Longevity and Health

One of the basic services provided by the environment is the support of human life. Changes in the life-support capacity of the environment brought about by (for example) pollution of the air or water can lead to increases in the incidence of disease, impairment of daily activities, and perhaps reduced life expectancy. Human alteration of the environment can affect health through a number of channels, including

- organic compounds, which may contaminate aquifers used as sources of drinking water;
- poorly treated sewage or septic tank leachate, which may spread disease-causing bacteria and viruses into drinking water supplies and among shellfish bound for human consumption;
- elevated levels of particulate matter air pollution, which can increase the risk of premature mortality;
- air emissions from manufacturing facilities, auto body repair and painting shops, and the like, which may include carcinogens; and
- depletion of the stratospheric ozone layer, which can cause increased ground-level ultraviolet radiation and increase the incidence of skin cancer.

This chapter describes and evaluates the currently available methods and techniques for estimating the monetary values of changes in human health that are associated with environmental changes.* Two links must be

*I am indebted to Maureen Cropper for her contributions to this chapter in the first edition of this book, especially the portions dealing with morbidity benefits and the life-cycle models for valuing changes in mortality risks.

established in estimating these values. The first is the link between the environmental change and the change in health status. The second is that between the change in health status and its monetary equivalent, willingness to pay (WTP) or willingness to accept compensation (WTA). There are two alternative strategies for using revealed preference methods to value environmental changes that affect human health. The first strategy is to develop a comprehensive model of individual behavior and choice in which environmental quality is one of the determining variables. Such models can provide a basis for measuring WTP directly as a function of the environmental change. The second strategy is to deal with the two links separately. Economic values of changes in health status or health risk would be derived first. They would then be combined with independently derived predictions of health changes or risk changes as a function of environmental change. An example of this approach would be to use measures of the value of risk reduction derived from studies of wage rates and occupational risk, combined with epidemiological studies of the relationship between air pollution and mortality rates, to estimate the economic value of an air-quality improvement that reduced the risk of premature mortality.

In the first section of this chapter I discuss models and measurement techniques for valuing changes in mortality as measured by the probability of dying. In the second section, I turn to morbidity or illness and review revealed preference models based on averting and mitigating behavior and household production. In the third section, I take up some broader issues concerning valuation of changes in health, including the use of alternative measures of health status such as quality-adjusted life years (QALYs) and the valuation of changes in the health of children.

Valuing Reduced Mortality Risks

Because some forms of pollution may increase mortality, economists have had to confront the question of the economic value of life and lifesaving. This is perhaps the most difficult and controversial aspect of valuing the health effects associated with environmental changes. To some, the idea of putting a price on human life is insensitive, crass, or even inhuman. However, this point of view does not give sufficient weight to the fact that individuals in their day-to-day actions and governments in their decisions about social policy make trade-offs between changes in the probability of death and other goods that have monetary values. These trade-offs make it possible to infer the implicit prices that people (or their governments) attach to changes in the probabilities of their deaths.

This section is about the economic theory of value as it is applied to the decisions made by individuals and governments concerning the trade-off between mortality risk and money. Here I present a brief description of the economic basis for assigning monetary values to certain kinds of lifesaving activity. I also identify and briefly discuss some of the important economic and ethical issues that have been identified in the literature. I then turn to a more detailed discussion of models and methods for measuring willingness to pay for reduced mortality risk. I conclude with a review of some of the issues that arise in using WTP measures in the evaluation of environmental policy.

Willingness To Pay

In keeping with the assumption that individuals' preferences provide a valid basis for making judgements concerning changes in their economic welfare, reductions in the probability of death due to accident or illness should be valued according to what an individual is willing to pay to achieve the reductions or is willing to accept in compensation to forgo the reductions. Schelling (1968) appears to have been the first to propose applying WTP concepts in this area. Other early contributors include Mishan (1971) and Jones-Lee (1974, 1976). The use of WTP concepts presupposes that individuals treat longevity more or less like any other good rather than as a hierarchical value. Perhaps surprisingly, this seems to be a reasonable assumption, at least for small changes in mortality risk. Individuals in many situations act as if their preference functions include life expectancy or the probability of survival as arguments. In their daily lives, they make a variety of choices that involve trading off changes in the risk of death for other economic goods whose values can be measured in monetary terms. For example, some people travel to work in cars rather than by bus or by walking because of the increased convenience and lower travel time of cars, even though these people thus increase their risk of dying prematurely. Also, some people accept jobs with known higher risks of accidental death because the jobs pay higher wages (Chapter 12). In such cases, people must perceive themselves as better off because of the alternatives they have chosen; otherwise they would have chosen some other alternative. When the thing that is given up (or gained) can be measured in dollars, the individual's WTP (or WTA) is revealed by these choices. These choices are the basis of measures of the economic value of reductions in the risk of death.

As the ordinary, everyday nature of these examples should make clear, this economic question is not about how much an individual would be willing to pay to avoid his or her certain death or how much compensation that individual would require to accept that death. In this respect, the term "value of life" is an unfortunate phrase that does not reflect the true nature of the question at hand. Most people would be willing to pay their total wealth to avoid certain death; there is probably no finite sum of money that could compensate an individual for the sure loss of life. Rather, the economic question is about how much the individual would be willing to pay to achieve a small reduction in the *probability* of death during a given period or how much compensation that individual would require to accept a small increase in that probability. This is an appropriate question to investigate because most environmental regulatory programs—even those aimed at fairly serious toxins—result in relatively small changes in individuals' mortality risks.

For this kind of situation, the theory of individual choice under uncertainty described in Chapter 8 provides a useful analytical framework. Individuals do not know which state of the world will exist at some specified date in the future, but they must make choices affecting their future utility before the future is revealed. Individuals are assumed to assign probabilities to alternative states of the world and to make their choices to maximize their expected utility. One aspect of uncertainty about the future is the date of one's death, or to put it differently, whether one will survive or succumb to some hazard during some time. Individuals can affect the probabilities of death during present and future periods by the choices they make. The value of a reduction in risk to an individual is the amount of money that person would be willing to pay to achieve it, other things being equal.

An individual's willingness to pay for changes in the probability of his or her death can be translated into a more convenient figure for evaluating policies that reduce the risk of death, namely, the *value of a statistical life* (VSL) or the *value of a statistical death avoided.* Suppose that there were a group of 10,000 people, each of whom has a probability of 0.0004 of dying during the next year. Suppose that a pollution control policy would reduce that probability to 0.0003, a change of 0.0001 (1 in 10,000). Furthermore, suppose that each individual in that group expresses a willingness to pay $500 for this policy. Because the policy would affect all the people equally, it is a form of collective good for the group. The total WTP of the group is $5 million. If the policy is adopted, there will be on average one fewer death during the year. Thus the total willingness to pay for the policy that results in one fewer death is $5 million. This is the value of statistical life.

As discussed thus far, the WTP approach focuses on the individualistic dimensions of human behavior, that is, an individual's willingness to pay to increase his or her own life expectancy. However, there is nothing in the logic of the WTP approach to prevent consideration of the effects of kinship and friendship, that is, an individual's altruistic willingness to pay to reduce the probability of death of close relatives and friends. Again, we should make it clear that the question is not the willingness to pay to prevent an imminent or highly probable death of another person (that is, the ransom for the kidnapped child or the search for the lost hiker or boater), but the willingness to pay for a small reduction in the probability of death for the group of which the friend or relative is a part. This would be a form of paternalistic altruism. See the discussion in Chapter 5.

Ex Ante versus Ex Post Perspectives: An Ethical Issue

One way of characterizing the economic approach is by saying that it avoids the issue of valuing life, per se, by recognizing that what people actually "buy" and "sell" through their choices and trade-offs is not life versus death but small changes in the probability of dying. Another way of characterizing the economic approach is by saying that the economic value is derived by focusing on choices ex ante, that is, before the uncertainty about the individual's death during a specified period is resolved. However, at some time, the uncertainty is resolved. Each individual will know if he or she is to die now or live a while longer. From this ex post perspective, those who would die would be willing to pay their total wealth to change the outcome or would require an infinite compensation to accept it. Critics of the economic approach to valuation, such as Broome (1978), have argued that this difference in perspective can have no ethical or moral significance and that therefore the WTP or compensation measures based on the ex ante perspective are morally unacceptable.

One defense of the economic perspective is based on the observation that people appear to be willing to make ex ante trade-offs involving risks of death. If people are rational and if their preferences are taken to be the basis of measures of economic value, then their willingness to consent to ex ante trade-offs must have some ethical significance. Furthermore, for many of the public policy issues where value of risk reduction information might be used, it will never be known ex post whose deaths were caused by failure to adopt a policy or whose lives were prolonged by a policy to reduce risks of death. For example, suppose a proposed regulation on drinking water contamination is predicted to reduce cancer mortality in the U.S. population by 100. If the policy is adopted, the identities of the 100 people whose lives were prolonged will never be known. If the policy is not adopted, it can never be known which 100 of the roughly 500,000 or so cancer deaths in any one year were "caused" by the failure to adopt the policy. Economists argue that it is consent and the veil of ignorance about who dies and who is saved that legitimize the ex ante perspective and its focus on the value of changes in risks rather than on the values of lives and deaths.

Human Capital: An Alternative Concept of the Value of Lifesaving

An alternative to using WTP for valuing lives is to assume that the value of an individual alive is what he or she produces and that this productivity is accurately measured by earnings from labor. Earnings before taxes are measured to reflect the government's, and therefore society's, interest in each individual's total productivity. With the death of the individual, that output is lost. This approach has a long tradition; in fact, Landefeld and Seskin (1982) trace the idea back almost 300 years. It has been the basis of some widely cited early estimates of the benefits of air pollution control, for example, by Lave and Seskin (1971, 1977).

The human capital approach is fundamentally at odds with the individualistic perspective of welfare economics and the theory of value. By in effect asking what the individual is worth to society, the human capital approach ignores the individual's own well-being, preferences, and WTP. It defines the social worth of the individual in a narrow way, that is, as the individual's market productivity, thereby ignoring the value of that person's health and well-being to loved ones. Although I think that the human capital approach is inappropriate for valuing reductions in the risk of death, it makes economic sense as the starting point for determining compensation for dependents in wrongful death settlements.

According to the human capital approach, the value of preventing the death of an individual who is presently of age t is the discounted present value of that individual's earnings over the remainder of his or her expected life:

$$\text{value} = \sum_{i=1}^{T-t} \frac{\pi_{t+i} \cdot E_{t+i}}{(1+r)^i} \tag{10-1}$$

where

π_{t+i} = probability of the individual surviving from age t to age $t+i$
E_{t+i} = expected earnings of the individual at age $t+i$
r = discount rate
T = age at retirement from the labor force

Several issues must be addressed concerning the implementation of the human capital approach. One question is, should an individual's productivity be measured net of or inclusive of that person's own consumption? Netting out consumption leaves a measure of the individual's worth as a producing asset to the rest of society. However, this measure is the antithesis of the individualistic premise of conventional welfare economics.

Another question concerns the role of nonmarket production in the measure of productivity and value. The omission of nonmarket productivity is particularly serious and troublesome in the case of stay-at-home spouses and nonworking parents. Some studies have attempted to correct for this omission by imputing earnings equal to the wages of domestic ser-

Table 10-1. Present Value of Future Earnings of Males by Selected Age Groups (2001 dollars)

Age group (years)	*Real discount rate*		
	2.5%	*6%*	*10%*
1 to 4	1,186,456	319,748	93,319
20 to 24	1,507,877	833,740	499,051
40 to 44	975,154	709,292	527,297
65 to 69	74,061	63,740	55,038

Note: Figures are based on the present value of both expected lifetime earnings and housekeeping services at 1977 wage rates and life expectancies and an annual increase in labor productivity of 1%.

Source: Landefeld and Seskin 1982.

vants. Others have argued that the average earnings of employed females provide a measure of the opportunity cost of working at home for women. If women are rational, the value of home production must be equal to or greater than this opportunity cost. However, even these adjustments do not capture the nonmarket productivity of many individuals.

A third issue is what discount rate to use in calculating the present value. The human capital value of children and young adults will be particularly sensitive to the choice of a discount rate. Rates of 6% (Cooper and Rice 1976) to 10% (Landefeld and Seskin 1982) have been used in the literature, but these rates are high relative to estimates of the real rates of return on riskless financial assets discussed in Chapter 7.

Some of the implications of the human capital approach are unsettling. Because of discounting and the time lag before children become productive participants in the economy, the human capital approach places a much lower value on saving children's lives compared with saving the lives of adults in their peak earnings years. Because of earnings differences by sex and race, the human capital approach values saving the lives of women and nonwhites less than it does saving the lives of adult white males. Also, the human capital approach assigns zero value to persons who are retired or totally disabled by illness or handicap. Human capital values for men of different ages, using alternative discount rates, are shown in Table 10-1. According to these calculations, men in the age group 20 to 24 are "worth more" than males in other age groups. The table also shows the dramatic effect of discounting on the human capital value of children. The human capital values of women and nonwhite males would be lower than those shown in the table.

Even if it is accepted that the human capital approach is flawed in principle, one could ask if it might be a reasonable approximation to the value of statistical life based on WTP. Both theoretical reasoning and empirical evidence (both presented below) suggest that human capital measures are a poor proxy for the desired WTP measure of value for small changes in the risk of death. Whereas it is likely that an individual's income and the consumption it allows are positively related to the utility that person derives from his or her own life, the human capital method does not reflect the probabilistic nature of death and death avoidance and individuals' differing attitudes toward risk. By definition an individual with no financial wealth could pay no more than the present value of his or her expected earnings stream to avoid certain death. However, his or her statistical value of life based on willingness to pay for small probability changes could be several times his or her discounted earnings stream.

Modeling Individual Choice and Willingness To Pay

I now describe some of the results generated by economic models of individual choice under uncertainty. We analyze the models here to derive hypotheses about the determinants of an individual's willingness to pay for a reduction in the risk of mortality and to examine the relationships between these predicted WTP measures and other economic variables, such as the individual's earnings and purchases of life insurance.

These models are based on the assumption that, as discussed earlier, individuals make choices that affect their risk of death so as to maximize the mathematical expectation of utility. A key assumption of the models is that individuals know the relevant probabilities of dying and know how they are changed by the choices they make. Because individuals must choose alternatives before the uncertainty is resolved, expected utility is an ex ante concept. These models are used to derive the willingness to pay for a reduction in the probability of death, defined as the maximum sum of money that can be taken from the individual ex ante without leading to a reduction in that person's expected utility. This is a compensating surplus measure of welfare change. Alternatively, these models could be used to derive the willingness to accept compensation for an increase in the probability of dying, defined as the sum of money that just compensates for the greater risk by increasing consumption sufficiently to equalize the expected utilities of the two alternatives.

Two aspects of the choice problem are of interest in this analysis: uncertainty and time. The uncertainty concerns the unknown timing of the individual's death and the effects of choices of occupation and consumption activity on the probabilities of surviving to any given date. The time aspect concerns the effects of choices made at any time on the probabilities and utilities associated with future periods. This is of particular relevance for environmental problems because many environmental policy issues are characterized by a substantial interval of time between exposure and the perceived effects on health. Most models of intertemporal choice have focused on identifying the optimal consumption stream given an income stream and opportunities for borrowing and lending. Our concern here is with choices made in the present that affect the probabilities of survival at future dates. The problem is to identify the marginal willingness to pay now for increases in the probabilities of survival during some future period.

Static Models. I begin with a simple one-period choice model to avoid the complications of intertemporal choice, identify the marginal willingness to pay for increases in survival probability, and then extend the model to the intertemporal case to see whether estimates of the value of statistical life based on intratemporal models can be applied to the intertemporal case.

Assume that an individual derives utility from the consumption of a composite good, X, with a price normalized to 1. The initial endowment of X and the probability of surviving to enjoy its consumption, π, are both given to the individual. Let X^0 and π^0 represent this initial endowment. Arbitrarily normalizing the utility function so that the utility of death is zero, expected utility is

$$E[u] = \pi^0 \cdot u(X^0) \tag{10-2}$$

We can obtain an expression for the individual's marginal willingness to pay (MWTP) in units of X for a reduction in π^0 by taking the total differential of equation 10-2 and setting it equal to zero:

$$\text{MWTP} = \frac{u(X^0)}{\pi^0 \cdot (\partial u / \partial X^0)} \tag{10-3}$$

As equation 10-3 shows, one of the fundamental results of models of this sort is that if individuals are not free to adjust their survival probabilities in the market, their marginal willingness to pay for enhanced survival will depend on their initial survival situation. Other things being equal, the higher the risk of death is (lower π^0), the higher will be the marginal willingness to pay to reduce that risk. Another result is that diminishing marginal utility of X implies a higher marginal willingness to pay for π with higher initial endowments of X, other things being equal. These propositions can be verified by taking the derivatives of equation 10-3 with respect to π and X.

This model, while providing useful insights, does not suggest a method for estimating WTP other than by direct questioning. To use indirect market methods we must link WTP to voluntary risk-taking behavior, that is, π must be made endogenous. Now suppose that the individual has the opportunity to rearrange his or her consumption and survival position through exchange, for example by giving up some X to improve his or her chances of surviving to enjoy the remainder. Let p_π represent the price at which consumption can be exchanged for enhanced survival probability. For the simple model, I assume that there are no opportunities for insurance. Comprehensive models encompassing bequest motivation (utility derived from unconsumed X remaining at death) and insurance behavior can be developed. For examples, see Conley 1976; Jones-Lee 1976; Thaler and Rosen 1976; and Cropper and Sussman 1988.

The individual chooses X and π to maximize expected utility, subject to the budget constraint

$$\max : E[u] = \pi \cdot u(X) + \lambda\left[X^0 - X + p_\pi \cdot \left(\pi^0 - \pi\right)\right] \tag{10-4}$$

The first-order conditions for a maximum of expected utility can be combined to obtain

$$\frac{u(X)}{\pi \cdot (\partial u / \partial X)} = p_\pi \tag{10-5}$$

This expression requires that the individual equate his or her marginal willingness to pay for enhanced survival (the left-hand side of the expression) with the given price of enhanced survival. This is similar to the expression for marginal willingness to pay for reductions in risk derived in Chapter 8 because here utility and the marginal utility of consumption are zero in the event of death.

The utility-maximizing individual may choose either to forgo consumption to enhance his or her survival probability or to take on increased risk (lower π) to enhance consumption opportunities. The actual choices depend on the initial endowment of X^0 and π^0, the price of π, and the individual's preferences. Whatever the final outcome, equation 10-5 allows us to infer the individual's marginal willingness to pay for enhanced survival because this value will be equated to the observable price of changes in π.

If the risk in question is an environmental risk, it may not be feasible to have a market for reductions in π. However, if there are other risks, the individual must be able to affect the level of one of these risks and view equal size reductions in any of these risks as equally valuable, or to put it differently, be indifferent as to the source of the risk of death. Suppose now that there are three sources of risk of death: one exogenous environmental risk; one job-related risk; and one related to the level of consumption of a private good x_i, an element in the vector of market goods $\mathbf{X}$. Assume that in addition to affecting the risk of death, this good conveys utility directly. The corresponding conditional probabilities of death are denoted ρ_e, ρ_j, and $\rho_x(x_i)$. Assuming that these causes of death are independent, the probability of surviving the current period is the product of the probabilities that the individual does not die from each of the three causes (Sussman 1984); that is,

$$\pi = (1 - \rho_e)(1 - \rho_j)(1 - \rho_x(x_i)) \tag{10-6}$$

In general, good x_i could either increase or decrease the risk of death. Some goods such as skydiving and cigarettes increase risk, while others such as smoke detectors decrease risk. Our interest is in risk-reducing goods, so that $\partial \rho_x / \partial x_i < 0$. Also, suppose that all types of jobs are alike in every respect except for the risk of accidental death and that more risky jobs have higher wage rates—in other words, the individual receives an annual wage, $M(\rho_j)$, where $\partial M / \partial \rho_j > 0$.

If I is exogenous income, then total income, M^*, is $I + M(\rho_j)$. Expected utility is given by

$$E[u] = (1 - \rho_e)(1 - \rho_j)[1 - \rho_x(x)] \cdot u\left[M^*(\rho_j) - p_x \cdot x_i, \mathbf{X}\right] \tag{10-7}$$

By total differentiation of equation 10-7, willingness to pay for a marginal change in exogenous risk of death, $dI/d\rho_e$, is given by

$$w_{\rho_e} = (1 - \rho_j)(1 - \rho_x)\left[\frac{u(\cdot)}{\pi \dfrac{\partial u}{\partial M^*}}\right] \tag{10-8}$$

This is the value of the utility lost if the individual dies, $u(\cdot)$, converted to dollars by dividing by the expected marginal utility of income $(\pi \cdot \partial u / \partial M^*)$ and multiplied by the probability that the individual does not die due to other causes $[(1 - \rho_j)(1 - \rho_x)]$.

The major question of interest is whether MWTP can be estimated by observing risk-taking behavior in consumption or in the labor market. To answer this, assume that the individual chooses x_i and a job with its associated risk ρ_j, so as to maximize equation 10-7 and derive the first-order conditions for choice of job risk and consumption of x_i. They are

$$\frac{\partial M}{\partial \rho_j} = (1 - \rho_e)(1 - \rho_x)\left[\frac{u(\cdot)}{\pi \dfrac{\partial u}{\partial M^*}}\right] \tag{10-9}$$

and

$$-\frac{p_x}{(\partial \rho_x / \partial x_i)} = (1-\rho_e)(1-\rho_j)\left\{\frac{u(\cdot)}{\pi\left[\frac{\partial u}{\partial M^*} - \frac{(\partial u / \partial x_i)}{p_x}\right]}\right\} \tag{10-10}$$

Equation 10-9 implies that the individual equates the marginal wage income forgone by moving to a safer job with the marginal benefit of a reduction in job risk. The latter is almost identical to the value of an exogenous risk change—equation 10-8—except that the probability of not dying due to other causes is now $(1 - \rho_j)(1 - \rho_x)$ instead of $(1 - \rho_e)(1 - \rho_x)$. If $(1 - \rho_e) \approx (1 - \rho_j)$, then the marginal price of risk reduction can be used as an estimate of the willingness to pay for a change in exogenous risk. This marginal price, which is an implicit price, can be estimated with a hedonic wage model. That method is described in Chapter 12.

The first-order condition for the choice of x_i differs from the expression for MWTP in part because it includes a term for the marginal utility of a dollar spent on x_i. This term is present because of the assumption that x_i conveys utility directly as well as through its effect on risk. Equation 10-10 can be used as an approximation for MWTP only if $\partial u/\partial x_i = 0$. Then, if $(1 - \rho_e) \approx (1 - \rho_x)$, the left-hand side of equation 10-10, which is the marginal cost of reducing risk through purchasing x_i, can be used to approximate marginal willingness to pay for an exogenous risk change.

To use data on "safety" goods to estimate the value of risk reduction, one must establish that the good's contribution to safety is known. What governs each individual's purchase decision is that person's perception of the risk-reduction capability of the good. In the absence of adequate information, individuals' perceptions might vary substantially and be difficult to observe. This would make the use of safety good purchases for estimating values of risk reduction problematic.

The revealed preference model of consumer choice is based on the assumption that the safety good is divisible. However, this assumption is often invalid. Some of the goods used in actual studies—for example, smoke detectors—are indivisible, that is, their purchase involves a 0–1 decision. The good is purchased if its marginal benefit is equal to or greater than its marginal cost. The equality of WTP and price occurs only for the marginal purchaser of the good. To estimate an average value of MWTP we must have data on the cost of the safety good and on its effect in reducing risk of death for a cross section of individuals. If marginal cost and marginal risk reduction vary across individuals in the sample, we can estimate the average value of MWTP by using a discrete choice model as described in Chapter 4.

Perhaps the most important conclusion to be drawn from this review of static models of individual choice and WTP is that each person is likely to attach a different value to a small reduction in the probability of dying because of differences in underlying preferences, degree of risk aversion, wealth, current level of risk exposure, age, number of dependents, and, perhaps, quality of the additional years of life expected to be gained from the reduced risk. This conclusion must be kept in mind when interpreting and using the results of empirical estimates of WTP that are based on averages of groups of perhaps heterogeneous people.

A second conclusion is that in the case of multiple risks of death where the individual can "purchase" reductions in some component of risk, the observed price or marginal cost of reducing that component of risk can be taken as a close approximation of the individual's willingness to pay for reductions in other components of risk, provided that the safety good or safer job does not also convey utility directly and that the individual values equal reductions in all components of risk equally.

Introducing Time. The static models discussed above are relevant to questions such as transportation and occupational safety, in which the individual's actions today affect the probabilities of dying today. However, many of the important environmental and occupational health questions involve actions taken today whose effects on the probability of dying are realized only at some time, perhaps 10 to 20 years, in the future. How much would an individual be willing to pay now to control current pollution when the effects of improved health might be realized only at some future time? The next step in the analysis is to develop a simple multiperiod model that allows us to investigate how willingness to pay now for a reduction in the probability of death is influenced by the time period to which the probability applies. To focus on the intertemporal aspect of the problem, assume only one cause of death, and one consumption activity.

As an extension of the model of intertemporal choice developed in Chapter 7, assume that individuals maximize their expected lifetime utility u^*, where u^* is an additively separable function of the consumption stream:

$$E[u^*] = \sum_{t=1}^{T} \pi_t \cdot D^{t-1} \cdot u(X_t) \tag{10-11}$$

where $D = 1/(1 + d)$ and

$d \equiv$ subjective rate of discount or own time preference
$\pi_t \equiv$ probability of surviving from period 1 through period t, that is, the probability of living for at least t years from now (year 1), with

$$\pi_t = \prod_{i=1}^{t} \pi_i = \prod_{i=1}^{t} (1 - \rho_i) \tag{10-12}$$

The term ρ_i is the conditional probability of dying during year i, that is, the probability of dying in year i, given that the individual has survived to the beginning of year i. Similarly, π_i is the probability of surviving the year given being alive at the beginning of the year. For now, assume that the stream of consumption X_t is given exogenously and cannot be altered by borrowing or lending.

The form of the intertemporal utility function deserves a brief comment. As explained in Chapter 7, the assumption that individuals maximize discounted utilities does not necessarily impose any restrictions on the nature of individuals' preferences regarding present versus future consumption. The subjective rate of discount d could be positive, negative, or zero. A variety of types of preferences and behavior can be encompassed in this model, depending on the value of d. In equilibrium, borrowing and lending behavior depends on market interest rates, present and future income levels, and the rate at which the marginal utility of income is diminishing, as well as time preference.

From equation 10-11 the marginal rate of substitution (MRS) between present consumption and present mortality reduction is

$$\text{MRS}_{\pi_1 X_1} = \frac{u(X_1) + \sum_{t=2}^{T} \frac{\pi_t}{\pi_1} D^{t-1} \cdot u(X_t)}{\pi_1 \cdot (\partial u / \partial X_1)} \tag{10-13}$$

The second term in the numerator shows the dependence of willingness to pay to reduce present risk on life expectancy and the associated stream of expected future utilities. Higher future consumption (X_t) and longer life expectancy (T and π_t) both increase the willingness to pay now for a higher current survival probability.

Suppose now that the marginal rate of substitution between present mortality reduction and present consumption is known, for example, from a study of the demand for occupational safety. Can this marginal rate of substitution be used to estimate the demand for reductions in future mortality that might be obtained through an environmental or occupational health program? The marginal rate of substitution between a mortality reduction in future period t' and present consumption is

$$\text{MRS}_{\pi_{t'} X_1} = \frac{\sum_{t=t'}^{T} \frac{\pi_t}{\pi_{t'}} D^{t-1} \cdot u(X_t)}{\pi_1 \cdot (\partial u / \partial X_1)} \tag{10-14}$$

The two marginal rates of substitution might differ for any of three reasons. First, equation 10-13 could be greater than equation 10-14 because it includes a term for first-period utility, $u(X_1)$. Also, the more distant that period t' is from the present, the smaller is the stream of future periods being summed in equation 10-14. However, equation 10-14 could be greater than equation 10-13 if $\pi_{t'}$ is sufficiently smaller than π_1. Thus marginal rates of substitution estimated from one type of probability choice problem—for example, the single-period problem—cannot in general be used as predictors of the marginal rates of substitution for other types of probability choice problems with different intertemporal dimensions.

In this model the individual takes the intertemporal pattern of consumption opportunities as given. There is no borrowing or lending. Cropper and Sussman (1990) extended this model by incorporating borrowing and lending at a riskless interest rate. This allows individuals to adjust their consumption streams over time. Measures of WTP will be different when this opportunity is available to people. To see this, we turn now to a full life-cycle model of intertemporal choice.

A Life-Cycle Model of Willingness To Pay. Several authors, including Usher (1973), Conley (1976), Cropper and Sussman (1990), and Rosen (1994), have used a life-cycle consumption-saving model with uncertain lifetime to analyze an individual's willingness to pay at age j for a change in the conditional probability of dying at age t, $t \geq j$. In this section, I follow the approach of Cropper and Sussman. This model can be used to examine

1. the relationship between willingness to pay for a change in current probability of death and age;
2. the relationship between the present value of expected lifetime earnings (the human capital measure) and WTP;
3. the relationship between WTP and the latency of the risk, that is, the interval between the exposure to the risk (a carcinogen, for example) and its manifestation (death due to cancer); and
4. the relationship between the willingness to pay in advance to reduce a given risk and the willingness to pay at the time the risk is experienced (a discounting question).

I will show, among other things, that WTP will generally decrease with age; that under plausible circumstances, the present value of lifetime earnings will be less than WTP; that latency reduces WTP; and that the WTP at year 1 for a reduction in risk in year t' is equal to the WTP in year t' for that probability change discounted back to year 1 by a discount factor that in general will be different from the market interest rate.

In the life-cycle model, an individual of any given age has a probability distribution over the date of his or her death. Let j denote the individual's current age, and let $\rho_{j,t}$ be the probability that the individual dies at the end of the year in which he or she attains age t, that is, the person lives exactly $t - j$ more years. Because the $\rho_{j,t}$ constitute a probability distribution, it must be true that

$$\rho_{j,t} \geq 0, t = j, j+1, \ldots, T, \quad \text{and that} \quad \sum_{t=j}^{T} \rho_{j,t} = 1 \tag{10-15}$$

where T is the maximum attainable age. The probability that the individual survives to his or her tth birthday, given that he or she is alive at age j, is $\pi_{j,t}$, which also is the probability that he or she dies at $t + 1$ or later. Formally,

$$\pi_{j,t} = \sum_{s=t+1}^{T} \rho_{j,s} \tag{10-16}$$

Let δ_t be the probability of dying at age t, conditional on being alive at the beginning of that year. Thus the conditional probability of surviving that year is $1 - \delta_t$. This term can also be derived from the survival probabilities

$$1 - \delta_t = \frac{\pi_{j,t+1}}{\pi_{j,t}} \tag{10-17}$$

Expected lifetime utility at age j is the sum of the utility of living exactly $t - j$ more years times the probability of doing so. As in the preceding section, assume that utility is additively separable. Also assume that there is no bequest motive. Then we can write expected lifetime utility at age j as

$$V_j = E[u^*] = \sum_{t=j}^{T} \pi_{j,t} \cdot D^{t-j} \cdot u(X_t) \tag{10-18}$$

The utility function for each period, $u(X_t)$, is assumed to be increasing in X_t, strictly concave, and bounded from below.

Two points about this formulation of the problem should be emphasized. First, the model is based on the assumption that the utility of living depends only on consumption and not on length of life per se. The concavity of the utility function implies that it is always desirable to spread a given amount of consumption over a longer time. Thus lifetime utility is an increasing function of life expectancy. However, this is only because of the effect on consumption, not because of the value of being alive per se. I return to this point below. Second, this model treats survival probabilities as exogenous to the individual. To keep things simple, I do not attempt to introduce opportunities for the individual to alter risk levels into this model; see Conley 1976 and Viscusi and Moore 1989 for examples of intertemporal models incorporating this additional element of choice.

The individual has to choose a time pattern of consumption, given initial wealth W_j, annual earnings M_t, $t = j, \ldots, T$, and capital market opportunities, so as to maximize expected lifetime utility as given by equation 10-18. Arthur (1981) and Shepard and Zeckhauser (1982, 1984) assume that the individual can save by purchasing actuarially fair annuities and borrowing via life-insured loans. If actuarially fair annuities are available, an individual who invests \$1 at the beginning of his *j*th year will receive $\$(1 + A_j)$ at the end of the year with probability $1 - \delta_j$ and nothing with probability δ_j. For the annuity to be fair, that is, to have an expected payout of $1 + r$ where r is the riskless rate of interest, there must be an annuity rate of interest a_j that satisfies

$$(1 + a_j)(1 - \delta_j) = 1 + r \tag{10-19}$$

If the individual borrows, he or she must cover the possibility that he or she might die before repaying the loan. Agreeing to pay $\$(1 + a_j)$ if he or she survives is equivalent to paying $\$(1 + r)$ on survival plus purchasing a life insurance policy in this amount at actuarially fair rates. So we can call a_j the actuarial rate of interest.

The individual's budget constraint can be expressed as the requirement that the present value of expected consumption equal initial wealth plus the present value of lifetime earnings, where discounting is done at the riskless rate, r:

$$\sum_{t=j}^{T} \pi_{j,t} \cdot (1+r)^{j-t} \cdot X_t = W_j + \sum_{t=j}^{T} \pi_{j,t} \cdot (1+r)^{j-t} \cdot M_t \tag{10-20}$$

Alternatively, making use of equation 10-19 and the fact that

$$\pi_{j,t} = \prod_{s=j}^{t=1} (1 - \delta_s) \tag{10-21}$$

the budget constraint can be expressed in terms of the actuarial rate of interest:

$$\sum_{t=j}^{T} \left[\prod_{s=j}^{t-1} (1 + a_s)^{-1} \right] \cdot X_t = W_j + \sum_{t=j}^{T} \left[\prod_{s=j}^{t-1} (1 + a_s)^{-1} \right] \cdot M_t \tag{10-22}$$

The pattern of consumption over the life cycle is chosen to maximize equation 10-18 subject to equation 10-20 or equation 10-22. Formally, it is the solution to the Lagrangian problem:

$$\max_{X_t} V_\lambda = \sum_{t=j}^{T} \pi_{j,t} \cdot (1+d)^{j-t} \cdot u(X_t) + \lambda \left[W_j + \sum_{t=j}^{T} \pi_{j,t} \cdot (1+r)^{j-t} \cdot (M_t - X_t) \right] \tag{10-23}$$

Now consider how a government health and safety regulation that reduces conditional probabilities of death affects lifetime utility. A government regulation can alter the probability that a person dies in any year only if that person is alive at the beginning of the year. Consider a regulation that reduces $\delta_{t'}$, the conditional probability of dying at age t', that is, the probability that the individual dies between his t'th and $t' + 1$st birthdays. Note that when the conditional probability of death is altered at age t', it affects the probabilities of surviving to ages $t' + 1$ and beyond (for example, $\pi_{j,t'+k}$) because, by repeated use of the definition of δ_t,

$$\pi_{j,t'} = (1 - \delta_j)(1 - \delta_{j+1}) \ldots (1 - \delta_{t'-1}) \tag{10-24}$$

Formally, let $w_{j,t'}$ be the individual's marginal willingness to pay at age j for a change in $\delta_{t'}$. It is measured by the wealth that must be taken away from that person at age j to keep his or her expected utility constant, given the reduced risk of death, or

$$w_{j,t'} \equiv \frac{dW_j}{d\delta_{t'}} = -\frac{(dV_j / d\delta_{t'})}{(dV_j / dW_j)} \tag{10-25}$$

The envelope theorem implies that

$$\frac{(dV_j / d\delta_{t'})}{(dV_j / dW_j)} = \frac{(\partial V_\lambda / \partial \delta_{t'})}{(\partial V_\lambda / \partial W_j)} \tag{10-26}$$

Thus, using equation 10-18, we can express MWTP as

$$w_{j,t'} = (1 - \delta_{t'})^{-1} \sum_{t=t'+1}^{T} \pi_{j,t} \cdot \left[(1+d)^{j-t} \cdot u(X_t) \cdot \lambda^{-1} + (1+r)^{j-t} \cdot (M_t - X_t) \right] \tag{10-27}$$

Willingness to pay at age j for a reduction in the conditional probability of death at age t' equals the gain in expected utility from year t' onward, converted to dollars by dividing by the marginal utility of wealth in year j (that is, λ) and discounted at the individual's own rate of time preference. Added to this is the effect of the change in $\delta_{t'}$ on the budget constraint. A reduction in $\delta_{t'}$ makes the individual wealthier by increasing the present value of his or her expected lifetime earnings from age $t' + 1$ onward. However, an increase in survival probabilities also decreases the consumption that the person can afford in each of the years $t' + 1$ through T from a given earnings stream. Thus his or her WTP is reduced by the change in the present value of the consumption stream.

Note that $w_{j,t'}$ is the rate at which the individual is willing to trade wealth for a unit change in risk. To compute the dollar value of a small change in risk, equation 10-27 must be multiplied by the magnitude of the risk change. For example, if $w_{j,t'} = \$2 \times 10^6$ but the change in risk is only 10^{-6}, willingness to pay for the risk change is $2.

Several points should be made about willingness to pay for reductions in the risk of death on the basis of equation 10-27. First, for a policy that affects the conditional probabilities for dying over a number of years, the total MWTP is the sum of the willingness to pay for the changes in each of the $\delta_{t'}$.

Second, a key assumption of models of this type is that expected lifetime utility depends only on expected lifetime consumption, as Linnerooth (1979) noted. So what is being calculated is the willingness to pay for the opportunity to continue consumption. As Bergstrom (1982) pointed out, if the intertemporal objective function is derived from preferences among alternative lotteries, it should include a term that values survival per se. If this term is an increasing function of the $(\pi_{j,t})$, any WTP measure derived from equation 10-18 must be regarded as a lower bound to true WTP. This condition was first noted in a static context by Conley (1976) and Cook (1978).

Third, even granting the assumption that utility depends only on consumption, as long as the individual's average utility of consumption exceeds his or her marginal utility, WTP exceeds human capital and human capital must be interpreted as a lower bound to WTP. By use of the first-order conditions for utility maximization, the term in brackets in equation 10-27 can be written as

$$(1+r)^{j-t}\left[\frac{u(X_t)}{(\partial u/\partial X_t)} - X_t + M_t\right] \tag{10-28}$$

This implies that if $u(X_t)/(\partial u/\partial X_t) - X_t > 0$ for all t, then each year's contribution to $w_{j,t'}$ as given by equation 10-28 exceeds that year's contribution to the present value of lifetime earnings, $(1 + r)^{j-t} \cdot M_t$, and WTP must exceed the present discounted value of lifetime earnings. As noted by Conley (1976) and Cook (1978), the condition that $u(X_t)/(\partial u/\partial X_t) - X_t > 0$ implies that the average utility of consumption exceeds its marginal utility, a condition that holds for all increasing, concave utility functions, provided consumption exceeds a subsistence level.

Blomquist (1981), expanding on Linnerooth's analysis, reviewed the results of nine empirical estimates of WTP and showed that the implied value of statistical life was typically much larger than the expected lifetime earnings of the members of the sample population used in each study. His analysis lends strong support to the assertion that values of life based on lifetime earnings are a poor proxy for the willingness to pay for reduced risk of mortality.

Fourth, equation 10-27 implies that as t' (the age at which risk of death changes) increases, $w_{j,t'}$ must decline, at least as long as the individual is above subsistence. This implies that the value of reducing a person's current probability of dying must always be greater than the value of reducing exposure at age j to a carcinogen with a latency period of $t' - j$ years. In the latter case fewer expected life years are saved. Furthermore, the longer the latency period, the smaller is WTP, with other things held constant. This means that WTP measures based on behavior toward contemporaneous risk (for example, from studies of wages and occupational accident mortality) will not be good proxies for willingness to pay to reduce latent risks.

Cropper and Sussman (1990) have shown that there is an alternative approach to dealing with the latency problem. Suppose that we have obtained a measure of $w_{t',t'}$ for a group at age t by examining, for example, the group members' trade-offs between wages and contemporaneous job risks. Cropper and Sussman have shown that $w_{j,t'}$ $(j < t')$ can be calculated from the following expression:

$$w_{j,t'} = \prod_{s=j}^{t'} \left(1 + a_s\right)^{-1} \cdot w_{t',t'} \tag{10-29}$$

That is, the willingness to pay in advance for a reduction in risk in t' is the contemporaneous WTP discounted back by a factor derived from the actuarial interest rates over the interval. Recall that these actuarial interest rates are a combination of the riskless interest rate and conditional probabilities of death. As such, they vary over time for the individual and across individuals, depending on age and other factors.

Fifth, it is interesting to see how $w_{j,j}$ varies over the life cycle. Shepard and Zeckhauser (1982, 1984) used an expression like equation 10-27 to examine this question. If consumption were constant for all t, as would be the case if the riskless rate of interest were equal to the subjective rate of time preference, $w_{j,j}$ would decline monotonically with age. Younger people would always have a higher willingness to pay to reduce current risk of death than older people because there would be more years of consumption and utility in the summation. If, however, consumption increases over some portion of the life cycle, $w_{j,j}$ may also increase with age up to some point and then decline. If, for example, the individual cannot be a net borrower but can lend at the riskless rate of interest, his or her consumption is likely to be constrained by income at the beginning of his or her life. This will cause the present value of the utility of consumption and, hence, $w_{j,j}$ to increase up to some point, and then to decline.

Toward Measurement

The most commonly used method for estimating the value of reduced risk of death is the hedonic wage–risk trade-off approach that was introduced above and is described in more detail in Chapter 12. Viscusi (1993) describes the wage–risk model and reviews the results of its application to primarily U.S. data. He finds that a majority of the studies in his survey report values of statistical lives (VSLs) in the range of $4.0 to 9.5 million (in 2001 dollars), where the VSL is the marginal change in annual wages for a small change in risk divided by the change in the probability of a fatal accident on the job. Mrozeck and Taylor (2002) use meta-analysis on a set of wage–risk studies that includes more recent work to investigate the relationship between estimated VSLs and features of study design. They suggest that what they call "best practice" studies yield estimated VSLs in the range of $1.6 to $2.7 million.

Values for risk reduction might also be revealed by choices regarding purchases of goods that reduce mortality risks. For example, smoke detectors and seat belts are goods whose primary purpose is to produce safety, that is, to reduce the risk of death for those who purchase them. Data on the purchase and usage of these goods have been used to estimate the values of reducing risk of death. Some results are reviewed in Viscusi 1993. Most of these studies yield estimates at the lower end of the range of wage–risk studies.

Safety could also be one of the characteristics of a differentiated product like an automobile. Different automobile models have measurable differences in accident rates (probability) and the severity of injury. If these differences are systematically related to the prices of different models of automobiles, then the hedonic price model described in Chapters 4 and 8 can be applied to estimate individuals' willingness to pay for reductions in the risk of accident or death. For an example of this approach based on U.S. data, see Atkinson and Halvorsen (1990), who estimate a VSL of about $5.4 million.

Stated preference methods have also been used to estimate VSLs. These methods permit researchers to investigate the relationships between WTP and such variables as age at risk, income, health status, cause of death, level of baseline risk, and size of risk reduction. In Viscusi's (1993) survey, stated preference studies showed estimates of VSLs in the range of $1.4 to $21.1 million. More recently, Hammitt and Graham (1999) reviewed all the stated preference studies they could find since 1980 to test their sensitivity to scope. They argue that for small changes in risk of death, stated WTP should be approximately proportional to the size of the postulated risk reduction. They find that for most of the studies where this proportionality condition can be tested, the data are not consistent with proportionality, and in some cases WTP does not vary at all with changes in the postulated risk reduction. This casts doubt on the validity of stated preference measures of the value of risk reduction. But Hammitt and Graham suggest that improved methods for presenting information to respondents on risk changes may improve the performance of stated preference surveys. See also Corso and others (2001), who present some evidence to support this suggestion.

Using the Value of Statistical Life in Policy Evaluation

In principle, the benefit of a policy that reduces the risk of premature death for a specified group of people is the sum of the individual WTPs for the reduction in risk of all the members of the group. These WTPs could vary across the group for many reasons, including differences in their age, income, health status, cause of death, and level of baseline risk. The practice of environmental and safety policy has been to ignore these differences and to use a VSL based on the WTP of the average individual in the group that provided the data for the WTP estimate. For example, if the source of data were a hedonic wage regression, the sample mean values for all of the independent variables would be used to compute the WTP of the sample mean individual. This in turn would provide the basis for computing the VSL for the sample.

In the United States, this practice has been sanctioned for all federal agencies by the Office of Management and Budget (U.S. Office Management and Budget 2000) and the Environmental Protection Agency (EPA) (U.S. EPA 2000a). EPA recommends the use of a VSL of $6.5 million in 2001 dollars. This figure is based on an analysis of 5 stated preference and 21 wage–risk studies in which a Weibul distribution was fitted to the VSLs and the mean of the Weibul distribution was calculated. There is no standard VSL for all federal agencies.

In the past five years or so, this practice has come under increasing criticism for failing to reflect adequately a variety of factors thought to influence individuals' willingness to pay for risk reduction. Perhaps the most important of these is the age of the population at risk. The wage–risk studies that figure so importantly in the VSLs used by most analysts reflect the WTPs of a group of healthy, mostly male individuals of working age. The mean age of the workers included is typically around 40. If the population affected by an environmental policy is mostly older and if WTP depends on age and years of life at risk, then the VSL based on wage–risk studies could be unrepresentative of the WTP of the affected population. Other factors leading to differences between estimated VSLs and VSLs of affected groups include differences in income, health status, cause of death, the nature of the risk (voluntary versus involuntary), and the level of baseline risk. For a recent review of these and related issues, see U.S. EPA 2000b. These issues represent important topics for further research.

Discounting Statistical Lives?

The practice of discounting statistical lives is common in the economic appraisal and evaluation of environmental and safety public health initiatives. The practice allows policymakers to compare policies that affect probabilities of premature death with different time patterns. However, the practice is also controversial from ethical perspectives. Discounting is justified on the basis of two factors. The first is the observation that people tend to prefer present consumption over future consumption (time preference). The second factor is the productivity of capital investments, meaning that $1 of resources invested today rather than consumed will make it possible to consume more than $1 at a future time.

The practice of discounting lives saved might be attractive to policymakers because it gives the appearance of making possible the comparison of policies that affect probabilities of premature death with different time patterns and different costs without engaging in the controversial practice of assigning monetary values to statistical lives saved. However, at a fundamental level, it is not possible to separate the thing being discounted (lives) from its economic value, especially in the realm of policymaking and choices among alternative policies with different costs. Thus if economic values should be discounted, then discounting should also apply to statistical lives saved. Discounting of statistical lives saved is a controversial practice. Here, the objection to the practice offered by moral philosophers will be outlined first. Then, two complementary arguments in favor of discounting from economics will be offered.

Speaking explicitly of discounting lives saved, the philosopher Douglas MacLean (1990) has argued that mere differences in the timing of events can have no moral significance. Similar people should not be treated differently solely on account of differences in the timing of their deaths. For example, the premature death of any 20-year-old person today can be considered as no worse than the premature death of a similar 20-year-old 10 years from now. Similarly, if it is better to save 100 lives today than to save 99 lives today, it must also be better save 100 lives 10 years from now than to save 99 lives today.

This is just a special case of a more general point that Broome (a philosopher and economist) attributes to the nineteenth-century philosopher and economist Henry Sidgwick. Broome says:

> *[F]rom a universal point of view* the time at which a man lives cannot affect the value of his happiness. A universal point of view must be impartial about time, and impartiality about time means that no time can count differently from any other. In overall good, judged from a universal point of view, good at one time cannot count differently from good at another. Nor can the good of a person born at one time count differently from the good of a person born at another. [italics in original] (Broome 1992, 92)

Thus, whether considering lives or well-being, the amount realized must be given the same moral weight independent of the time of its realization. This is equivalent to saying that in moral terms the discount rate is zero.

Looking at deaths alone, the ethical perspective has validity. However, it abstracts from two important considerations. The first consideration is individuals' preferences concerning the timing of the benefits and costs that they experience and that contribute to their well-being. The second is the opportunity cost of resources committed to lifesaving policies. Each consideration will be discussed in turn.

As shown in Chapter 7, if individuals wish to maximize their lifetime utility, they must equate their marginal rate of substitution between present and future consumption with the market rate of interest. This is true whether they have a positive, zero, or negative time preference. Furthermore, as long as market goods and nonmarket goods such as health and risk reduction are substitutes in preferences, the marginal utilities of future nonmarket goods should also be discounted. It follows that if policy choices should be based on (or at least reflect) individuals' preferences regarding the benefits and costs they receive, then the benefits and costs of policies—including policies that affect the risk of premature death—should be discounted.

For example, consider the question of choosing between policy A, which prevents x immediate deaths in a group now (e.g., reduces the risks of fatal accidents) and policy B, which reduces the same group's exposure to a carcinogen thereby preventing x deaths in, say, 30 years. The group would undoubtedly prefer policy A because of the larger number of life years saved. This would be reflected in their higher willingness to pay now for policy A than for policy B. One way for policymakers to reflect this preference would be to discount the avoided deaths of policy B.

Turning to the second consideration, discounting concerns choices among policies that involve commitments of resources that have opportunity costs; these opportunity costs can include forgoing other opportunities to reduce mortality at other times. The practice of discounting is a way of bringing these opportunity costs into the decision process. Consider a policy involving costs now that avoids x deaths per year in perpetuity. The undiscounted sum of lives saved is infinite. Without discounting, this would justify an infinite commitment of resources. However, surely there are limits on the cost that the present generation is obliged to incur to save future lives. The discounted present value of this stream of lives saved converges on x/r. The present generation can (and must) decide what cost it is willing to incur to save a finite sum of discounted lives.

What about policy A, which prevents x deaths in a group of 40-year-old people now versus policy B, which prevents y ($> x$) deaths in a group of otherwise similar 40-year-old people in 30 years? The ethical perspective claims that policy B is preferred because of the larger number of deaths avoided. However, if the resources used in this policy were instead invested to earn r% for 1 year and then used in a policy similar to B (call it policy C), then y' ($> y$) deaths would be avoided. Thus, C should be preferred to B, and so forth for policies D, E, etc., involving further postponement of the lifesaving policy. None of the policies would be undertaken because an additional postponement can always increase the undiscounted number of deaths avoided. Discounting avoids this counterintuitive result.

These two lines of argument (time preference and opportunity cost) come together in the following way. To identify policies that are potential Pareto improvements, both the benefits and costs of lifesaving policies must be discounted at a rate that reflects the interaction of time preference and opportunity cost: the market interest rate in simple models.

Valuing Reduced Morbidity

Morbidity is a general term that refers to cases of disease or being in less than "good" health. Morbidity can be classified in a variety of ways, among them duration of the condition (chronic or acute), degree of impairment of activity, or type of symptom. An episode of acute morbidity would last only a matter of days and would have a well-defined beginning and end. Chronic morbidity refers to cases of a longer term illness of indefinite duration. The degree of impairment could be defined in terms of, for example, "restricted activity days," on which a person is able to undertake some, but not all, normal activities; "bed disability days," on which a person is confined to bed, either at home or in an institution, for all or most of a day; or "work loss days," on which a person is unable to engage in ordinary gainful employment. However, these measures of morbidity reflect responses to ill health rather than the health condition itself. Whether a given clinical manifestation of ill health results in any restriction on activity, bed disability, or work loss depends upon a number of socioeconomic variables, such as employment and labor-force status, nonlabor sources of income, the presence of other income-earners in the household, and so forth. Morbidity can also be measured by "symptom days," that is, by the occurrence of specific symptoms such as an asthma attack, a headache, a cough, throat irritation, or diarrhea.

The choice of symptoms for defining and measuring morbidity has implications for the economic valuation of health effects. The economic perspective on health focuses attention on effects that people are aware of and wish to avoid, that is, effects that would reduce their utility. Yet some biomedical clinical research focuses on effects of questionable significance to individuals and measures health effects that are difficult to relate to individual perceptions and behavior. The question of how to define morbidity is tied to a legal and policy issue that vexes the U.S. Environmental Protection Agency: What constitutes an adverse health effect? The Clean Air Act of 1970 calls for setting air-quality standards that protect individuals from *adverse* health effects. The question of whether or not an effect is adverse arises when clinical studies reveal that exposure to an air pollutant under controlled conditions leads to detectable changes in organ structure or function without necessarily causing pain, impeding people's activities, or reducing life expectancy. Are these changes adverse? From an economic perspective based on the WTP definition of value, the answer depends on whether the changes are perceived by the individual and whether the individual reveals or expresses a willingness to pay to avoid the effect or requires compensation to experience it.

Measures of morbidity must also take into account the fact that, unlike mortality, morbidity is not a discrete event but a process involving time. Cases observed during a period of time may fall into one of four categories:

1. onset of morbidity occurs before the period and morbidity terminates by either recovery or death during the period;
2. onset of morbidity occurs before the period and morbidity terminates after the period;
3. onset of morbidity occurs during the period and morbidity terminates during the period; and
4. onset of morbidity occurs during the period and morbidity terminates after the period.

The *prevalence rate* encompasses all four categories; it is defined as the total number of cases in the period as a percentage of the average number of persons in the population during the period. The *incidence rate* covers only the third and fourth categories; it is defined as the number of new cases during the period as a percentage of the average number of persons in the population. Incidence rate data would be more appropriate for investigating causal relationships between changes in exposure and changes in health status. Because willingness to pay to reduce morbidity is likely to depend on the length as well as the number of cases, the prevalence rate and measures incorporating data on duration would be more appropriate for analyzing the social costs of morbidity.

Broadly speaking, estimates of the monetary value of reduced morbidity take one of two forms, those based on individual preferences (WTP or WTA), and those based on the resource and opportunity costs associated with illness. The latter form is typically referred to as a cost-of-illness or sometimes a damage cost measure. These are examples of the damage function method described in Chapter 2. They seek to identify the real costs of illness in the form of lost productivity and output and the increase in resources devoted to medical care. Costs per case of illness or per day are multiplied by the number of cases or days sick to determine an aggregate value. As I will show formally in this section, the damage cost approach to valuation is unsatisfactory because it fails to capture the variety of behavioral responses to illness and the threat of illness.

To see this in a simple way, consider the case of an individual who experiences one fewer day of asthma attacks because of an improvement in air quality. The benefit to that person might include avoiding the lost wages associated with being unable to work one day and the reduction in costs for medicine and treatment. In addition, though the individual avoids the discomfort associated with the attack itself. The first two components are captured by the cost-of-illness approach. Only a comprehensive WTP measure would capture the discomfort component as well.

Now consider an individual who experiences no asthma attacks at present levels of air pollution because he or she spends money operating an air purifier and stays at home indoors on high-pollution days to prevent the attacks that would be associated with exposure to the outdoor air. If air quality is improved, this individual benefits by being able to reduce the monetary expenditures and the lost wages and opportunities for leisure activities that are associated with these defensive activities. This individual will benefit from reduced air pollution even though there is no observed reduction in the actual incidence of asthma attacks related to air pollution or in the associated cost of illness.

As these two examples show, air pollution that affects human health can reduce people's well-being through four channels: the medical expenses associated with treating disease induced by air pollution (including the opportunity cost of time spent in obtaining such treatment); the lost wages resulting from the inability to work; the defensive or averting expenditures and activities associated with attempts to prevent disease induced by air pollution (including the opportunity cost of time); and the disutility associated with the symptoms and lost opportunities for leisure activities caused by the illness. Improving environmental quality can yield benefits to individuals by reducing some or all of these adverse effects. Portions of the first three of these effects have readily identifiable monetary counterparts, but effects of the fourth kind may not. A truly comprehensive benefit measure should be able to capture all these relevant effects. Measures based solely on decreases in medical costs or lost wages are not comprehensive because they omit major categories of beneficial effects.

Although individual WTP is the correct starting point for analyzing health-related values, there is one important respect in which society's valuation of changes in health might diverge from that of the affected individual. Society has developed several mechanisms for shifting some of the costs of illness away from the individual who is ill and onto society at large. These mechanisms include medical insurance, which spreads the costs of treatment among all policyholders, and sick leave policies, which shift at least part of the cost of lost workdays onto the employer and ultimately onto the consumers of the employer's products. An individual's expressed willingness to pay to avoid illness would not reflect those components of the costs of his or her illness that are borne by or shifted to others. However, the value to society of avoiding his or her illness includes these components. Empirical measures of the value of reducing illness must take account of these mechanisms for shifting costs. This will be discussed in more detail below.

In this section, I develop a general model of individual choice that captures the principal ways by which individuals can affect their health status. I use this model to derive measures of the value of decreases in pollution that affect health and of improvements in health per se. I also describe several extensions of the basic model and consider the problems that arise when this approach to valuation is applied to reductions in the incidence of chronic disease.

A Basic Model of Health Production and Choice

Most of the formal models used for deriving the value of reduced morbidity use some variant of the health production function first developed by Grossman (1972). Cropper (1981) introduced a pollution variable into the health production function. Harrington and Portney (1987) extended the model to examine explicitly the relationships among willingness to pay for a reduction in pollution, reductions in the cost of illness, and changes in defensive expenditures. In this section I develop an expanded version of the Harrington and Portney model. Dickie (2003) presents an excellent exposition of these models.

The health production function relates exogenous variables (including environmental variables such as air pollution) and choice variables (such as preventive medicine and treatment costs) to some measure of health status. We assume that individuals know their health production function, choose the output level optimally, and choose inputs to minimize the cost of production of any level of health. Of course, these are strong assumptions.

As originally formulated by Grossman (1972), the health production function was dynamic, allowing for "investments" in "health capital" that yielded benefits in the form of reduced illness over several time periods. The simple model presented here will abstract from this intertemporal dimension of the problem. Intertemporal models are briefly discussed in a later section.

Let health in any time period be measured by the number of days sick, represented by s. This is a simplification because it makes no distinction between one episode of illness of two days' duration and two separate illnesses of one day each and because differences in the types of symptoms and the severity of illness are ignored. Some of the implications of richer specifications of the health variable are discussed in a later section.

Assume that one of the determinants of health status is the level of exposure to or dose of some environmental contaminant. Dose is represented by the scalar variable d, which depends on the concentration of pollution, c, and the amount of an averting activity, a, undertaken to avoid or reduce exposure to pollution. Examples of averting activities include filtering tap water before drinking and staying indoors on days of high air pollutant levels. If the contaminant is an air pollutant, c could be interpreted as the number of days during which some measure of air pollution exceeds a stated standard, the mean value of the pollutant averaged over the relevant time period, or the highest value recorded for the pollutant during that period. Because a change in pollutant emissions is likely to change all these measures in a predictable way, the choice of a measure for d should be based on whatever is the best predictor of changes in health status. Similar questions arise for other forms of environmental contamination, such as chemicals in drinking water and pesticide residues in food.

Assume also that there is a set of mitigating activities and treatments that can be undertaken to reduce the health effect of any given exposure to pollution, represented by b. Examples of mitigating activities include taking antihistamines and visiting a doctor to obtain relief from a sinus headache. In this model, assume that the level of mitigating activities can be chosen by the individual to maximize utility.

The health production function for an individual can be written as

$$s = s(d, b) \tag{10-30}$$

$$d = d(c, a) \tag{10-31}$$

and by substitution

$$s = s(c, a, b) \tag{10-32}$$

with

$$\partial s/\partial c > 0 \tag{10-33}$$

$$\partial s/\partial b, \partial s/\partial a < 0 \tag{10-34}$$

The health production function can be estimated from cross-section data on illness, pollution, and averting and mitigating activities. It would also be necessary to control for other determinants of health status such as physical and socioeconomic characteristics of individuals, for example, age, sex, use of tobacco, income, and education.

The individual derives utility from the consumption of a numeraire good, X, normalized with a price of 1, and leisure, f. Illness causes disutility, so

$$u = u\,(X, f, s) \tag{10-35}$$

with

$$\partial u/\partial X, \partial u/\partial f > 0 \tag{10-36}$$

$$\partial u/\partial s < 0 \tag{10-37}$$

The individual chooses X, f, a, and b to maximize utility subject to the budget constraint of

$$I + p_w(T - f - s) = X + (p_a \cdot a) + (p_b \cdot b) \tag{10-38}$$

where

I = nonlabor income
p_w = the wage rate
T = total time available
p_a = the price of averting activities
p_b = the price of mitigating activities

Where there are time costs associated with the averting or mitigating activities, they should be incorporated into the full income budget constraint. For an example, see Gerking and Stanley 1986.

The first-order conditions for a maximum include

$$\partial u/\partial X = \lambda \tag{10-39}$$

$$\partial u/\partial f = \lambda \cdot p_w \tag{10-40}$$

and

$$\lambda \cdot \frac{p_b}{(\partial s/\partial b)} = \frac{\partial u}{\partial s} - \lambda \cdot p_w = \lambda \cdot \frac{p_a}{(\partial s/\partial a)} \tag{10-41}$$

where λ is the Lagrangian multiplier and can be interpreted as the marginal utility of income.

The Marginal Value of Reduced Pollution. This model of choice can be used to derive an observable measure of the individual's marginal willingness to pay to reduce pollution. An individual's willingness to pay for a small reduction in ambient pollution is the largest amount of money that can be taken away from that person without reducing his or her utility. If pollution enters the utility function directly because, for example, of aesthetic disamenities associated with pollution, then there are additional benefits that are not associated with health. This point is discussed further in the section called Complex Models.

In the health production model, in which pollution affects utility only through health, WTP is the reduction in the cost of achieving the optimal level of health made possible by the decrease in pollution. For example, if a reduction in ozone levels from 0.16 to 0.11 parts per million (ppm) reduces the number of days of respiratory symptoms from 6 to 4, and if an expenditure of $20 on averting activities or on medicine has the same effect, then (all else being equal) the individual should be willing to pay no more than $20 for the ozone reduction.

Formally, marginal willingness to pay for a reduction in pollution (w_c) is given by the reduction in sick time associated with the reduction in pollution times the marginal cost of reducing sick time. The latter is given by the cost of an additional mitigating input divided by the reduction in sick time that input produces, or, alternatively, by the cost of averting behavior divided by the reduction in sick time that averting behavior produces. To see this, first totally differentiate the indirect utility function, $v(I, p_w, p_a, p_b, c)$, and solve for WTP (w_c) defined as dI/dc to obtain

$$\frac{dI}{dc} = -\frac{(\partial v / \partial c)}{(\partial v / \partial I)} = -\frac{(\partial v / \partial c)}{\lambda} \tag{10-42}$$

Next note that the effect of c on utility consists of two components: the direct loss of utility due to illness and the opportunity cost of time spent sick valued at the wage rate, or

$$\frac{\partial v}{\partial c} = \left(\frac{\partial u}{\partial s} \cdot \frac{\partial s}{\partial c}\right) - \left(\lambda \cdot p_w \cdot \frac{\partial s}{\partial c}\right) = \left[\frac{\partial u}{\partial s} - (\lambda \cdot p_w)\right]\frac{\partial s}{\partial c} \tag{10-43}$$

Then, substitute the first order condition 10-41 for the term in square brackets in equation 10-43:

$$\frac{\partial v}{\partial c} = \left[\lambda \cdot \frac{p_b}{(\partial s / \partial b)}\right]\frac{\partial s}{\partial c} \tag{10-44}$$

Finally, substitute equation 10-44 into equation 10-42 to obtain

$$w_c = -p_a \frac{(\partial s / \partial c)}{(\partial s / \partial a)} = p_a \frac{\partial a}{\partial c} \tag{10-45}$$

A similar procedure leads to

$$w_c = -p_b \frac{(\partial s / \partial c)}{(\partial s / \partial b)} = p_b \frac{\partial b}{\partial c} \tag{10-46}$$

The right-hand terms in equations 10-45 and 10-46 follow from application of the implicit function rule.

Most of these results can also be derived from the expenditure function, as follows:

$$e \equiv \min\left[X + (p_a \cdot a) + (p_b \cdot b) + (p_w \cdot s)\right] + \mu\left[u^0 - u(X, f, s)\right] \tag{10-47}$$

where $s = s(c, a, b)$. The first-order conditions include

$$\frac{p_a}{(\partial s / \partial a)} = \mu \cdot \frac{\partial u}{\partial s} - p_w \tag{10-48}$$

So

$$w_c = -\frac{\partial e}{\partial c} = \left(\mu \cdot \frac{\partial u}{\partial s} - p_w\right)\frac{\partial s}{\partial c} = p_a \cdot \frac{(\partial s/\partial c)}{(\partial s/\partial a)} \tag{10-49}$$

which is the same as equation 10-45. Equation 10-46 can be found by similar manipulations.

There are several things to note about these expressions for MWTP. First, the ratios

$$\frac{(\partial s/\partial c)}{(\partial s/\partial b)} \quad \text{and} \quad \frac{(\partial s/\partial c)}{(\partial s/\partial a)} \tag{10-50}$$

can be interpreted as marginal rates of technical substitution (MRTSs) between pollution and the other variable in producing a constant level of sickness. MWTP can be expressed in terms of any of the MRTSs between pollution and another input in the production of health because to minimize the cost of producing health, the values of marginal products of all inputs must be equal at the margin. Second, all of the measures are functions of observable variables that can be calculated given knowledge of the health production function. Third, as shown by the right-hand terms, MWTP can be calculated from the reductions in expenditures on either mitigating or averting behavior that are required to attain the original health status, holding all else constant. As noted in Chapter 4, this will not in general be equal to the observed reduction in mitigating or averting behavior associated with the reduction in pollution.

To compute equations 10-45, 10-46, or 10-49, it is necessary to estimate a production function for the health outcome of interest and evaluate the numerator and denominator of the equation at current levels of all inputs. In practice this has proven to be a difficult task. For one effort to implement this basic model, see Gerking and Stanley 1986. In addition to having data on the relevant health outcome and on ambient pollution levels, one must identify averting and mitigating behaviors and measure their costs. In practice the most effective method of reducing exposure, given ambient pollution levels, is to spend time indoors. Although the amount of time spent indoors could be measured, determining its cost would be difficult. Devices that reduce indoor pollution concentrations (such as air conditioners and air filters) have costs that can be measured; however, they produce other services, such as reducing indoor temperature, so that it is inappropriate to allocate all these costs to pollution avoidance. The implications of this possibility are discussed in the section called Complex Models.

Because of the difficulties of implementing these measures, it is useful to consider an alternative expression that shows the relationship between the observable cost of illness and MWTP. The first step in deriving this expression is to obtain the demand functions for a and b: $a^*(I, p_w, p_a, p_b, c)$ and $b^*(I, p_w, p_a, p_b, c)$. These functions give the optimal quantities of a and b as functions of income, prices, and pollution. The second step is to take the total derivative of the health production function:

$$\frac{ds}{dc} = \left(\frac{\partial s}{\partial a} \cdot \frac{\partial a^*}{\partial c}\right) + \left(\frac{\partial s}{\partial b} \cdot \frac{\partial b^*}{\partial c}\right) + \frac{\partial s}{\partial c} \tag{10-51}$$

This gives the effect of a change in pollution on illness after taking account of the optimal adjustments of a and b to the pollution change. This expression can be rearranged as follows:

$$\frac{\partial s}{\partial c} = \frac{ds}{dc} - \left(\frac{\partial s}{\partial a} \cdot \frac{\partial a^*}{\partial c}\right) - \left(\frac{\partial s}{\partial b} \cdot \frac{\partial b^*}{\partial c}\right) \tag{10-52}$$

and multiplied by the first-order conditions of equation 10-41

$$-\frac{p_a}{(\partial s/\partial a)} = p_w - \frac{(\partial u/\partial s)}{\lambda} \tag{10-53}$$

to obtain

$$\begin{aligned} -p_a \frac{(\partial s/\partial c)}{(\partial s/\partial a)} = \left[p_w - \frac{(\partial u/\partial s)}{\lambda}\right] \cdot \frac{ds}{dc} - \left[p_w - \frac{(\partial u/\partial s)}{\lambda}\right] \cdot \frac{\partial s}{\partial a} \cdot \frac{\partial a^*}{\partial c} \\ - \left[p_w - \frac{(\partial u/\partial s)}{\lambda}\right] \cdot \frac{\partial s}{\partial b} \cdot \frac{\partial b^*}{\partial c} \end{aligned} \tag{10-54}$$

or, after rearranging,

$$w_c = \left(p_w \cdot \frac{ds}{dc}\right) + \left(p_b \cdot \frac{\partial b^*}{\partial c}\right) + \left(p_a \cdot \frac{\partial a^*}{\partial c}\right) - \left[\frac{(\partial u/\partial s)}{\lambda} \cdot \frac{ds}{dc}\right] \tag{10-55}$$

This expression says that MWTP is the sum of the observable reductions in the cost of illness and averting activities and the monetary equivalent of the disutility of illness. The change in the cost of illness consists of the economic value of reductions in sick time and mitigating expenditures. The term $p_w \cdot (ds/dc)$ includes both actual lost wages and lost leisure time valued at the wage rate.

To compute ds/dc we do not need to estimate a health production function but can instead estimate a dose–response function, a reduced form relationship between illness and ambient pollution controlling for other variables that affect health status. In the health production framework a dose–response function is obtained by substituting the demand functions for b and a into the health production function. Full implementation of equation 10-55 as a measure of value therefore requires estimation of these demand functions.

As a practical matter, the first three terms on the right-hand side of equation 10-55 can be approximated after the fact by using the observed changes in illness and averting and mitigating expenditures. In this way, equation 10-55 can be used to derive a lower bound to individual WTP. Because the last term in the equation is negative $(\partial u/\partial s < 0)$, the first two terms—the value of lost time plus the change in averting and mitigating expenditures—give a lower bound to WTP. In the health literature, the term "cost of illness" typically refers only to the social cost of lost earnings plus the medical expenditures associated with illness. This term therefore ignores two components of the social cost of illness—the social value of averting expenditures and the cost of lost leisure time that results from illness.

The Marginal Value of Reduced Illness. The marginal willingness to pay for an exogenous reduction in illness falls out of the above model as a special case. Suppose that averting behavior is not possible and that mitigation (b) reduces realized s from its exogenous level s^* according to

$$s = f(s^*, b) = s^* - s(b) \tag{10-56}$$

The analogs of equations 10-42 to 10-45 are

$$\frac{dI}{ds^*} = -\frac{(\partial v/\partial s^*)}{(\partial v/\partial I)} = -\frac{(\partial v/\partial s^*)}{\lambda} \tag{10-42$'$}$$

$$\frac{\partial v}{\partial s^*} = \frac{\partial u}{\partial s^*} - \lambda \cdot p_w \tag{10-43$'$}$$

$$\frac{\partial v}{\partial s^*} = \lambda \cdot \frac{p_b}{(\partial s/\partial b)} \tag{10-44$'$}$$

$$w_{s^*} = -p_b \cdot \frac{\partial b}{\partial s^*} \tag{10-45$'$}$$

Similarly, the analog to equation 10-55 is

$$w_{s^*} = p_w + p_b \cdot \frac{\partial b^*}{\partial s^*} - \frac{(\partial u/\partial s^*)}{\lambda} \tag{10-55$'$}$$

or, MWTP is the sum of the cost of illness (lost wages and mitigation costs) and the monetary equivalent of the lost utility (pain and suffering).

Valuing Nonmarginal Changes in Pollution. In the preceding subsection I defined a measure of value for a small change in health status. This value measure is likely to be a function of health status itself. Specifically, for reasons analogous to the standard assumption of diminishing marginal utility, the marginal willingness to pay for further decreases in morbidity is likely to decrease as health status increases. In principle, this functional dependence of MWTP on health status should be taken into account whenever large (that is, nonmarginal) changes in morbidity are being valued. The proper way to do this is straightforward in principle but may be hard to implement in practice.

If the function relating MWTP to health status is known, the value of the nonmarginal change is simply the integral of this function over the relevant range. Typically, however, empirical methods produce a point estimate of MWTP. If it can be assumed that MWTP is (approximately) constant over the relevant range, the value of the nonmarginal change can be calculated by multiplying the MWTP times the change in health status. The larger is the change in morbidity and the more rapidly MWTP decreases with increases in health status, the larger is the error in using this simple approach. Alternatively, one could compute a lower bound to MWTP. As shown in Chapter 4, the change in averting and mitigating expenditures, holding illness constant, constitutes a lower bound to willingness to pay for pollution improvement.

Extensions of the Health Production Model

The basic model of choice and value described in the preceding section involves a considerable amount of simplification to obtain some insights about the relationship between observed behavior and economic value. In this section I extend the model in several directions toward greater realism to see to what extent the conclusions about values are preserved.

Multiple Symptoms. The model can be generalized to the case of many symptoms and various forms of averting and mitigating behavior. Consider the special case in which each mitigating activity enters the separate health production function for one symptom, $s_i(c, a, b_i)$. Following procedures similar to those of the basic model, it can be shown that

$$w_c = \Sigma p_i \cdot \frac{(\partial s_i / \partial c)}{(\partial s_i / \partial b_i)} \tag{10-57}$$

where p_i is the price of b_i. Estimation of w_c in this case involves estimating production functions for all symptoms. Similarly, it can be shown that

$$w_c = p_a \cdot \frac{(\partial s_i / \partial c)}{(\partial s_i / \partial a)} \tag{10-58}$$

for any symptom.

Unfortunately, simple expressions such as these cannot be obtained when mitigating activities affect more than one symptom. Suppose that there are two symptoms, s_1 and s_2, neither of which restricts activities, and two mitigating activities, b_1 and b_2, with health production functions:

$$s_1 = s_1(c, a, b_1, b_2) \tag{10-59}$$

and

$$s_2 = s_2(c, a, b_1, b_2) \tag{10-60}$$

For example, b_1 could be a multisymptom cold remedy. The first-order conditions are

$$\left[\frac{(\partial u / \partial s_1)}{\lambda} \cdot \frac{\partial s_1}{\partial b_1}\right] + \left[\frac{(\partial u / \partial s_2)}{\lambda} \cdot \frac{\partial s_2}{\partial b_1}\right] - p_1 = 0 \tag{10-61}$$

$$\left[\frac{(\partial u / \partial s_1)}{\lambda} \cdot \frac{\partial s_1}{\partial b_2}\right] + \left[\frac{(\partial u / \partial s_2)}{\lambda} \cdot \frac{\partial s_2}{\partial b_2}\right] - p_2 = 0 \tag{10-62}$$

$$\left[\frac{(\partial u / \partial s_1)}{\lambda} \cdot \frac{\partial s_1}{\partial a}\right] + \left[\frac{(\partial u / \partial s_2)}{\lambda} \cdot \frac{\partial s_2}{\partial a}\right] - p_a = 0 \tag{10-63}$$

As Dickie and Gerking (1991) show, as long as the derivatives of the health production function are known and the number of mitigating and averting activities (three here) equals or exceeds the number of symptoms (two), this system of equations can be used to solve for the unobserved utility terms, for example,

$$\frac{(\partial u / \partial s_1)}{\lambda} \tag{10-64}$$

The solution can be used to substitute into the following expression to obtain a measure of WTP:

$$w_c = -\frac{(\partial v / \partial c)}{\lambda} = -\left[\frac{(\partial u / \partial s_1)}{\lambda} \cdot \frac{\partial s_1}{\partial c}\right] - \left[\frac{(\partial u / \partial s_2)}{\lambda} \cdot \frac{\partial s_2}{\partial c}\right] \tag{10-65}$$

However, if the number of symptoms exceeds the number of mitigating and averting activities, then the expression for MWTP will necessarily include some unobservable marginal utility terms.

Shifting the Cost of Illness. The measures derived above give each individual's willingness to pay to avoid pollution-induced illness. However, where social mechanisms have been developed for spreading some of the costs of illness, the social value of reducing pollution may exceed the aggregate of the individuals' WTP. For example, where employers grant a certain number of days of paid sick leave, the individual does not incur any financial cost for the loss of a workday. Thus if a truthful person were asked a contingent valuation question about the value of reducing pollution, the response would not include the lost wages component of the first term on the right of equation 10-55. Nevertheless, society loses that person's output for that day.

The burden of this lost output is borne initially by the employer but ultimately shows up as some combination of higher prices for outputs and lower equilibrium wages. A nonmarginal reduction in pollution that resulted in fewer lost workdays would reduce the costs of paid sick leave plans. In principle this would result in changes in general equilibrium wages and prices.

Similarly, medical insurance has the effect of shifting the marginal cost of treatment away from the sick individual. A 100% coverage plan reduces the price of medical treatment to zero. Thus MWTP measures based on the health production function would have to focus on changes in averting expenditures. Again, changes in the cost of treatment covered by insurance should be added to the aggregate of individuals' willingness to pay to obtain the social benefit of reduced pollution-induced illness.

As Harrington and Portney (1987) point out, the introduction of sick leave and medical insurance plans changes incentives by changing the prices of these activities and thus could alter individuals' choices of mitigating and averting activities and days of illness. If the effects of these plans on prices has been taken into account in estimating the models of individual behavior, then the measures of social values described here will be correct.

Health Capital and the Health Production Function. The basic model outlined above ignores any dynamic aspects of health and illness. In that model, averting or mitigating behavior and pollution in any one period affect only illness in that time. It would be more in the spirit of Grossman's (1972) original treatment of the health production function to formulate the choice problem as one of optimizing investment in health capital over time. Health capital can be thought of as a measure of health status that depreciates over time unless it is maintained or augmented with some kind of investment. To formalize this, let h_t be the stock of health capital at t. Then h_t can be augmented by investment of time, y_t, in activities such as exercising and by the purchase of "health" goods, z_t. Health capital is also subject to depreciation at the rate of g% per year. The health stock changes over time according to

$$dh_t / dt = I_t\ (z_t, y_t) - g \cdot h_t \tag{10-66}$$

The rate of depreciation itself can be made a function of age and other characteristics.

One approach to deriving WTP is to assume that health capital is an argument in the symptom production functions. Then, measures of WTP similar to those of the basic model can be derived. Cropper (1977, 1981) used a different approach. She made the rate of depreciation a function of pollution, among other things. She also assumed that illness was not a direct argument in the utility function. She derived a WTP measure that involved two terms. The first term is the value of the reduction in sick time. The second is the change in the rate of investment in health capital caused by the change in pollution. The sign of this term is ambiguous because reducing pollution both lowers the price of an investment in health capital and increases the desired quantity of health capital. This second term is analogous to the expressions involving averting and mitigating activities. Once again, the conclusion is that true WTP is different from changes in the cost of illness because of the induced changes in other activities and expenditures that affect health.

Complex Models. The models outlined so far have been simple in the sense that pollution affects only sick days and averting behavior serves only to reduce exposure to pollution. However, if either pollution or averting activities also affect utility directly, then the model becomes more complex in the sense that it is no longer possible to derive expressions for MWTP that are functions only of observable data. Rather, the expressions for MWTP become contaminated with unobservable marginal utility terms, marginal disutility terms, or both.

Suppose that in addition to causing illness, pollution causes disutility directly, for example, by impairing the view from one's house. Then the expression for WTP in this modified version of the basic model is

$$w_c = \frac{(\partial v / \partial c)}{\lambda} = -\frac{\left[(\partial u / \partial s) - (\lambda \cdot p_w)\right](\partial s / \partial c) + (\partial u / \partial c)}{\lambda} \qquad (10\text{-}67)$$

The first-order condition can be used to eliminate $(\partial u/\partial s)/\lambda$, but the unobservable $(\partial u/\partial c)/\lambda$ remains. Observations on averting and mitigating behavior are not sufficient to measure MWTP. However, if pollution varies across housing sites, then the hedonic housing price model provides a way around this problem (see Chapter 11). Similarly, suppose a generates utility in addition to reducing exposure. For example, a could be air conditioning or water filters that reduce unpleasant tastes in tap water. Then the first-order condition for a becomes

$$\frac{\partial u}{\partial s} - \lambda \cdot p_w = \frac{\left(\lambda \cdot p_a - \dfrac{\partial u}{\partial a}\right)}{\left(\dfrac{\partial s}{\partial a}\right)} \qquad (10\text{-}68)$$

When this is substituted into the indirect utility function, the unobservable $(\partial u/\partial a)/\lambda$ remains.

Chronic Morbidity

I turn now to some of the problems that arise when these models of choice and value are applied to the case of chronic morbidity. One problem is how to quantify and measure chronic morbidity. Chronic morbidity is more like a state of being than like an illness where incidence and duration are the primary characteristics of concern. It may be more useful to model chronic morbidity in terms of state-dependent preferences; for example,

$$u = u(X, s) \qquad (10\text{-}69)$$

with $s = 0$ for health and $s = 1$ for chronic disease.

A second problem concerns the choice between an ex ante and an ex post perspective. We have already discussed some of these issues in the context of mortality risks. Suppose that air pollution increases the risk of chronic lung disease. The ex ante perspective asks for a healthy individual's willingness to pay to reduce the risk of incurring chronic lung disease at some time in the future. This is essentially the approach used by Viscusi and others (1991). In a variation on the contingent ranking method, they asked individuals to choose between different bundles of chronic bronchitis and other attributes to estimate their risk–risk and risk–dollar marginal rates of substitution. The ex post perspective would be applicable to individuals already suffering from chronic lung disease, who would be asked their willingness to pay to be restored to a healthy state. As discussed in Chapter 8, the ex ante and ex post perspectives might give different values to the same chronic health effect, even leaving aside the likelihood that an individual who actually experiences a chronic health effect might learn something about it that would alter his or her WTP. Which perspective is appropriate would depend, at least in part, on the nature of the policy question being asked. If a potential cure for the chronic condition is available, the ex post perspective yields the correct answer to the valuation question. If the policy in question is one that would reduce exposure to, say, air pollution leading to reduced risk of chronic disease, then the ex ante perspective is appropriate.

Toward Measurement

To implement measures based on the health production function, such as equations 10-45 and 10-55, we would require the following data for a cross section of individuals over some time period:

1. frequency, duration, and severity of pollution-related symptoms;
2. ambient pollution levels to which the individual is exposed;
3. actions the individual takes to avoid or mitigate the effects of air pollution;
4. costs of avoidance and mitigating activities; and
5. other variables affecting health outcomes (age, general health status, presence of chronic conditions, and so forth).

These data would be used to estimate health production and input demand functions, which, in turn, would be used to calculate the expression for MWTP. The same data could be used to implement a cost-of-illness measure. Items 1, 2, 4, and 5 would be used to estimate a dose–response function. The predicted effect of a change in pollution on illness would be multiplied by the value of lost time plus the observed change in medical and avoidance expenditures to calculate the cost of illness.

There is an important difference, however, between the data requirements of the averting behavior approach and the cost-of-illness approach. To infer MWTP using data on avoidance and mitigating behavior, we should know each individual's beliefs about the efficacy of these behaviors because these beliefs motivate observed behavior. Individuals must believe that air pollution affects illness for them to take actions to reduce their exposure. The spirit of the cost-of-illness approach is to measure observable costs associated with actual changes in illness induced by air pollution. In practice, the objective effect of avoidance and mitigating behavior is measured under the averting behavior approach, with the tacit assumption that individuals correctly perceive the effects of their actions.

An alternative approach to the revealed preference method based on the health production function is to ask some form of stated preference question. For example, people could be asked directly what they would be willing to pay to reduce pollution. Alternatively, people could be asked to value a reduction in the number of days with a specified set of symptoms. Then the value per symptom day could be multiplied by the predicted change in the number of symptom days based on an estimate of the dose–response function. In theory these responses could be collected without obtaining any of the information in the five data categories listed above. However, collecting information on the age, health status, and income of respondents is necessary if values from hypothetical responses from a given study are to be used in other contexts.

If willingness to pay for morbidity benefits varies with health status, age, and income, pollution control policies may have different benefits depending on the characteristics of the target population. To value the morbidity benefits of different policies, we must therefore know the distribution of key variables in the population affected by the policy and how WTP (or the cost of illness) varies with these characteristics.

In principle, each of the valuation techniques presented above is capable of describing how the value of reduced morbidity varies with the characteristics of the respondent. The averting behavior models suggest that WTP depends on any variables that affect the marginal product of pollution, mitigating activities, or avoidance activities. In practice these variables would include health status (whether or not the respondent has a chronic respiratory condition), age, and perhaps education. WTP will also vary with factors such as earned income or education that affect the cost of averting activities. The effect of these variables on WTP could be calculated from the health production function or from equations describing the unit cost of averting activities.

When WTP is estimated by a stated preference method, data are typically gathered on variables that would enter the health production function or affect the level of averting (or mitigating) behavior undertaken. WTP responses can then be regressed on these variables. A similar approach can be taken with the cost of illness, provided that cost-of-illness data are gathered as part of a survey that also elicits socioeconomic and demographic data.

What are the prospects for measuring WTP using the averting behavior approach? It is usually difficult to measure the cost of activities that reduce exposure to pollution or that prevent symptoms from occurring altogether. The mere presence of an air conditioner in a home or in a car is an imperfect measure of reduced exposure to air pollution. The cost of averting behavior is inherently difficult to measure. For example, to determine the cost of spending leisure time indoors rather than outdoors, we cannot rely on observed prices but must question individuals directly. Also, measuring the cost of averting behavior is complicated by the fact that many avoidance activities produce joint products.

Another possibility is to use mitigating behavior, in the form of medication purchased, to measure WTP. Medication has an advantage in that it is a pure averting good with a cost that is easily measurable. If we can find a product that is appropriate for each symptom of interest and that affects only that symptom, then we can sum the WTP estimates produced by equation 10-55 across symptoms.

Special Topics

Subtle Health Effects

In addition to the mortality and morbidity effects discussed in this chapter, there is now substantial evidence that chemical contaminants can produce subtle physical and mental changes in people. For example, relatively low levels of lead in blood have been implicated in decrements to IQ in children, in low birth weight, and in lower body growth rates and, presumably, height at maturity (U.S. EPA 1986). To the extent that subtle effects such as low birth weight are associated with other adverse health effects and illness, they pose no special methodological problems from an economic perspective. The models and techniques discussed in this chapter can be used to estimate the monetary values of avoiding these adverse health effects. However, problems arise in estimating monetary values for avoiding more subtle effects such as decrements in IQ and reduced growth rates. Some studies have estimated the costs of special education and lost productivity of individuals with impaired intellectual development due to elevated levels of lead in blood (for example, U.S. EPA 1997). From a conceptual perspective, though, these cost-of-illness and lost-wage measures are only partial and incomplete measures of the total willingness to pay to avoid adverse health effects. Surely there is a loss of utility or well-being associated with intellectual impairment, but basing estimates of the value of this loss on an individual's WTP is problematic.

Valuing Health Effects on Children

The economic valuation of policies that protect or improve the health of children raises interesting ethical as well as empirical questions, especially what normative perspective to adopt in valuation. For a discussion of some of these issues, see U.S. EPA 1999a. The standard theory of welfare economics is based on the assumptions that each individual is the best judge of how well off he or she is in a given situation and that individuals have well-defined preferences over alternative states and choose rationally among alternatives subject to the usual constraints. These assumptions define what might be called the consumer sovereignty normative perspective of welfare economics. When we turn to the welfare economics of children, there appear to be three alternative ethical or normative perspectives. The first is a natural extension of consumer sovereignty to children or what I will call children's sovereignty. The second perspective, what I will call parental sovereignty, would use the parents' values for changes in the health of their children. According to the third or "child as adult" perspective, we would base values on what the adult would have chosen for him or herself in childhood. None of these ethical perspectives is entirely satisfactory. I briefly discuss each in turn.

Children's Sovereignty. This perspective is consistent with the individualistic basis for welfare economics, but it is not ethically attractive. Children are immature and lack the cognitive ability to make choices about health and safety. They may not have well-defined preferences over the full range of alternatives necessary to make reasoned choices. Also, they do not control the financial resources that are required to make trade-offs between money and health or safety.

Parental Sovereignty. How is this perspective justified on ethical grounds? One possibility is to assert guardianship or stewardship. Another possibility is parental altruism of some form. Some authors simply presume altruism without much discussion. Others note that parents do not always seem to be the best judges of what is good for their children and sometimes engage in activities such as smoking and drinking that actually harm their children.

Parental sovereignty also has some ethically unattractive implications at a more fundamental level. The economic analysis of fertility choice emphasizes the utility that children convey to parents and the potential economic benefits they bring through providing labor for household production, and in the long-term economic security of their parents. If the marginal utility of a child (or its marginal productivity) is decreasing in the number of children, then the value to the parent of reducing the risk of death to the child or preventing disease depends on the number of children in the family and, perhaps, on its birth order.

Child as Adult. This is the perspective most consistent with the basic welfare economics principles, but it is difficult to implement. We cannot observe the relevant choices. Therefore, stated preference methods would have to be used, and they would impose difficult cognitive tasks on respondents.

In conclusion, the choice of an ethical or normative perspective involves some difficult questions with no easy answers. However, each perspective has implications for how we measure values for children. Perhaps it is more accurate to say that each of the methods discussed in this chapter map back to one of these perspectives. The household production function and averting behavior methods are consistent with parental sovereignty because we observe the choices of parents. When we attempt to use benefits transfer where the values come from adults and are adjusted on some basis, this implies the child-as-adult perspective. Finally, cost-of-illness measures still have a role as lower bound estimates of value from either of the perspectives.

Quality-Adjusted Life Years

The models described above take as measures of the outcome of policy the change in the probability of dying and the reduction in days spent sick. However, as noted above, these outcome measures are simplifications of a more complex reality. I will briefly describe an alternative outcome measure that has been used extensively in the health economics literature—the quality-adjusted life year (QALY)—and I will evaluate its potential for use in valuing the outcomes of environmental health and safety policies. The QALY is a measure of the performance of medical treatments and interventions. It captures in a single metric two important dimensions of medical outcomes: the degree of improvement in health and the time interval over which the improvement occurs, including any increase in the duration of life itself.* Duration is measured in years of life, and quality is indexed by a number between 0, representing death, and 1, representing perfect health. So a treatment that is expected to increase the duration of life by 1 year of perfect health is said to produce 1 QALY. A treatment that improves health status from an index number of 0.25 to 0.75 for two years also produces 1 QALY. A treatment that extends a life by 5 years at a quality level of 0.4 produces 2 QALYs. In the evaluation of alternative health policies or treatment programs, the numbers of QALYs produced for each patient or recipient of treatment are simply added to obtain an aggregate measure of program effectiveness. As a summary measure with simple intuitive appeal, can the QALY be used in benefit–cost analysis to describe the outcomes of environmental policies? I will attempt to shed some light on this question.

*Most of what is said here about QALYs applies also, with appropriate modification, to disability-adjusted life years, health-adjusted life years, and related concepts.

Many of the earliest papers on this topic relied on the judgements of medical professionals to provide the quality weights for different health states (Torrance 1986). However, the preferred approach has been to obtain the weights by some form of questioning of a sample of individuals representative of the population of interest. There are three principal ways of asking questions to elicit values or weights. They all start with a description of a health state including symptoms, degree or level of pain, degree of impairment of activity or function, and so forth (Fabian 1994; Gold et al. 1996).

In the simplest form of questioning, respondents are simply asked to assign a weight or numerical value between one and zero that reflects the utility they assign to the health state relative to states of perfect health and of death, respectively. Often, respondents are provided with a visual aid such as a horizontal line with a scale between zero and one marked on it.

The second approach is known as the "time-tradeoff" approach. Respondents are asked to choose between two options: living in a given state of less than perfect health for a fixed period of time (T), for example, five years, and living a shorter time (N) in perfect health. The number of years of perfect health is varied until the individual expresses indifference between the two options. The value or weight attached to the impaired health state is simply N/T.

The third approach to questioning is the standard gamble question derived from the method first outlined by von Neumann and Morgenstern for eliciting cardinal utilities. Respondents are asked to choose between two options, where option A is living with the impaired health state with certainty for the rest of one's normal life span and option B is a gamble in which one outcome is living for the same period of time in perfect health (with a probability of p) and the other outcome is immediate death (with a probability of $1 - p$). The probability is then varied until the individual expresses indifference between the gamble and the given health state with certainty. This means that the expected utilities of the two choices are equal. The quality weight for the given health state is simply the probability (p) that makes the individual indifferent between the two choices.

Advocates of the use of QALYs in policy evaluation cite as one advantage of the concept that QALYs are based on individuals' preferences. To the extent that QALYs are derived from the responses of representative individuals, there is something to this claim. However, as several authors have shown, given the way that QALYs are used in policy evaluation, they are consistent with utility theory and the economic theory of individual preferences only if individuals' utility functions and preference structures satisfy some quite restrictive conditions. I will use examples to illustrate the nature of some of these restrictions. This discussion is based on Freeman and others (2002). For further discussion and illustration, the reader should consult Broome 1993; Fabian 1994; Johansson 1995; Garber et al. 1996; Hammitt 2000a; and Johnson and Lievense 2000.

For QALYs to represent an individual's preferences over risks of changes in health and longevity, his or her preferences must satisfy several restrictive conditions (Pliskin et al. 1980; Bleichrodt et al. 1997). The first is "risk neutrality" over longevity, which means that an individual is indifferent to patterns of mortality risks that have the same life expectancy. For example, an individual must be indifferent between living 25 more years for certain and a gamble offering a 50% chance of living 50 more years and a 50% chance of dying immediately.

The second condition is "constant proportional trade-off" (of longevity for health), which implies that the fraction of remaining longevity an individual would trade to improve his or her health from one state to another (for the rest of his or her life) does not depend on his or her longevity. For example, if he or she is willing to give up 10 of 50 remaining years to improve his or her health from "fair" to "excellent," he or she would also be willing to give up 1 of 5 remaining years for the same health improvement. Alternatively, if future QALYs are discounted (as is recommended practice; Gold et al. 1996), then the fraction of discounted longevity an individual is willing to give up must remain constant (Johannesson et al. 1994). Johnson and Lievense (2000) cite evidence suggesting that the assumption of constant proportional trade-off is unduly restrictive.

An additional condition is that an individual's preferences for health and longevity are "utility independent" of his or her wealth and future income, which means that his or her preferences for risks that affect health or longevity do not depend on income. This assumption implies that the effect of income on utility is positively related to the quality weight for each health state. There is little empirical information available on this point, although the notion that the marginal utility of income is smaller in impaired health than in full health is consistent with one study (Sloan et al. 1998).

Because QALYs impose restrictive assumptions on preferences, the ranking of health interventions using QALYs may differ systematically from the ranking using WTP. For example, given these restrictions, the QALY value of reducing mortality risk within the current year to different people should be proportional to life expectancy. This implies that the value of reducing risk to a 20-year-old is about three times larger than the value of reducing risk to a 65-year-old. In contrast, individual willingness to pay to reduce mortality risk does not fall as sharply with decreasing life expectancy and may even increase as life expectancy declines over some range of ages. Under the WTP approach, it is not necessarily more valuable to reduce mortality risk to a younger person, and in any case the differential value assigned is likely to be smaller than under the QALY approach. Willingness to pay to reduce mortality risk does not fall in proportion to life expectancy because the opportunity cost of spending on risk reduction also falls with decreasing life expectancy, as the individual has less to save for (Hammitt 2000b).

For another example, consider two alternative treatments or interventions, one of which yields an improvement in health status for next year of 0.5 while the other extends life by six months of perfect health. They both produce a half a QALY. If the improvement in health status lasts for 4 years and the increase in healthy life duration is 2 years, they both produce 2 QALYs. Independent of the discounting issues discussed above, this implies that the marginal valuations of health status and of longevity are both constant (i.e., risk neutrality). It also implies that the marginal valuation of health status is independent of longevity and vice versa (i.e., mutual utility independence). This means that, for example, a year of perfect health for an individual at age 80 has the same QALY value as a year of perfect health for that person at age 40, despite the more restricted range of activities available to the 80-year-old.

This discussion suggests that, although QALYs are consistent with economic theory, they require that preferences satisfy a number of quite restrictive conditions. This means that policy choices based on QALYs might at least in some cases be different from policy choices based on WTP measures. Despite this, do QALYs offer any special advantages in the evaluation of environmental and consumer safety policies that affect human health, especially as a way of characterizing changes in health status?

On the positive side, QALYs provide a way of quantifying some of the complications in valuing reductions in mortality risk that the simple wage–risk based VSL measures miss. These include the age at risk or life years saved and the health status enjoyed in these years. However, as noted above, the profile of QALY values and age is not consistent with the predictions of standard utility theory. Also, QALYs are unable to take into account factors such as cause of death and involuntariness of the risk that some authors argue are important (Revesz 1999). Regarding morbidity, QALYs capture the pain and suffering (lost utility) associated with illness, but they do not capture the "cost-of-illness" dimension of the value of reducing morbidity. If the objective is to value the health benefits of reducing exposures to an environmental agent, QALYs are likely to underestimate benefits because at least some of the benefit is in the form of reduced averting and defensive expenditures.

Another issue to consider is the validity of QALY weights as indicators of individuals' preferences over alternative health states. Because QALY weights are based on responses to hypothetical questions, they are susceptible to the same kinds of criticisms as value measures based on contingent valuation or other stated preference methods. Relatively little effort has been devoted to tests of the validity of QALY weights, and validity tests are difficult to design (Nord 1992). QALY weights are also known to be sensitive to framing effects and to question format (Nord 1992; Gold et al. 1996; Kaplan 1996). If respondents are actually risk averse or have positive time preference, this can independently affect responses to different kinds of standard gamble and time trade-off questions (Pliskin et al. 1980; Nord 1992).

For policymaking, QALYs may be perceived as more equitable than WTP because they treat a gain in health or longevity equally, regardless of the individual's wealth or income. In contrast, WTP often increases with ability to pay. In practice, however, this difference might not be a major factor because the health effects of policies can be (and usually are) valued using some population-average WTP.

Despite these reservations about QALYs, is there a potential role for them in the economic assessment of environmental and consumer safety policies affecting health? In some circumstances, there may be a limited potential role. Suppose that an environmental policy is predicted to reduce the number of days individuals experience acute respiratory symptoms. Further suppose that there are estimates of the willingness to pay to avoid asthma attacks but not to avoid acute respiratory symptoms. If it were known, the ratio of QALY weights for a day of acute respiratory symptoms and for an asthma attack could be used to scale the known WTP in a form of benefits transfer exercise. Some of the restrictive assumptions about preferences discussed above would not seem to be serious where the two endpoints do not entail major differences in life expectancy or duration of symptoms. Johnson and colleagues (1997) propose a similar approach to calculating WTP based on the quality of well-being index.

However, most of the QALY weights are for more serious diseases of longer duration and for various degrees of long-term disability or impairment. One could propose to do the primary research to estimate the QALYs for the relevant health endpoints. However, this is unlikely to be any easier or any better than doing the primary research to estimate the relevant WTP values directly. There is some question about whether QALYs, which are measured in years, can be reliably applied to short-term acute diseases that have durations of a few days at most (U.S. EPA 1999b, 22, A-18).

In summary, QALYs are not a panacea for the conceptual and empirical difficulties in assessing and evaluating environmental policies that affect health. Where WTP estimates of health values are available, they are likely to be superior reflections of individuals' preferences. Even though a limited role for QALYs has been identified in some kinds of benefits transfer exercises where WTP values are not available, QALY weights are not available in the literature for the most relevant health endpoints. I think that using limited research resources to obtain QALY weights for these endpoints would not be wise because the cost of that line of research is the lost opportunity to obtain better WTP values for the same endpoints.

Summary

In this chapter I have described the economic framework for valuing reductions in illness and risk of death on the basis of what an individual is willing to pay. The aggregate benefit of a reduction in the risk of death and incidence of illness is the sum of what each of the affected people is willing to pay to reduce his or her own risks of death and illness.

To assert that an individual has a willingness to pay for a reduction in risk of death or illness implies that the individual can perceive and is aware of changes in his or her health status. It does not require that the individual know that the reduction is attributable to a specific environmental change. If the value of reduced risk or symptom days is known, policy analysts can calculate benefits if they can predict the magnitude of the reduction caused by pollution control. Assumptions about individuals' knowledge play important roles in the empirical estimation of WTP in some circumstances, however. For example, if willingness to pay for risk reductions is to be inferred from wage–risk premiums, then it must be assumed that individuals know the relative risk levels of different jobs. Also, if willingness to pay for reduced illness is to be inferred from an individual's averting or mitigating behavior in response to changes in pollution, then it must be assumed that the individual knows the relationship between pollution and his or her illness experience.

In valuing risks to life, two forms of the indirect approach have been widely used. One is based on compensating wage differentials received by workers in risky occupations; the other is based on the cost of goods that increase safety, such as seat belts and smoke detectors. Safety goods are typically 0–1 activities whose benefits exceed their costs for most people. Thus, unless the data permit estimation of the parameters of a discrete choice model, basing risk valuations on these activities is likely to understate the value of a risk reduction. On the other hand, if the safety good also conveys utility directly, then value of risk reduction alone will be overestimated. The compensating wage approach may also understate the value of a risk reduction because persons who are willing to be paid to accept increased risk (such as structural ironworkers) may have lower values for risk reduction than the average person.

A serious shortcoming of studies of wage–risk premiums and use of safety equipment is that they value only voluntarily assumed risks of accidental death (for example, risk of death on the job or risk of dying in an auto accident). By contrast, environmental risks are largely involuntary and may lead to a painful illness (such as cancer) before death occurs. These considerations suggest that individuals might be willing to pay more to reduce environmental risks than to reduce risk of death in an auto accident. On the other hand, to the extent that environmental risks may not occur until after a long latency period, fewer years of life will be lost than are risked from death in an auto accident during the current year. This suggests that estimates of willingness to pay for a reduction in current risk of death may overstate willingness to pay to reduce environmental risks.

One advantage of stated preference methods is that the risks being valued can be tailored to circumstances relevant to environmental health policy. For example, people can be asked their willingness to pay to reduce exposure to a pollutant that increases their risk of dying of cancer in 20 years. However, as the risk being valued becomes smaller in magnitude and more distant in time, individuals may have trouble understanding what they are valuing.

It may not be appropriate to use the value of a reduction in risk of accidental death during the coming year to value the reduction in risk of death associated with exposure to an environmental carcinogen. Exposure to a carcinogen may not increase risk of death until 20 years hence, which suggests that the value of reducing risk of death today overstates the value of reducing exposure to the carcinogen. On the other hand, death due to cancer may be feared more than accidental death. It is therefore unclear what the relationship between the two risk valuations should be.

We have also reviewed models for inferring this value of reduced morbidity from observable data. The value of a reduction in symptoms can be inferred either from activities that reduce exposure to air pollution (e.g., running an air purifier or staying indoors) or that mitigate the severity or duration of symptoms (e.g., taking medication or visiting a doctor).

The principal advantage of these revealed preference methods is that they are based on observed behavior. However, to implement them requires a considerable amount of data. We must be able to measure and estimate the relationships among the health outcome of interest, the amount of mitigating or averting behavior undertaken, and its price. In addition, we must either know how effective the individual thought his or her averting behavior was, or we must know its actual effectiveness and assume that the individual correctly perceived this effectiveness.

References

Arthur, W.B. 1981. The Economics of Risks to Life. *American Economic Review* 71(1): 54–64.

Atkinson, Scott E., and Robert Halvorsen. 1990. Evaluation of Risks to Life: Evidence from the Market for Automobiles. *Review of Economics and Statistics* 72(1): 133–136.

Bergstrom, Theodore C. 1982. When Is a Man's Life Worth More than His Human Capital? In *The Value of Life and Safety*, edited by Michael W Jones-Lee. Amsterdam: North-Holland.

Bleichrodt, Han, Peter Wakker, and Magnus Johannesson. 1997. Characterizing QALYs by Risk Neutrality. *Journal of Risk and Uncertainty* 15(2): 107–114.

Blomquist, Glenn. 1981. The Value of Human Life: An Empirical Perspective. *Economic Inquiry* 19(1): 157–164.

Broome, John. 1978. Trying To Value a Life. *Journal of Public Economics* 9(1): 91–100.

———. 1992. *Counting the Cost of Global Warming*. Cambridge, U.K.: The White Horse Press.

———. 1993. Qalys. *Journal of Public Economics* 50(2): 149–167.

Conley, B.C. 1976. The Value of Human Life in the Demand for Safety. *American Economic Review* 66(1): 45–55.

Cook, Philip J. 1978. The Value of Human Life in the Demand for Safety: Comment. *American Economic Review* 68(4): 710–711.

Cooper, B.S., and Dorothy P. Rice. 1976. The Economic Cost of Illness Revisited. *Social Security Bulletin* 39(2): 21–36.

Corso, Phaedra S., James K. Hammitt, and John D. Graham. 2001. Valuing Mortality-Risk Reduction: Using Visual Aids To Improve the Validity of Contingent Valuation. *Journal of Risk and Uncertainty* 23(2): 165–184.

Cropper, Maureen L. 1977. Health, Investment in Health, and Occupational Choice. *Journal of Political Economy* 85(6): 1273–1294.

———. 1981. Measuring the Benefits from Reduced Morbidity. *American Economic Review* 71(2): 235–240.

Cropper, Maureen L., and Frances G. Sussman. 1988. Families and the Economics of Risks to Life. *American Economic Review* 78(1): 255–260.

———. 1990. Valuing Future Risks to Life. *Journal of Environmental Economics and Management* 19(2): 160–174.

Dickie, Mark. 2003. Defensive Behavior and Damage Cost Methods. In *A Primer on Non-market Valuation*, edited by Kevin J. Boyle and Patricia A. Champ. Boston: Kluwer Academic Publishers.

Dickie, Mark, and Shelby Gerking. 1991. Valuing Reduced Morbidity: A Household Production Approach. *Southern Economic Journal* 57(3): 690–702.

Fabian, Robert. 1994. The Qualy Approach. In *Valuing Health for Policy: An Economic Approach*, edited by George Tolley, Donald Kenkel, and Robert Fabian. Chicago: University of Chicago Press.

Freeman, A. Myrick, III, James K. Hammitt, and Paul De Civita. 2002. On Quality Adjusted Life Years (QALYs) and Environmental/Consumer Safety Valuation. *AERE Newsletter* 22(1): 7–12.

Garber, A.M., M.C. Weinstein, G.W. Torrance, and M.S. Kamlet. 1996. Theoretical Foundations of Cost Effectiveness Analysis. In *Cost-Effectiveness in Health and Medicine: Report to U.S. Public Health Service by the Panel on Cost-Effectiveness in Health and Medicine*, edited by Marthe R. Gold, Dennis G. Fryback, Alan M. Garber, David Hadorn, and others. Washington, DC: U.S. Department of Health and Human Services.

Gerking, Shelby, and Linda R. Stanley. 1986. An Economic Analysis of Air Pollution and Health: The Case of St. Louis. *Review of Economics and Statistics* 68(1): 115–121.

Gold, M.R., D.L. Patrick, G.W. Torrance, D.G. Fryback, and others. 1996. Identifying and Valuing Outcomes. In *Cost-Effectiveness in Health and Medicine: Report to U.S. Public Health Service by the Panel on Cost-Effectiveness in Health and Medicine,* edited by Marthe R. Gold, Dennis G. Fryback, Alan M. Garber, David Hadorn, and others. Washington, DC: U.S. Department of Health and Human Services

Grossman, Michael. 1972. On the Concept of Health Capital and the Demand for Health. *Journal of Political Economy* 80(2): 223–255.

Hammitt, James K. 2000a. QALYs v. WTP. Paper presented at the Society for Risk Analysis annual meeting, Washington, DC.

———. 2000b. Valuing Mortality Risk: Theory and Practice. *Environmental Science & Technology* 34(8): 1396–1400.

Hammitt, James K., and John D. Graham. 1999. Willingness To Pay for Health Protection: Inadequate Sensitivity to Probability? *Journal of Risk and Uncertainty* 18(1): 33–62

Harrington, Winston, and Paul R. Portney. 1987. Valuing the Benefits of Health and Safety Regulations. *Journal of Urban Economics* 22(1): 101–112.

Johannesson, Magnus, Joseph S. Pliskin, and Milton C. Weinstein. 1994. A Note on QALYs, Time Tradeoff, and Discounting. *Medical Decision Making* 14(2): 188–193.

Johansson, Per-Olov. 1995. *Evaluating Health Risks: An Economic Approach.* Cambridge, U.K.: Cambridge University Press.

Johnson, F. Reed, Erin E. Fries, and H. Spencer Banzhaf. 1997. Valuing Morbidity: An Integration of the Willingness-To-Pay and Health-Status Index Literatures. *Journal of Health Economics* 16(6): 641–665.

Johnson, F. Reed, and Katharine Lievense. 2000. *Stated-Preference Indirect Utility and Quality-Adjusted Life Years.* Prepared for Health Canada. Durham, NC: Triangle Economic Research.

Jones-Lee, Michael W. 1974. The Value of Changes in the Probability of Death or Injury. *Journal of Political Economy* 99(4): 835–849.

———. 1976. *The Value of Life: An Economic Analysis.* Chicago: University of Chicago Press.

Kaplan, Robert M. 1996. Utility Assessment for Estimating Quality-Adjusted Life Years. In *Valuing Health Care: Costs, Benefits, and Effectiveness of Pharmaceuticals and Other Medical Technologies,* edited by Frank A. Sloan. Cambridge, U.K.: Cambridge University Press.

Landefeld, J. Steven, and Eugene P. Seskin. 1982. The Economic Value of Life: Linking Theory to Practice. *American Journal of Public Health* 72(6): 555–566.

Lave, Lester B., and Eugene P. Seskin. 1971. Air Pollution and Human Health. *Science* 169 (August 21): 723–731.

———. 1977. *Air Pollution and Human Health.* Baltimore, MD: Johns Hopkins University Press for Resources for the Future.

Linnerooth, Joanne. 1979. The Value of Human Life: A Review of the Models. *Economic Inquiry* 27(1): 52–74.

MacLean, Douglas. 1990. Comparing Values in Environmental Policies: Moral Issues and Moral Arguments. In *Valuing Health Risks, Costs, and Benefits for Environmental Decision Making*, edited by P. Brett Hammond and Rob Coppock. Washington, DC: National Academy Press.

Mishan, Ezra J. 1971. Evaluation of Life and Limb: A Theoretical Approach. *Journal of Political Economy* 79(4): 687–705.

Mrozeck, Janusz R., and Laura O. Taylor. 2002. What Determines the Value of Life? A Meta-Analysis. *Journal of Policy Analysis and Management* 21(2): 269–286.

Nord, Erik. 1992. Methods for Quality Adjustment of Life Years. *Social Science Medicine* 34(5): 559–569.

Pliskin, Joseph S., Donald S. Shepard, and Milton C. Weinstein. 1980. Utility Functions for Life Years and Health Status. *Operations Research* 28: 206–224.

Revesz, Richard L. 1999. Environmental Regulation, Cost–Benefit Analysis, and the Discounting of Human Lives. *Columbia Law Review* 99(4): 941–1017.

Rosen, Sherwin. 1994. The Quantity and Quality of Life: A Conceptual Framework. In *Valuing Health for Policy: An Economic Approach*, edited by George Tolley, Donald Kenkel, and Robert Fabian. Chicago: University of Chicago Press.

Schelling, Thomas C. 1968. The Life You Save May Be Your Own. In *Problems in Public Expenditure Analysis*, edited by Samuel B. Chase, Jr. Washington, DC: Brookings Institution.

Shepard, Donald S., and Richard J. Zeckhauser. 1982. Life Cycle Consumption and Willingness To Pay for Increased Survival. In *The Value of Life and Safety*, edited by Michael W. Jones-Lee. Amsterdam: North-Holland.

———. 1984. Survival vs. Consumption. *Management Science* 30: 423–439.

Sloan, Frank A., W. Kip Viscusi, Harrell W. Chesson, Christopher J. Conover, and Kathryn Whetten-Goldstein. 1998. Alternative Approaches to Valuing Intangible Health Losses: The Evidence for Multiple Sclerosis. *Journal of Health Economics* 17(4): 475–497.

Sussman, Frances G. 1984. A Note on the Willingness To Pay Approach to the Valuation of Longevity. *Journal of Environmental Economics and Management* 11(1): 84–89.

Thaler, Richard G., and Sherwin Rosen. 1976. The Value of Life Savings. In *Household Production and Consumption*, edited by Nester Terleckyj. New York: Columbia University Press.

Torrance, George W. 1986. Measurement of Health State Utilities for Economic Appraisal: A Review. *Journal of Health Economics* 5(1): 1–30.

U.S. EPA (U.S. Environmental Protection Agency). 1986. *Air Quality Criteria for Lead*. Research Triangle Park, NC: U.S. Environmental Protection Agency.

———. 1997. *The Benefits and Costs of the Clean Air Act: 1970 to 1990*. Prepared for the U.S. Congress, Washington, DC: U.S. Environmental Protection Agency.

———. 1999a. *Valuing Health for Environmental Policy with Special Emphasis on Children's Health Protection*. Proceedings of the Second Workshop in the Environmental Policy and Economics Workshop Series. http: //yosemite1.epa.gov/ee/epa/eerm.nsf/vwSER/D811C47399E896C48525679D0074ED76?OpenDocument (accessed March 11, 2002).

———. 1999b. *An SAB Report on the National Center for Environmental Assessment's Comparative Risk Framework Methodology: A Review by the Drinking Water Committee.* Washington, DC: U.S. Environmental Protection Agency.

———. 2000a. *Guidelines for Preparing Economic Analyses.* Washington, DC: U.S. Environmental Protection Agency.

———. 2000b. *An SAB Report on EPA's White Paper Valuing the Benefits of Fatal Cancer Risk Reductions.* EPA–SAB–EEAC–00–013. Washington, DC: U.S. Environmental Protection Agency, http: //www.epa.gov/sab/eeacf013.pdf (accessed February 19, 2002).

U.S. Office of Management and Budget. 2000. *Guidelines To Standardize measures of Costs and Benefits and the Format of Accounting Statements.* Memorandum from Jacob J. Lew, March 22, 2000, Washington, DC: U.S. Office of Management and Budget, http: //www.whitehouse.gov/omb/memoranda/m00-08.pdf (accessed February 19, 2002).

Usher, Dan. 1973. An Imputation to the Measure of Economic Growth for Changes in Life Expectancy. In *The Measurement of Economic and Social Performance.* Studies in Income and Wealth, vol. 38, edited by Milton Moss. New York: National Bureau of Economic Research.

Viscusi, W. Kip. 1993. The Value of Risks to Life and Health. *Journal of Economic Literature* 31(4): 1912–1946.

Viscusi, W. Kip, Wesley A. Magat, and Joel Huber. 1991. Pricing Health Risks: Survey Assessments of Risk–Risk and Risk–Dollar Tradeoffs. *Journal of Environmental Economics and Management* 21(1): 32–51.

Viscusi, W. Kip, and Michael J. Moore. 1989. Rates of Time Preference and Valuation of the Duration of Life. *Journal of Public Economics* 38(3): 297–317.

CHAPTER

11

Property Value Models

Economists were documenting the relationship between the prices of housing units and quantities of environmental amenities even before this relationship had been recognized as an application of the newly developed theory of hedonic prices (for example, Ridker and Henning 1967). Examples of the statistical analysis of the relationship between farmland prices and the characteristics of the land can be found as early as 1922. (See Colwell and Dilmore 1999 for a review.) The past 25 years have seen an explosion of both theoretical and empirical studies of the monetary values of nonmarket amenities and disamenities based on hedonic price theory. Now it is well accepted that housing price differentials reflect differences in the quantities of various characteristics of housing and that these differentials have significance for applied welfare analysis. For example, Smith and Huang (1995) conducted a meta-analysis of hedonic studies of air pollution and housing prices. They reported finding 37 studies and more than 160 separate estimates of the effect of some air-quality variable on housing prices.

The two issues that have been explored most intensively in this literature are the proper specification and estimation of the model relating housing prices to amenities and the development of measures of welfare change that make the best use of the available data and that are consistent with the underlying economic theory. This chapter reviews the current state of knowledge concerning these two issues.

After a brief review of the evolution of economic thinking about property prices and environmental amenities, I present a more detailed exposition of the hedonic property value model. This presentation includes discussions of problems in estimating the hedonic price function; approaches to recovering information on preferences and the demands for characteristics from the hedonic price function; and the measurement of welfare change. Also included are discussions of the relationship between property prices and annual rental values, the relationship between the hedonic model and the averting behavior model, and the application of the hedonic model to markets for agricultural and commercial land. I then describe two alternative approaches to interpreting property price information: the repeat-sales model and discrete choice models. All these models have their foundations in individual choice theory and a model of the equilibrating processes of the housing market.

Historical Background

The theory of rents holds that the equilibrium price for a parcel of land will be the present value of the stream of rents produced by the land. Economic theory has long recognized that the productivity of land differs across sites. These productivity differentials will yield differential rents to land and therefore differential land values. Where land is a producer's good, competition and free entry are sufficient to ensure that productivity differentials are fully reflected in the land rent structure. For any property where the land rent is less than the productivity, the activity occupying that land must earn a profit. Some potential entrant will be willing to bid above the going rent to occupy that site and reap the rewards of a superior productivity. This competition bids up land rents and eliminates profit. Rent differentials will be equal to productivity differentials. Because the price at which a unit of land sells in the market is the present value of the stream of future rents, productivity differentials will also be reflected in land prices.

Some environmental characteristics such as air or water quality may affect the productivity of land either as a producer good or a consumer good. Where this is so, the structure of land rents and prices will reflect these environmentally determined productivity differentials. These results from classical rent theory aroused considerable interest among economists about the possibility of using data on land rent or land value for residential properties to measure the benefits to households brought about by improvements in environmental characteristics such as air or water quality. Ridker (1967) was the first economist to attempt to use residential property value data as the basis for estimating the benefits of changes in measures of environmental quality such as air pollution. He reasoned as follows:

> If the land market were to work perfectly, the price of a plot of land would equal the sum of the present discounted streams of benefits and costs derivable from it. If some of its costs rise (e.g., if additional maintenance and cleaning costs are required) or if some of its bene-

fits fall (e.g., if one cannot see the mountains from the terrace) the property will be discounted in the market to reflect people's evaluation of these changes. Since air pollution is specific to locations and the supply of location is fixed, there is less likelihood that the negative effects of pollution can be significantly shifted on to other markets. We should therefore expect to find the majority of effects reflected in this market, and we can measure them by observing associated changes in property values (Ridker 1967, 25).

The last sentence of this passage raises three questions. The first is whether environmental variables such as air pollution systematically affect land prices. Assuming an affirmative answer to this question, the second is whether knowledge of this relationship is sufficient to predict changes in land prices when, say, air pollution levels change. And the third question is whether changes in land prices accurately measure the underlying welfare changes.

Ridker (1967) and Ridker and Henning (1967) provided the first empirical evidence that air pollution affects property values by regressing median census tract property values in an urban area on a measure of sulfate air pollution. They then asserted positive answers to the second and third questions. Specifically, they argued that the coefficient on the air pollution variable in the regression equation could be used to predict the change in the price of any residence conditioned on a change in its air pollution level and that the sum of all such changes could be taken as a measure of the benefit of improving air quality in an urban area (Ridker 1967, 1, 6–137; Ridker and Henning 1967, 254).

This work stimulated a now large literature on the proper theoretical interpretation of the observed air pollution–property value relationship. Early contributions included those by Freeman (1971, 1974a, 1974b) and Anderson and Crocker (1972). Subsequent efforts to provide a sound theoretical basis for interpreting the air pollution–property value relationship have taken one of two paths. The first has been the development of models of the urban land market to determine whether and under what circumstances changes in aggregate land values accurately measure the benefits associated with environmental improvements. Early efforts in this direction included those by Strotz (1968), Lind (1973), Pines and Weiss (1976), Polinsky and Shavell (1976), and Kanemoto (1988). Although some of these models can be given an interpretation in the context of hedonic price theory, they do not lend themselves to empirical application, so they are not covered in this book. For further discussion of this branch of the literature, see Bartik and Smith 1987 and Palmquist 1991.

The second path drew on hedonic price theory to interpret the derivative of the cross-section regression equation with respect to air pollution as a marginal implicit price and therefore a marginal value for air-quality improvement (see Freeman 1974b and Rosen 1974). In this chapter I describe this approach to welfare measurement, along with more recent developments, such as the repeat-sales hedonic model and discrete choice models of housing choice. My primary emphasis is on model specification and interpretation rather than econometric estimation. My objectives are to provide an overview of the methods of welfare measurement based on property values and to identify the major conceptual issues.

The Hedonic Property Value Model

The hedonic price theory first introduced in Chapter 4 provides the basis for deriving welfare measures from observed differences in prices of houses. In this section I first review the basic theory of hedonic prices in the context of the housing market. I then discuss a number of topics involving the estimation of the model. For other current treatments of these matters with greater emphasis on econometric and estimation issues, see Palmquist (forthcoming) and Taylor 2003.

The Basic Model

Assume that each individual's utility is a function of that person's consumption of a composite commodity X, a vector of location-specific environmental amenities $\boldsymbol{Q}$, a vector of structural characteristics of the house the person occupies (such as size, number of rooms, age, and type of construction) denoted by $\boldsymbol{S}$, and a vector of characteristics of the neighborhood in which the house is located (such as quality of local schools, crime rates, and accessibility to parks, stores, and work places) denoted by $\boldsymbol{N}$.

Any large area contains a variety of sizes and types of housing with different locational, neighborhood, and environmental characteristics. An important assumption of the hedonic technique is that the urban area as a whole can be treated as a single market for housing services. Individuals must have information on all alternatives and must be free to choose a house anywhere in the urban market. It is as if the urban area were one huge supermarket offering a wide selection of goods. Individuals cannot move their shopping carts through this supermarket. Rather, their selections of residential locations fix for them the whole bundle of housing services. It is much as if shoppers were forced to make their choices from an array of already filled shopping carts. Individuals can increase the quantity of any characteristic by finding an alternative location alike in all other respects but offering more of the desired characteristic.

Because our interest is in the values of characteristics to buyers of houses, there is no need to model formally the supply side of this market.

Let us simply assume that the housing market is in equilibrium, that is, that all individuals have made their utility-maximizing residential choices given the prices of alternative housing locations and that these prices just clear the market given the existing stock of housing and its characteristics. Under these assumptions, the rental price of the *j*th residential location can be taken to be a function of the structural, neighborhood, and environmental characteristics of that location. In other words,

$$R_{h_j} = R_h\left(\boldsymbol{S}_j, \boldsymbol{N}_j, \boldsymbol{Q}_j\right) \tag{11-1}$$

As explained in Chapter 4, this relationship can be linear in a characteristic if repackaging of that characteristic is possible, but in general this need not be the case. Two living rooms with 6-foot ceilings are not equal to one living room with a 12-foot ceiling. Where repackaging is not possible, equation 11-1 may be nonlinear.

To model the problem more formally, consider an individual who occupies house *j*. Her utility is given by

$$u = u(X, \boldsymbol{Q}_j, \boldsymbol{S}_j, \boldsymbol{N}_j) \tag{11-2}$$

where X is a Hicksian composite good with a price of 1. This assumption makes the demands for characteristics independent of the prices of other goods, a convenient property for empirical work. The individual maximizes $u(\cdot)$ subject to the budget constraint:

$$M - R_{h_j} - X = 0 \tag{11-3}$$

A typical first-order condition for the choice of environmental amenity q (an element in $\boldsymbol{Q}$) is

$$\frac{(\partial u / \partial q)}{(\partial u / \partial X)} = \frac{\partial R_h}{\partial q} \tag{11-4}$$

Assume now that the hedonic price function $R_h(\cdot)$ has been estimated for an urban area. Its partial derivative with respect to any of its arguments, for example q, gives the implicit marginal price of that characteristic, that is, the additional amount that must be paid by any household to move to a housing bundle with a higher level of that characteristic, other things being equal. If this function is nonlinear, the marginal implicit price of a characteristic is not constant but depends on its level and perhaps the levels of other characteristics as well. If the individual is assumed to be a price

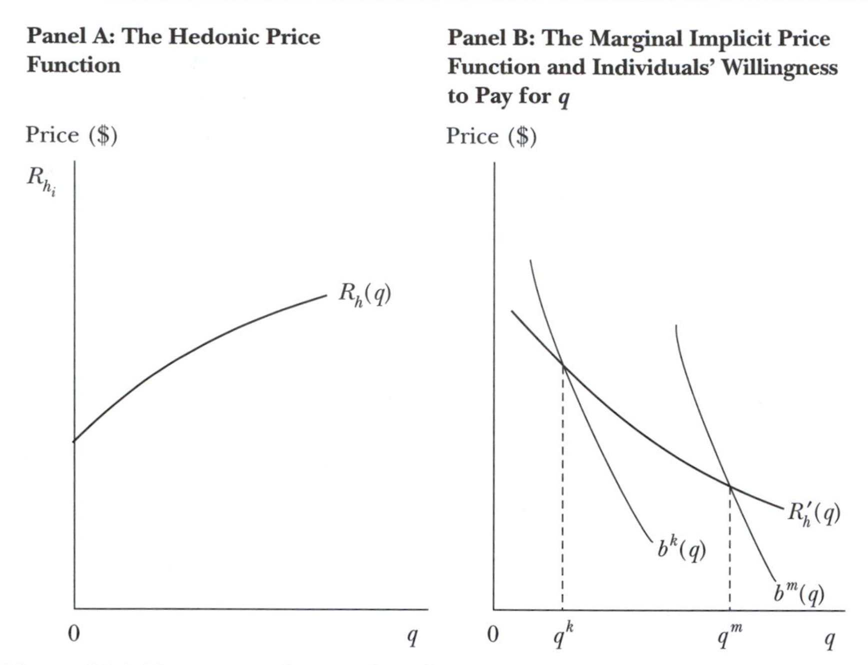

Figure 11-1. Property prices and environmental quality

taker in the housing market, that person can be viewed as facing an array of implicit marginal price schedules for various characteristics. An individual maximizes utility by simultaneously moving along each marginal price schedule until she reaches a point where her marginal willingness to pay for an additional unit of that characteristic just equals the marginal implicit price of that characteristic. If an individual is in equilibrium, the marginal implicit prices associated with the housing bundle actually chosen must be equal to the corresponding marginal willingnesses to pay for those characteristics.

Panel A in Figure 11-1 shows the partial relationship between R_h and q as estimated from equation 11-1. Panel B shows the marginal implicit price of q, $\partial R_h / \partial q$. It also shows the marginal willingness-to-pay (MWTP) curves for two individuals, k and m, who have chosen utility-maximizing bundles of housing characteristics; these are labeled $b^k(q)$ and $b^m(q)$. These curves show each individual's marginal willingness to pay for changes in the characteristic, holding utility constant at the level achieved by maximizing equation 11-2 subject to equation 11-3. Let this level be u^*. Both individuals in Figure 11-1 have chosen locations where their marginal willingnesses to pay for q are equated with its marginal implicit price.

The analysis described here results in a measure of the price of and the marginal willingness to pay for q but does not directly reveal the MWTP function. The second stage of the hedonic technique is to combine the quantity and implicit price information in an effort to identify the MWTP function for q. The individual's demand price or willingness to pay for q is a function of the level of q. Because there may be substitute and complementary relationships among characteristics, the willingness to pay for q may also depend on the levels (or marginal implicit prices) of other characteristics. It is convenient to assume that the utility function is weakly separable in housing so that prices of other goods can be omitted in the specification of the MWTP function. Also, it is convenient to assume that each individual purchases only one housing bundle. If more than one were purchased, it would be necessary that the bundles be identical or that the hedonic price function be linear in all characteristics. This is because there can be only one marginal implicit price recorded for each individual for each characteristic.

Given these assumptions, for the ith individual we can derive an MWTP function for q by differentiating the expenditure function, as shown in Chapter 3. The result is

$$b^i = b^i\left(q, \boldsymbol{Q}^*, \boldsymbol{S}, \boldsymbol{N}, u^*\right) \tag{11-5}$$

where $\boldsymbol{Q}^*$ is all amenities except q. If equation 11-5 can be estimated, it can be used to estimate the welfare change of an individual associated with changes q, assuming that other things are held equal. Specifically, if the quantities of other characteristics and amenities do not change, the welfare change can be found by integrating b^i over the relevant range of the change in q. However, a change in the quantity of one characteristic can result in changes in the quantities of other characteristics the individual chooses and in changes in the hedonic price function itself. The task of welfare measurement when individuals can fully adjust to the new supply of amenities and characteristics is discussed below, in the section called Measuring Welfare Changes.

Estimating the Hedonic Price Function

The Dependent Variable. In the discussion so far, the hedonic price has been the annual rental price of the property. What is observed usually is the purchase price of the house, which I will denote as P_h for the rest of this chapter. P_h can be interpreted as the discounted present value of the stream of expected rental values. This leads to two complications. First, when housing price differentials are used to estimate welfare changes that are usually expressed as annual flows, care must be taken to convert the house price measures into the appropriate temporal dimension; this topic is discussed in a later section. Second, it might be necessary to take account of expected changes in the characteristics of a house, especially environmental changes, when estimating and interpreting the hedonic price function. For example, if air pollution at a given location is expected to improve over time, the present price of the house should be bid up to reflect not only the current conditions but the expected improvement as well.

One question to be asked is whether the dependent variable to be explained should be a pure site or land value or the full price of the house and land together. Because the environmental amenities of interest are location specific but not a part of the structure, the values of the environmental amenities should be reflected in the price of land alone. However, at least in the United States land is not usually traded separately from the structures placed upon it, so the observed prices reflect the values of both the land and its structural improvements. This causes no problems at the theoretical level, but it does require that the hedonic price equation adequately control for structural characteristics.

Another question concerns the source of data on housing prices. Data on actual market transactions are preferable. For rental housing there is a regular monthly "market transaction," from which fairly accurate data on housing rents could be gathered. However, the majority of residential housing is owner-occupied, and only a small percentage of the total owner-occupied housing stock is exchanged through the market each year. The most preferred source of data is systematically collected information on actual sales prices of individual dwellings, along with relevant characteristics. Fortunately, in many parts of the country these data are collected by multiple-listing services and tax assessing agencies. In recent years, virtually all published hedonic property value studies have used data on individual transaction prices.

However, other sources of data can be used. The earliest hedonic property value studies used data from the U.S. Census of Population and Housing. The census asks each owner to estimate the value of his or her property. The census also gathers other data on structural characteristics, as well as socioeconomic data on occupants. These data are aggregated by census tracts and reported as means or medians. Although the census tract observations represent a convenient source of data for property value studies, there are two kinds of problems with them.

The first concerns the degree of accuracy of estimates of values by individual owners. Kiel and Zabel (1999) recently compared individual home owners' estimates of market values with the house's most recent sale price. They found a modest tendency for owners to overstate values by about 5%. They conclude, however, "… that hedonic equations based on owners' valuations will provide unbiased estimates of … the prices of house and neighborhood characteristics" (Kiel and Zabel 1999, 264).

The second problem with census tract data is the loss of detail and reduced ability to control for relevant housing and location characteristics, both because of the limited number of variables reported and because of aggregation of individual data by census tract. Census tract boundaries are chosen in an effort to construct relatively homogeneous units in terms of housing and socioeconomic characteristics. If within-tract variation is relatively small compared with the variation among tracts, then relatively little is lost by aggregating a given set of observations to census tract units before undertaking the statistical analysis. However, even within generally homogeneous communities there may be substantial variation in relevant characteristics, such as number of rooms. The effect of these variables on individual property values would be masked by aggregation.

An alternative source of property value data would be professional appraisals of individual properties for taxation or other purposes. Some jurisdictions have developed computer-based systems of appraisals, made for tax purposes, which include data not only on appraised values but also on a variety of structural and site characteristics. As these systems are developed and extended, they can provide a valuable data source for further property value studies. For citations to studies examining the accuracy of appraisals and tax assessments, see Kiel and Zabel 1999, Table 1.

Where the data are based on assessments, appraisals, or self-reporting, they may not correspond to actual market prices. The errors in measuring the dependent variable will tend to obscure any underlying relationship between true property value measures and environmental amenities. However, estimates of the relationship will not be biased unless the errors themselves are correlated with other variables in the model.

The presumed superiority of individual transaction data over nonmarket estimates provided by experts is based on the assumption that the housing market is in an equilibrium in which all opportunities for possible gains from further trade at the revealed set of prices have been exhausted. This is a heroic assumption. Buyers and sellers often operate with substantial ignorance about the true willingness to pay (WTP) and willingness-to-accept offers of other potential buyers and sellers. Sellers typically state an asking price that effectively truncates potential offers at that price. Sellers must choose to accept or reject offers more or less at the time they are received, without knowing when (or even if) a higher offer might come along. Buyers lack information on possible prior bids made by others for a given property (Horowitz 1986). Horowitz developed an alternative model of the bidding and acceptance strategies of buyers and sellers and estimated both a standard hedonic model and his alternative model with the same data set. He found that the statistical performance of his bidding model was substantially superior to the standard model in predicting sales prices. However, because he used principal components in his estimation procedure rather than actual attributes, it is not possible to analyze the impact of his alternative modeling strategy on marginal implicit prices for environmental attributes.

Explanatory Variables. In choosing the appropriate explanatory variables, the first question to be addressed is the way in which environmental amenities and location characteristics enter the hedonic price function. The typical practice has been to enter a simple scalar measure of an amenity—for example, parts per million of an air pollutant, or distance to a park. However, Parsons (1990) showed that this practice is not consistent with a restriction imposed on the hedonic price function by profit-maximizing behavior on the supply side of the housing market. The implication of profit maximization is that the effect of the environmental amenity can only be captured without bias by weighting the amenity by the area of the lot on which the house sits. The restriction is that if an area of land of given q is developed and sold in two or more different lot sizes, the prices of the lots must be such that the return per acre is independent of the sizes of the lots. For example, if two one-acre lots sell for \$$X$ apiece, one two-acre lot must sell for \$$2X$. This means that the premium on lots with higher levels of q must be twice as high for the two-acre lot compared with the one-acre lots. The higher premium on the larger lot is necessary to compensate the landowner for the forgone opportunity to capture two premiums with the smaller lots.

Although the argument is correct in principle, there is some question about its relevance in practice. As Parsons points out, once lots are developed, the cost of changing the size of lots on which houses sit may be too high to force amenity premiums to take the weighted form in the second-hand market for houses. Although Parsons shows that biased estimates of implicit prices for characteristics are possible, few empirical researchers have used his proposed weighted amenity values.

The levels of some environmental amenities are fixed by location, whereas the levels of others, especially those related to air quality, vary over time with changes in emissions and meteorological conditions. With time-varying amenities, there is the question of how best to represent the level of the amenity in the regression equation. The typical practice in air pollution–property value studies has been to use the annual mean as a summary statistic. However, Murdoch and Thayer (1988) showed that in the case of visibility, using more information on the probability distribution of visual range improves the statistical performance of the hedonic price function.

Several other conceptual and practical issues must be resolved in the course of selecting a set of explanatory variables for a hedonic price function. These include the following: Which measures of environmental quality should be used to characterize environmental amenities? Is it possible to separate the effects of different amenities on property values when measures of the amenities are correlated? What objective data best capture neighborhood characteristics? Does the spatial scale of the socioeconomic data often used in these studies correspond to peoples' perceptions of these characteristics? (For example, is there sufficient homogeneity within census tracts so that census tract means or medians adequately measure "neighborhood"?) Is there sufficiently close correspondence between peoples' perceptions of amenity levels (which presumably govern the choices reflected in property prices) and the objective measures of amenity levels that are available to the researcher?

Because the objective of the hedonic analysis is to determine the effect of one amenity on property values, other things being equal, a key issue is the control for structural, neighborhood, and other environmental variables. The issue is made more difficult by the likelihood of multicollinearity among housing characteristics. This raises the troublesome question of the trade-off between increasing bias through the omission of variables that are correlated with the variable of concern and increasing the variance or imprecision of coefficient estimates when collinear variables are included. Theory does not provide any hard-and-fast answers to this question. Work by Atkinson and Crocker (1987) and Graves and coauthors (1988) suggests the value of approaching this question systematically, using Bayesian principles. These authors have also examined the effects of errors in the measurement of other explanatory variables on the estimates of the coefficients of the environmental variables of concern.

Functional Form. Functional forms for the hedonic price function that have been proposed or used in the literature include the linear, the quadratic, the log–log, the semi-log, the inverse semi-log, the exponential, and the Box–Cox transformation. The first step in choosing a functional form is to see what theory can tell us. According to theory, a hedonic price function is an equilibrium relationship derived from the interaction of individuals' preferences and suppliers' cost or profit functions. The only obvious general restriction on the form of the hedonic price function is that its first derivative with respect to an environmental characteristic be positive (negative) if the characteristic is a good (bad).

Rosen (1974) and Epple (1987) showed that it is possible to solve for the hedonic price function analytically after making specific assumptions about the form of individual utility functions and the distribution of suppliers' characteristics (Rosen) or the exogenous supply of housing characteristics (Epple). These analytical solutions are only possible for a limited set of assumed forms of preferences and supply. For example, Rosen assumed that individuals' utility functions were linear. This is not an attractive assumption, especially if the ultimate objective of the analysis is to measure welfare values for changes in supplies of environmental characteristics.

Early researchers tried alternative functional forms for the hedonic price function (typically the semi-log, inverse semi-log, and log-linear) and selected one on the basis of goodness of fit. Goodman (1978) was one of the first to experiment with a flexible functional form. He employed a Box–Cox transformation of the dependent variable:

$$P_h^{[\lambda]} = \frac{P_h^{[\lambda]} - 1}{\lambda} \tag{11-6}$$

For $\lambda = 1$, this is a simple linear function. As λ approaches zero, this becomes the semi-log form. Some authors have found estimates of λ that were significantly different from both zero and one, indicating that this more complicated form fit the data better than either the linear or semi-log forms.

Transforming only the dependent variable still produces a very limited range of possibilities. Halvorsen and Pollakowski (1981) proposed that what they called a quadratic Box–Cox functional form be estimated. It would have the form

$$P_h^{[\lambda]} = a_0 + \sum_{i=1}^{n} a_i \cdot z_i^{[\beta]} + \tfrac{1}{2} \sum_{i=1}^{n} \sum_{j=1}^{m} c_{ij} \cdot z_i^{[\beta]} \cdot z_j^{[\beta]} \tag{11-7}$$

for all i, j and where i and j index the characteristics, z, and λ and β are estimated from the data.

Cassel and Mendelsohn (1985) pointed out that for welfare analysis, it is not the goodness of fit of the hedonic price function that matters; rather, it is the estimate of the marginal implicit price of the environmental attribute. In the regression equation for housing price, the environmental variable is likely to have relatively little influence in determining the estimated magnitude of β. But the estimate of β would have a major impact on the estimated marginal implicit price of the environmental characteristic.

The most general flexible form would be an extension of the quadratic Box–Cox of equation 11-7) to:

$$P_h^{[\lambda]} = a_0 + \sum_{i=1}^{n} a_i \cdot z_i^{[\beta_i]} + \tfrac{1}{2} \sum_{i=1}^{n} \sum_{j=1}^{m} c_{ij} \cdot z_i^{[\beta_i]} \cdot z_j^{[\beta_j]} \tag{11-8}$$

for all i, j. This functional form allows for different transformations for each independent variable. But it is not feasible to estimate this form for any realistic number of characteristics. A compromise, proposed by Palmquist (1991), would be to set $\beta_i = \beta_j$ for all $i, j = 1, \ldots, n - 1$, where n indexes the environmental attribute of interest. This would be responsive to the point raised by Cassel and Mendelsohn. Allowing for a separate transformation of the environmental amenity should give better results.

One question about functional form is whether the form chosen allows the marginal implicit price of the environmental characteristic to depend on the levels of the other attributes of houses. Of the commonly used functional forms, only the log and the Box–Cox transformation make the implicit prices of characteristics depend on the levels of other characteristics; the other forms impose independence. But this is a question that should be answered by the data, not by assumption.

A study by Cropper and others (1988) sheds some light on the question of functional form. These authors simulated the performance of a housing market using real data on buyer and housing characteristics drawn from the Baltimore, Maryland, area. After specifying the functional form and parameters of individuals' utility functions and the distribution of characteristics that reflect taste differences across individuals, they solved an assignment problem that produces a housing market equilibrium with each house being sold to the individual with the highest willingness to pay for its bundle of characteristics. With knowledge of the utility function parameters, it was then possible to calculate the true marginal implicit price for each individual and for the mean across all individuals. The authors used the equilibrium prices to estimate six alternative functional forms for the hedonic price function. They were then able to compare the mean true marginal bids with the bids calculated from each hedonic price function. They found that when all the housing characteristics were included in the hedonic price function, the linear and quadratic versions of the Box–Cox transformation provided the most accurate estimates of marginal implicit prices. However, when they experimented with various forms of misspecified hedonic price functions (by omitting variables or using proxy variables), they found that the linear version of the Box–Cox transformation was consistently superior in generating marginal implicit prices. This was apparently because with the quadratic version of the function there are more coefficients to estimate for each characteristic and omitting a variable leads to bias in more of the coefficients. Because there is no reason to expect these biases to be offsetting, the net result is a larger bias in the estimated marginal implicit prices (which are functions of all the relevant coefficients).

Spatial Interdependence. Recently attention has turned to the spatial interdependence of property prices. This means that the prices of nearby properties might help to explain the prices of any specific property of interest. The problems caused by spatial interdependence are primarily econometric, so they will not be covered in this book. A good introductory discussion and references to the literature can be found in Palmquist (forthcoming).

The Hedonic Price Function as a Market Equilibrium. Interpreting the marginal implicit prices as measures of households' MWTPs requires the assumption that each household is in equilibrium with respect to a given vector of housing prices and that the vector of housing prices is the one that just clears the market for a given stock of housing. These conditions ensure that the hedonic price function is the price vector that makes all participants in the market in aggregate just willing to hold the existing stock of housing. For these two aspects of equilibrium to be fully achieved, we require first that households have full information on all housing prices and attributes and that their transactions and moving costs be zero; and second, that the price vector adjust instantaneously to changes in either demand or supply. The market for housing can be viewed as a stock-flow model where the flow (change in stock) is a function of prices but the prices at any point are determined only by the stock at that time.

This idealized model is clearly not an accurate representation of real-world housing markets, but in evaluating the strength of this criticism of the hedonic price model, we must focus on several distinct issues. One issue concerns the speed of adjustment of the market to changed conditions of supply and demand. If adjustment is not complete, observed marginal implicit prices will not accurately measure household MWTPs. A major question is whether imperfect adjustment will lead to systematic biases in estimates of WTP.

Consider households' imperfect adjustment to changing prices. First, an increase in housing prices need not affect the marginal implicit prices of attributes, in which case no adjustment of the attribute bundles is necessary. Even if marginal implicit prices change, households will not move unless the potential utility gain to returning to full equilibrium exceeds the information costs, transactions costs, and moving costs associated with the change. These costs help define a band within which observed marginal implicit prices can diverge from household marginal willingnesses to pay for housing attributes. If housing prices change so that the marginal implicit price schedule for an attribute moves consistently in one direction, households will consistently lag in their adjustment to that change, and the MWTPs will be overstated or understated according to whether the marginal implicit price is rising or falling.

A second issue concerns expectations about future environmental amenity levels. Market prices for long-lived assets such as housing reflect the discounted present value of the stream of expected future services from that asset. A change in expectations about future environmental amenity levels can affect housing prices and marginal implicit prices independently of the present level of these amenities. For example, if there are widespread expectations of an improvement in air quality and the market adjusts reasonably quickly to these expectations, the price differential between currently dirty houses and clean houses should decrease. Correlating these prices with existing levels of air pollution would lead to an underestimate of the marginal implicit price of air quality.

Divergences from full equilibrium of the housing market in many circumstances will only introduce random errors into the estimates of MWTPs. However, where market forces are moving continuously in one direction or are expected to move in one direction, incomplete market adjustment or full adjustment to changing expectations, or both, can introduce biases in both directions. We should be cautious about using the cross-section hedonic price model in those cities and at those times during which market forces and environmental quality levels are changing rapidly (granted that "rapidly" is an imprecise term). However, it is also possible in these circumstances to determine the direction of bias. Thus, estimates of MWTP derived from such studies can be labeled as an upper bound or a lower bound on the basis of that analysis.

Another issue concerns the possibility of corner solutions. If there is not a sufficiently wide variety of housing models available, corner solutions are likely. The hedonic price function defines an opportunity locus across attribute space. A household chooses a housing model such that its indifference surface is tangent to the given opportunity locus, provided that a model with that precise set of attributes is available. If the optimum model is not available, the household must pick the nearby housing model that gives the highest utility level, but then the first-order conditions for utility maximization are not satisfied as equalities (Mäler 1977, 361–362).

The hedonic model is based on an assumption that the implicit price function is differentiable and continuous. However, this is an artifact of the statistical and mathematical technique. If this assumption is not satisfied in practice, two problems can arise. The first problem is that the statistically fitted hedonic price function is a good approximation only when the number of housing units is large and there is more continuous variation in characteristics among units. A small number of distinctly different types of housing units might be better analyzed with one of the discrete choice models described later in this chapter. The second problem arises if there are no units available with particular combinations of attributes. If there are substantial gaps in the opportunity locus, some households will not be able to satisfy the first-order conditions as equalities. This could be a problem for certain subsets of the urban population.

Market Segmentation. Straszheim (1974) was the first to raise the question of market segmentation in the context of estimating hedonic price functions for housing within an urban area. He argued that the urban housing market really consisted of a series of separate, compartmentalized markets with different hedonic price functions in each. As evidence in support of the segmentation hypothesis, Straszheim showed that estimating separate hedonic price functions for different geographic areas of the San Francisco Bay area reduced the sum of squared errors for the sample as a whole.

For different hedonic price functions to exist in an urban area, two conditions must be met. The first is that the structure of demand, the structure of supply, or both, must be different across segments. Either buyers in separate submarkets must have different structures of demands, or the structure of characteristics of the housing stocks must be different. The second condition is that purchasers in one market segment must not participate significantly in other market segments. In other words, there must be some barrier to mobility of buyers among market segments that prevents arbitrage from occurring in response to differences in marginal implicit prices. Such barriers could be due to geography, discrimination, lack of information, or a desire for ethnically homogeneous neighborhoods. Even with buyer immobility, if demand and supply structures are the same they will produce similar structures of hedonic prices. Perfect mobility and information on the part of buyers will eliminate differences in the implicit prices for any characteristic across market segments.

If market segmentation does exist, the hedonic price function estimated for the urban area as a whole will provide faulty estimates of the implicit prices facing subsets of buyers in different market segments. Thus estimates of benefits and estimates of demand functions based on faulty price data will also be faulty. If market segmentation does exist, separate hedonic price functions must be estimated for each segment, and benefit and demand functions must be separately estimated for each segment with a different set of implicit prices.

It is not clear how significant the problem of market segmentation is for air pollution–property value studies within single urban areas, although there are enough positive results in the literature to suggest that it is not a problem that can be dismissed out of hand. Some authors have found evidence of different hedonic price functions for submarkets within larger urban areas, suggesting segmentation, but this could be due to misspecification of the model. Others have not found evidence of segmentation in their data.

The existence of market segmentation does not render the hedonic price technique invalid; rather, it makes application of the technique more difficult. If the appropriate basis for segmentation can be identified, it is conceptually possible to estimate separate implicit price functions for each submarket. Although these functions would be different across markets, they each would accurately reflect the outcome of the market processes in each submarket. Thus the functions could be used to estimate equilibrium MWTPs.

Estimating Characteristics Demands

The attractiveness of the hedonic price model for applied welfare analysis lies in the potential for using estimates of individuals' marginal implicit prices for a characteristic to recover the uncompensated inverse demand function for q or information on the underlying structure of preferences. Rosen (1974) had argued that the inverse demand and marginal supply price functions could be estimated from the information contained in the hedonic price function in the following manner:

> ... compute a set of implicit marginal prices ... for each buyer and seller evaluated at the amounts of characteristics ... actually bought or sold, as the case may be. Finally, use estimated marginal prices ... as endogenous variables in the second-stage simultaneous estimation of [the inverse demand and supply price functions]. Estimation of marginal prices plays the same role here as do direct observations on prices in the standard theory and converts the second-stage estimation into a garden variety identification problem. (Rosen, 1974, 50)

This suggestion has been the source of a large literature for more than two decades. Because the emphasis in this book is on models and basic economic method rather than on econometric issues, I provide here only an overview of the sources of problems in estimating and identifying demand functions for characteristics and alternative approaches to solving them.*

The difficulties in estimating the inverse demand functions from hedonic price data come in two forms. The first arises from the fact that the source of data for the dependent variable in the MWTP function is not direct observation of the inverse demand prices; rather, it is the calculation of the marginal implicit price $\partial P_h/\partial q$ from the estimated hedonic price function. However, this variable is itself computed as the function of the same characteristics that are explanatory variables in the MWTP function. Brown and Rosen (1982) and Mendelsohn (1987) show that at least in some cases this procedure leads to parameter estimates for the MWTP function that are identical to the estimated coefficients in the hedonic price function. As Brown and Rosen put it,

*Readers interested in more technical discussion, especially from an econometric perspective, should consult Brown and Rosen 1982; Epple 1987; Bartik 1987; and McConnell and Phipps 1987. Bartik and Smith 1987 and Palmquist 1991 also provide useful reviews.

> Contrary to Rosen's original statement, we claim that marginal attribute prices constructed as above will not necessarily play the same role in estimation that direct observation on prices would play if they were available. Because such constructed prices are created only from observed sample quantities, any new information that they may provide (i.e., any information beyond that already provided directly by observed sample quantities) can only come from a priori restrictions placed on the functional form of the price function $P_h(\cdot)$. In the absence of such additional restrictions, second stage "structural" estimation of the sort suggested by Rosen may only reproduce the information already provided by the first-stage estimation of the $P_h(\cdot)$ function. (1982, 176; notation changed by the author)

In other words, because the second-stage estimation procedure uses no additional data beyond that already contained in the hedonic price function, it can do no more than reproduce the coefficients estimated from the hedonic price function.

The nature of the problem can be illustrated with Figure 11-2. This example is due to Bartik (1987, 84–85). Consider two individuals with the same income and uncompensated inverse demand functions of the form

$$b^* = b^*(q, M, \boldsymbol{T}) \tag{11-9}$$

where M is income, and $\boldsymbol{T}$ is a vector of unobserved determinants of tastes. For the first individual, we observe point A on her inverse demand function, but we have no information on the demand price for other levels of q. If the two individuals choose different levels of q, for example, q^1 and q^2, it must be because of differences in tastes. This means that the demand-shifter, which is unobserved in the data of Figure 11-2, is correlated with the observed choices of q. As Palmquist (1991) put it, "the other marginal prices [on the individual's MWTP function] are only observed for other individuals with other socioeconomic characteristics and provide no information on the original consumer's bid for different quantities of the characteristic" (1991, 96). This makes it difficult to separate the effects of demand-shifters from the price–quantity relationship itself.

One approach to solving the identification problem is to impose sufficient structure on the problem by assumption to ensure that the conditions for identification of the inverse demand function are met. An early example is provided by Quigley (1982), who assumed a functional form for preferences that included homotheticity as a property. See Chattopadhyay 1999 for a more recent example of this approach. By specifying the relationship between income and demand, this assumption made it possible to separate the effects of income and quantity change on the marginal will-

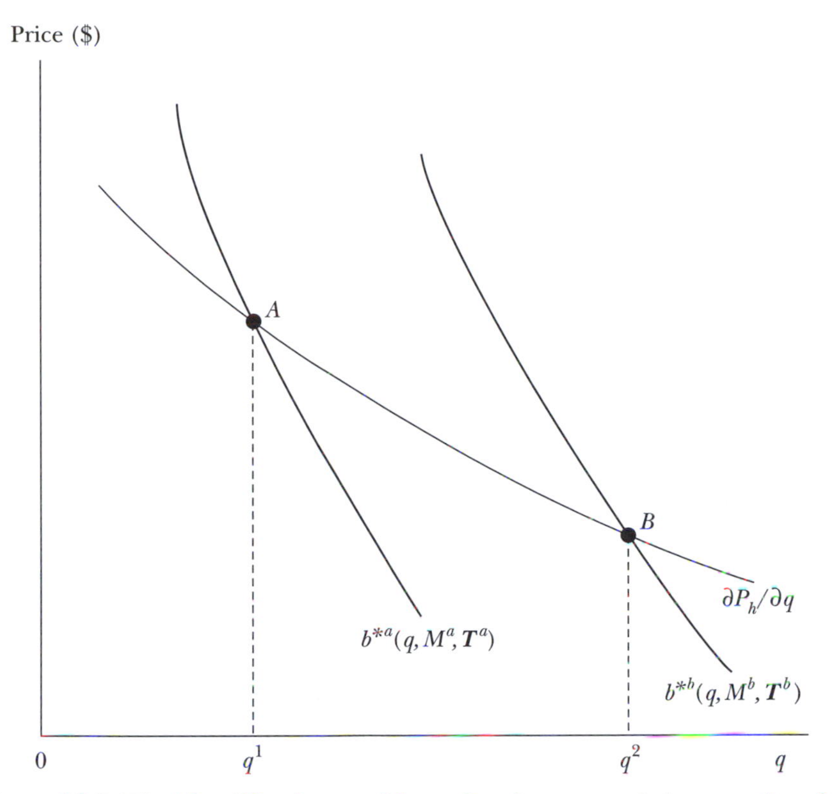

Figure 11-2. The identification problem when income and the quantity of the characteristic are correlated

ingness to pay for characteristics. Note, however, that the assumptions about functional form are not testable.

Probably the most reliable approach to solving the identification problem is to find cases where the marginal implicit prices of characteristics vary independently of the other demand-shift variables. Specifically, this means finding cases where individuals with the same preferences, income, and other traits face different marginal implicit prices. This can only occur if similar individuals must choose in markets with different hedonic price functions, which in turn implies either segmented markets within a city or observations taken from several different housing markets (as, for example, in different cities).

The first step in implementing this approach is to estimate separate hedonic price functions for each housing market, using the same specification. The second step is to compute the marginal implicit price faced by each individual from the hedonic price function in that market. Then the computed marginal implicit prices can be regressed on the observed quantities of the characteristics and the exogenous demand-shifters to obtain the uncompensated bid function. Assuming sufficient independent variation across markets and assuming that there are no unobserved differences in preferences across individuals, this approach will lead to reliable and properly identified bid functions. For examples of this approach, see Palmquist 1984; Bartik 1987; and Zabel and Kiel 2000.

The second source of difficulty in estimating inverse demands for attributes lies in the fact that both the quantity of the characteristic and its marginal implicit price are endogenous in the hedonic price model. Unlike the standard market model in which an individual faces an exogenously determined price and chooses a quantity, and unlike a quantity-rationed market in which an individual faces an exogenously determined quantity and reveals an MWTP, the individual chooses both a point on the hedonic price schedule and its associated quantity. The choice of that point simultaneously determines the MWTP and the quantity of the characteristic.

One approach to solving this problem is to find truly exogenous variables to be used as instruments. This appears to be a difficult task, however; for some of the suggested possibilities and their problems, see Mendelsohn 1984, 1985; Bartik 1987; Bartik and Smith 1987; and Palmquist 1991. Another possibility is to assume that there is a characteristic in the marginal implicit price function that is not an argument in the MWTP function for another characteristic. This makes it possible to use the omitted characteristic as an instrument. Recently many authors have been critical of this approach because the results are only as good as the assumptions imposed to obtain them. See Palmquist 1984; Bartik and Smith 1987; Horowitz 1987; and Mendelsohn 1987. Unfortunately, such assumptions are not testable.

Measuring Welfare Changes

In an equilibrium of the housing market, utility-maximizing individuals equate their marginal willingness to pay for housing characteristics with the marginal implicit prices of these characteristics. In some circumstances it may be possible to estimate inverse demand functions on the basis of this information. We now ask the question, how can we use the information on prices and preferences that can be extracted from the hedonic housing market to calculate measures of welfare change for changes in environmental amenities? The basic concepts of welfare measurement at the level of the individual are straightforward and were introduced in Chapter 4. However, measurement of aggregate welfare changes based on hedonic prices is made difficult by the adjustments that people are likely to make in response to changes in environmental attributes and the possibility that the hedonic price function will change. Also, in principle, it is necessary to consider possible changes in the supply side of the hedonic property market.

In this section I first define the basic welfare measure for marginal changes in a characteristic or environmental amenity, holding other things constant, in particular, individuals' choices of housing bundles. I next consider the benefits of a nonmarginal change in an amenity, assuming that individuals cannot adjust their housing bundles by moving. This measure looks only at benefits to purchasers of housing bundles. I then consider a fully general measure of welfare change that includes possible changes in profits on the supply side of the hedonic market as well as the consequences of individuals' adjustments on the demand side of the market. Because this conceptually correct measure is not implementable in practice, I go on to describe some methods that can provide lower or upper bound approximations of the correct measure, at least in certain circumstances. Finally I consider localized changes in environmental quality that lead to benefits to some people without changing the hedonic price function.

Because a change in an environmental amenity in an urban area is nonexcludable and nondepletable, it is, in effect, a public good. The marginal value of the change, then, is simply the sum of the MWTPs of each of the n affected individuals evaluated at the existing housing equilibrium. In other words, for the amenity q

$$w_q = \sum_{i=1}^{n} b_i = \sum_{i=1}^{n} \left(\frac{\partial P_h}{\partial q} \right)_i \qquad (11\text{-}10)$$

where w_q is the aggregate marginal welfare change and b_i is the ith individual's MWTP. Although most proposed environmental policy changes are nonmarginal in magnitude, the ease of calculating equation 11-10 may make it useful for indicating whether some improvement is desirable, by comparing this measure with an estimate of the marginal cost of the improvement.

Welfare Changes without Adjustments. Recall from Chapter 4 that the equilibrium hedonic price function is given by the double envelope of the bid and offer curves for all of the characteristics. A change in the level of q will place at least some individuals out of equilibrium, given the existing set of marginal implicit prices. Their efforts to restore their equilibria will result in changes in the hedonic price function and marginal implicit prices. Also, in principle, changes in the hedonic price function could trigger changes in the supplies of houses with different bundles of characteristics. In the next section, I will discuss welfare measurement in the context of these changes, but for now let us assume that all individuals are constrained to stay at their original location, as might be the case with high transactions and moving costs or if a short-run perspective is taken. Also, assume that there is no supply response to the change in q.

Given these assumptions, the welfare value of the change in q from q^0 to q^1 is given by the sum of the areas under each individual's MWTP curve over the change in q, or

$$W_q = \sum_{i=1}^{n} \int_{q^0}^{q^1} b_i\left(q_i, \mathbf{Q}_i^*, u_i^*\right) dq \tag{11-11}$$

where W_q is the aggregate benefit. Notice that this measure requires knowledge of the MWTP functions of individuals. If the uncompensated bid functions from the second stage of the hedonic price estimation are used, the welfare gain from an increase in q will be overestimated. The magnitude of the error will depend upon the factors discussed in Chapter 3.

There is a method for calculating exact welfare measures for nonmarginal changes in a characteristic, holding all other things constant. It is based on an adaptation by Horowitz (1984) of Hausman's technique for exact welfare measurement for price changes (see Chapter 3). Suppose that the ith individual's uncompensated inverse demand function for q

$$b_i^* = b_i^*\left(q, \mathbf{Q}_i^*, M - P_h\right) \tag{11-12}$$

has been identified. Using the indirect utility function, in equilibrium,

$$\frac{(\partial v / \partial q)}{(\partial v / \partial M)} = b_i^*(\cdot) \tag{11-13}$$

and for individual i at location j

$$b_i^* = \frac{\partial P_h}{\partial q} \tag{11-14}$$

The left-hand side of equation 11-13 is the slope of the indifference curve between the numeraire, M, and q. So, in equilibrium

$$\frac{dM}{dq} = b_i^*(\cdot) = \left(\frac{\partial P_h}{\partial q}\right)_i \tag{11-15}$$

This expression can be solved for

$$M = f(q, \mathbf{Q}^*, C) \tag{11-16}$$

where C is a constant of integration. The benefit of an increase in q is

$$W_q = f\left(q^0, \mathbf{Q}^*, C\right) - f\left(q^1, \mathbf{Q}^*, C\right) \tag{11-17}$$

See Horowitz 1984 and Palmquist (forthcoming) for examples of this approach with various forms of the utility function.

In those cases where neither the uncompensated or compensated inverse demand functions are available, welfare changes could be estimated by making some assumption as to the shape of the MWTP function through the original equilibrium point. Three alternative assumptions can be used to establish bounds on the true measure. One is to assume that the MWTP for each individual is constant, that is, that the MWTP function for each individual is a horizontal line through the known point. In this case, each individual's benefit for the postulated improvement in the amenity is approximated by the product of the (assumed) constant MWTP and the change in the amenity. The aggregate benefit is obtained by summing over all individuals. This assumption leads to an estimate of aggregate benefits that is biased upward.

A second convenient assumption would be that each individual's MWTP curve decreases linearly from its observed point to the point of the highest attainable level of the amenity. MWTP would be zero at this point. It is not clear whether this approximation would lead to an overestimate or an underestimate of true benefits. The third assumption would be that all individuals' MWTP functions are identical. Then, as discussed in Chapter 4, the marginal implicit price curve is identified as the MWTP curve for the representative individual.

Welfare Changes with Full Adjustment. Bartik and Smith (1987, 1223) have presented a welfare measure that takes account of all the adjustments that individuals make in response to the nonmarginal change in q. At any location, the value of a nonmarginal change can be taken to be the integral of the values of a series of infinitesimal changes in the amenity. The value of each small change is taken to be the WTP of the occupant of that site at that point in the sequence of changes. The measures for all sites together is the sum of the values for each site. It is given by

$$W_q = \sum_{j=1}^{m} \int_{q_j^0}^{q_j^1} \left[\frac{\partial P_h(q, \mathbf{Z})}{\partial q} \right] dq_j \tag{11-18}$$

where $\mathbf{Z}$ is the vector of all other site characteristics (which are held constant by assumption), where j indexes locations, and where the hedonic prices change in response to the adjustments that people make.

In principle this measure allows individuals to relocate in response to changes in the quantity and price of the amenity because it sums individuals' marginal values as the amenity changes at each site. This is important because a major limitation of some of the measures described below is their inability to account for individual relocation decisions. Furthermore, this measure does not require knowledge of either the MWTP or the bid function. It relies on the fact that at each point in the sequence of changes, each individual's marginal bid is revealed by the marginal implicit price of the characteristic.

However, because the hedonic price function is shifting as a consequence of the change in the amenity level, it is necessary to know how the hedonic price function and the marginal implicit prices at each location change as the levels of the amenities at each location change along the path of integration. As a practical matter, this is a major limitation of the measure.

This limitation has forced researchers to look for practical measures that can be interpreted as approximations or upper or lower bounds on the true welfare change. Following the analysis by Bartik (1988), suppose that there are increases in several environmental amenities in an urban area. These increases need not be uniform across the area. Specifically, let us consider the case where the vector $\boldsymbol{Q}$ increases from $\boldsymbol{Q}^0$ to $\boldsymbol{Q}^1$. We will first look at the immediate welfare effects of this change on those occupying sites where amenities have changed. Then we will examine how the welfare gains are magnified and redistributed by the rational adjustments of individuals and ensuing price changes in the hedonic market.

First, assuming that individuals cannot move to new locations and that the hedonic price function does not change, the benefit to individuals is given by an expanded version of equation 11-11:

$$W_q = \sum_{i=1}^{n} \int_{Q_i^0}^{Q_i^1} \int \boldsymbol{B}_i \left(\boldsymbol{Q}, \boldsymbol{S}, u_i^* \right) d\boldsymbol{Q}_i \tag{11-19}$$

where each individual's welfare gain is computed from a path-independent line integral over the changes in the individual elements in $\boldsymbol{Q}$, $\boldsymbol{B}_i(\cdot)$ is the vector of individual MWTP functions for the characteristics, and $\boldsymbol{S}$ is the vector of structural characteristics of the house.

Now, at the existing hedonic price function, some people may wish to choose different bundles of characteristics. If they do change, it must be because they perceive themselves to be better off after the adjustment. This welfare gain is in addition to that given by equation 11-19. Thus equation 11-19 can be interpreted as a lower bound on the true measure. It requires knowledge of only the bid or compensated inverse demand functions for the characteristics that change. Furthermore, the effort to adjust to different characteristics bundles is likely to affect the hedonic price function, unless the number of people wishing to do so is quite small relative to the market. Also, it is possible that the suppliers of housing will respond to changes in the hedonic price function by offering different bundles of housing characteristics, $\mathbf{S}$. This could have further repercussions on the hedonic price function, and it will increase the profits of housing suppliers.

When all of these repercussions have worked themselves out, the aggregate benefit to individuals can be defined in terms of each individual's total willingness to pay for a housing unit with given characteristics, holding utility constant. Let this total WTP be given by

$$WH_i\left(\mathbf{Q}_i^{*j}, \mathbf{S}_i^{*j}, u_i^*\right) \tag{11-20}$$

where $\mathbf{Q}^{*j}$ and $\mathbf{S}^{*j}$ $(j = 0, 1)$ indicate the vectors of environmental and other characteristics actually chosen by the individual in the original and new equilibrium. Each individual's total benefit is the increase in total willingness to pay for the characteristics actually chosen, holding utility constant, minus any increase in actual expenditure on housing. Summing across all individuals, we obtain

$$\begin{aligned} W_q = & \sum_{i=1}^{n}\left[WH_i\left(\mathbf{Q}_i^{*1}, \mathbf{S}_i^{*1}, u_i^*\right) - WH_i\left(\mathbf{Q}_i^{*0}, \mathbf{S}_i^{*0}, u_i^*\right)\right] \\ & - \sum_{i=1}^{n}\left[P_h^1\left(\mathbf{Q}_i^{*1}, \mathbf{S}_i^{*1}\right) - P_h^0\left(\mathbf{Q}_i^{*0}, \mathbf{S}_i^{*0}\right)\right] \end{aligned} \tag{11-21}$$

Turning to the supply side of the market, there are m $(k = 1, \ldots, k, \ldots, m)$ housing producers. In aggregate, they realize a change in aggregate profits given by the increase in expenditures on housing net of any change in their costs. This is given by

$$\begin{aligned} \Delta\text{Prof} = & \sum_{i=1}^{n}\left[P_h^1\left(\mathbf{Q}_i^{*1}, \mathbf{S}_i^{*1}\right) - P_h^0\left(\mathbf{Q}_i^{*0}, \mathbf{S}_i^{*0}\right)\right] \\ & - \sum_{k=1}^{m}\left[C_k^1\left(\mathbf{Q}_i^{*1}, \mathbf{S}_i^{*1}\right) - C_k^0\left(\mathbf{Q}_i^{*0}, \mathbf{S}_i^{*0}\right)\right] \end{aligned} \tag{11-22}$$

where $C(\cdot)$ is the cost function for producers.

The welfare change for society as a whole is the sum of equations 11-21 and 11-22. One component of this sum is simply a transfer of revenue from buyers to sellers, so it nets out. The total welfare change is the sum of the increase in total WTP of individuals minus any cost increase on the part of producers of housing. Full implementation of this welfare measure would require enormous amounts of information. However, it should be noted that this measure reduces to equation 11-19 if the hedonic price function does not change and if the change in environmental amenities does not affect the costs of supplying housing amenities for producers.

Even if this set of conditions is not satisfied, equation 11-19 can be interpreted as a lower bound on the true measure of benefits. This can be seen by decomposing the true benefit measure into a three-step sequence of changes and adjustments. Consider first the change in amenity levels without any adjustment on the part of individuals or suppliers. The welfare change associated with this step is given by equation 11-19 plus any reduction in the costs of supplying existing houses at the affected locations. Second, suppose hypothetically that the hedonic price function is shifted to its new equilibrium position but that no adjustments to the new price function by individuals or suppliers are permitted. At this stage, although some individuals and suppliers may gain while others lose, on net all the price changes sum to zero. At this stage, there is no net change in welfare.

Finally, allow individuals and suppliers to respond to the new hedonic price function. Any adjustments that take place at this stage must represent welfare improvements for those responding. The total welfare change is the sum of equation 11-19, any cost reduction to suppliers, and the benefits of adjusting to the price change. The latter two components are either zero or positive. Thus, equation 11-19 represents a lower bound on the true measure of benefits. The smaller the adjustment to the changes in the hedonic price function is, the smaller is the error involved in using equation 11-19.

Localized Amenity Changes: A Special Case. If the hedonic price function does not shift, then exact welfare measurement may be a relatively easy task. Palmquist (1992a) has identified one situation in which the hedonic price function could be assumed to be constant. That is when the number of sites at which there is a change in the amenity level is small relative to the total urban market. If this is the case, and if individuals can move without cost from one site to another in response to the change in environmental amenity levels, then exact welfare measurement is straightforward. The hedonic price function can be used to predict the changes in the prices of affected properties. Benefits are exactly measured by the increase in the values of the affected properties, and knowledge of the marginal bid functions is not required.

Consider the case of an improvement from q^0 to q^1 at just one site, as shown in Figure 11-3. Assume that moving costs for occupants who choose

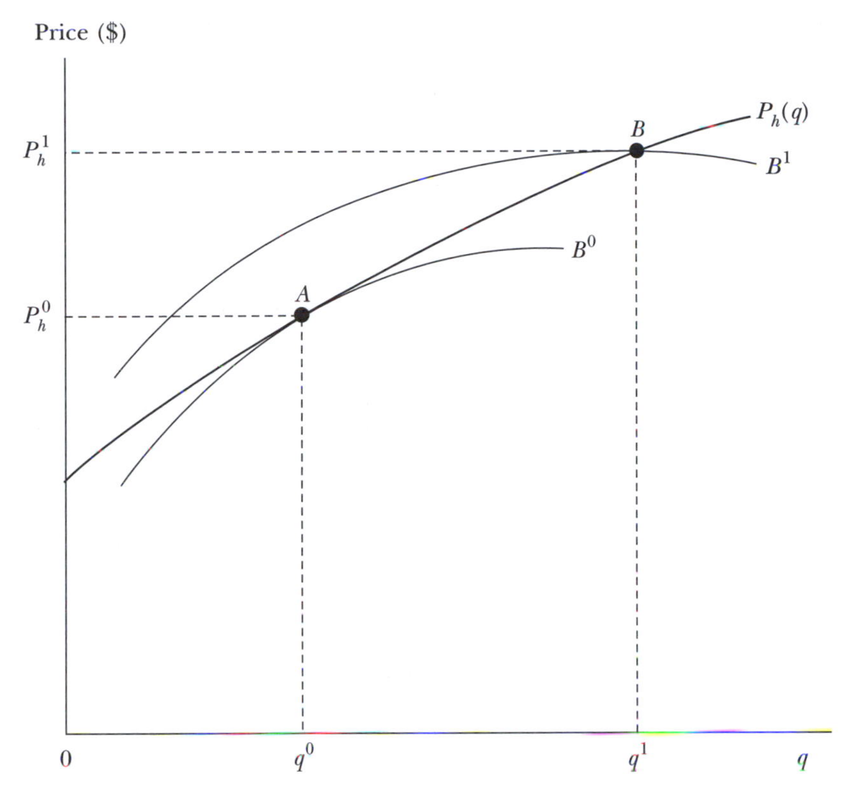

Figure 11-3. The benefit of an amenity improvement at one site when the hedonic price function is unchanged

to relocate are zero (the impact of positive moving costs is discussed below). The change in the amenity level results in an increase in the price of this house from P_h^0 to P_h^1. The owner of the property is made better off by this increase in wealth. Even though the occupant of the property experiences the increase in amenity level, he or she is made worse off because of the increase in the cost of occupying the property. The occupant is shifted from point *A* on the curve B^0 to point *B* on the curve B^1. However, with costless moving, the occupant can relocate to his or her original equilibrium position, so the net welfare change is the increase in wealth to the owner. If the owner and occupant are the same person, the result is still the same. This individual might choose to move to a property with an amenity level somewhat greater than q^0 because of wealth. However, the increase in wealth fully captures the benefit of the amenity improvement to this individual.

Now consider the case where the number of affected sites is small, so that the hedonic price function does not change, but where moving costs are positive. The renter either loses the moving costs involved in adjusting his or her housing bundle or bears a loss of utility associated with staying at a less preferred location after the amenity change. In either case, the increase in property prices is an upper bound on the total benefit (Palmquist 1992b).

Rents, Taxes, and Property Prices

In the development and exposition of the theoretical model, I have ignored the temporal dimension of housing prices and how welfare measures based on property prices might be converted to the annualized form usually used in welfare evaluation. Typically the market price of a property is observed. Inferences about the streams of rents and of benefits are drawn by converting observed present values to annual streams. The institutions of income and real property taxation affect how the market capitalizes rents (and changes and differentials in rents) into market prices for properties. These effects must be properly understood if the process of retracing these steps to infer rents from property value observations is to be successful. In this section, I develop and expand on some ideas first presented by Niskanen and Hanke (1977).

In the simplest case of a stream in perpetuity and with no taxes, the conversion of property value to rent is given by

$$R = P_h \cdot r \tag{11-23}$$

where r is the appropriate discount rate. I proceed by examining first the effects of the two forms of taxation separately, then their combined effects. I show that the effects of these two kinds of taxation on the relationship between observed differences in property values and welfare measures depends of the specific features of the tax system and parameter values such as interest rates and tax rates.

Ad valorem taxation of property can be viewed as a device for capturing some of the rent of land for the government. Because taxation affects the net return to the property owner, it should affect the market value of property as an asset. An individual would purchase a property as an asset only if its market price, P_h, is equal to or less than the discounted present value of the rental stream net of property taxation. Market forces would establish the following relationship between property values and rents:

$$P_h = \frac{R - t \cdot P_h}{r} \tag{11-24}$$

where t is the ad valorem tax rate.

If property values are known, the rental stream they represent can be computed by rearranging equation 11-24:

$$R = P_h \cdot (r + t) \tag{11-25}$$

Assume that the property value–amenity relationship, $P_h(q)$, has been estimated. The marginal benefit of a change in q at a site is

$$w_q = \frac{\partial R}{\partial q} = (r + t) \cdot \frac{\partial P_h}{\partial q} \tag{11-26}$$

The present value of this stream of benefits is

$$\frac{w_q}{r} = \left(1 + \frac{t}{r}\right)\left(\frac{\partial P_h}{\partial q}\right) \tag{11-27}$$

In other words, when the hedonic price function is defined in terms of property value, ignoring the effect of property taxation on the capitalization of rents can lead to the underestimation of benefits. The term (t/r) is a measure of the percentage error resulting from omitting the tax term in the calculation of benefits. For an interest rate of 10% and an effective tax rate of 10–20 mils per dollar of market value (1–2%), the error is between 10% and 20%.

However, this is not the whole story. The income tax code treats the imputed rental income of home owners differently than it does the rental income of landlords. The absence of a tax liability for imputed rent further complicates the task of inferring annual rents and benefits from observations of (capitalized) market prices for housing assets. This is because the market will place different values on two assets with the same rental stream if one is subject to income taxation while the other is not.

Assume two perpetual assets indexed as a and b with equal annual returns of R per year. The return to the first asset is taxable at the rate g%, and the return to the second asset incurs no tax liability. If r represents the market rate of return on assets with taxable returns, the two assets will be priced to equalize the after-tax rate of return:

$$P_a = R_a / r \tag{11-28}$$

and

$$P_b = \frac{1}{(1-g)} \frac{R_a}{r} > P_a \tag{11-29}$$

If P_b is observed, the tax-free rental stream can be computed as

$$R_b = r \cdot (1 - g) \cdot P_b \tag{11-30}$$

Taking account of this adjustment and using equations 11-26 and 11-27, the marginal annual benefits of amenity changes to home owners and their present value are

$$w_q = r \cdot (1 - g) \cdot \left(\frac{\partial P_b}{\partial q} \right) \tag{11-31}$$

$$\frac{w_q}{r} = (1 - g) \cdot \left(\frac{\partial P_b}{\partial q} \right) \tag{11-32}$$

Ignoring the effects of income taxation leads to an overestimation of benefits. The discount factor $r(1 - g)$ is analogous to the municipal bond rate, and it arises for the same reason. However, where the marginal tax rate is itself a function of income, g varies across individuals and equations 11-31 and 11-32 must be computed separately for each individual.

The tax code confers additional benefits on home owners by exempting from taxation capital gains realized on the sale of a primary residence and by allowing them to deduct property tax payments in calculating taxable income. This latter provision lowers the real cost of the property tax by $g\%$. Combining these effects (ad valorem taxation, deductibility, and exemption of imputed rental income), we have

$$P_b = [R_b - (1 - g) \cdot t \cdot P_b] r \cdot (1 - g) \tag{11-33}$$

Solving for R_b gives

$$R_b = (r + t) \cdot (1 - g) P_b \tag{11-34}$$

Marginal benefits are calculated by

$$w_q = (r + t) \cdot (1 - g) \cdot \left(\frac{\partial P_b}{\partial q} \right) \tag{11-35}$$

and

$$\frac{w_q}{r} = \left(1 + \frac{t}{r} \right) \cdot (1 - g) \cdot \left(\frac{\partial P_b}{\partial q} \right) \tag{11-36}$$

The effects of ignoring taxation in calculating benefits depend on the magnitudes of g and t/r. The higher the marginal income tax rate, the more likely benefits would be overstated if taxes were ignored. For an example, suppose the marginal income tax rate is 30%, the opportunity cost of capital is 10%, and the property tax rate is 2%. Then the terms in parentheses come to 0.84. Ignoring tax effects would lead to an overstatement of benefits by almost 20%, but lower income tax and discount rates can reverse this conclusion.

An alternative approach to dealing with taxation and discounting is to base the hedonic equation on measures of user cost (called gross rent by Sonstelie and Portney 1980). This variable captures the full cost of owning (and using) an asset such as a house. User cost would include property taxes and the opportunity cost of capital plus any change in the market price of the asset over the interval, say a year. It would be calculated as follows:

$$u = (r + t + m)P_h \tag{11-37}$$

where m is the percentage rate of change in market value.

The user cost approach differs from that outlined above in two respects. The first is the inclusion of the change in market value over time. P_h could be changing because of physical depreciation of the house, general price inflation, changes in the price of housing relative to other goods, and changes in the variables determining P_h. Only the latter changes have relevance for benefit estimation, and they would be captured by modified versions of equations 11-35 and 11-36, which generalized from the assumption of constant streams in perpetuity. However, the depreciation term might be useful in empirical work as an approximation of expected changes in these variables, provided that it were adjusted to net out general price level effects.

The second difference arises in considering the effects of some provisions of the income tax code on user cost. For one thing, the tax exemption for imputed rent does not affect the user cost of holding a house because user cost is an opportunity cost. For another, the tax deductibility of mortgage interest does affect user cost, but it does not affect the market capitalization of streams of benefits. If user cost is used to compute benefits, the net result of these two effects is to overstate benefits in comparison with equations 11-35 and 11-36.

The Hedonic Model with Averting Behavior

The use of the hedonic property value model for welfare measurement described in this section requires the assumption that preferences are separable in housing characteristics and those environmental amenities specific to housing locations. This is probably not always a reasonable assumption. If the assumption is not valid, it is necessary to explore the nature of the interdependence between housing characteristics and other arguments in the individual's utility function and how the welfare measures derived from hedonic theory are related to those derived from other models of behavior. In this section I explore some implications of one kind of interdependence between a location-specific amenity, say air quality, and other arguments of an individual's utility function.

Suppose that air quality as measured by q produces an amenity such as a better view of the distant mountains. Suppose that q also reduces the incidence of illness and disease related to air pollution. I examine here the relationship between the marginal value of improved health and the marginal implicit price of q as revealed from a hedonic property value analysis. (Models for valuing reduced morbidity have been explored in Chapter 10.)

As in Chapter 10, assume that the number of days spent sick (by whatever measure) is inversely related both to q and to the quantity of an averting activity, a. This relationship is $s = s(q, a)$. The averting activity is purchased at a price of p_a. The individual derives utility from the consumption of a numeraire good X, leisure f, a vector of housing attributes $\mathbf{Z}$, and a location-specific amenity v; v is itself a function of the level of air quality, $v(q)$, $\partial v/\partial q > 0$. The individual also experiences disutility from the days spent sick. In addition to disutility, illness results in lost workdays and reduction in wage income. The individual's choice problem can be represented as

$$\text{max: } u[X, f, \mathbf{Z}, v(q), s(q, a)] + \lambda\{I + p_w[T - f - s(q, a)] - X - (p_a \cdot a) - P_h(q, \mathbf{Z})\} \tag{11-38}$$

where $P_h(q, \mathbf{Z})$ is the hedonic property value function. The relevant first-order conditions include

$$\frac{\left[\dfrac{\partial u}{\partial v} \cdot \dfrac{\partial v}{\partial q}\right]}{\lambda} + \left[\frac{(\partial u/\partial s)}{\lambda} - p_w\right] \cdot \frac{\partial s}{\partial q} = \frac{\partial P_h}{\partial q} \tag{11-39}$$

and

$$\left[\frac{(\partial u/\partial s)}{\lambda} - p_w\right] = \frac{p_a}{(\partial s/\partial a)} \tag{11-40}$$

The left-hand side of equation 11-39 is the individual's marginal willingness to pay for an improvement in q. This marginal value can be inferred from observations of the marginal implicit price of q, $\partial P_h/\partial q$. The marginal willingness to pay for q has two components, one related to the amenity and the other related to the disutility and income loss associated with illness. Substituting equation 11-40 into equation 11-39 yields another way of looking at marginal value:

$$w_q = \frac{\left[\frac{\partial u}{\partial v} \cdot \frac{\partial v}{\partial q}\right]}{\lambda} + p_a \cdot \frac{(\partial s / \partial q)}{(\partial s / \partial a)} \tag{11-41}$$

or

$$w_q = \frac{\left[\frac{\partial u}{\partial v} \cdot \frac{\partial v}{\partial q}\right]}{\lambda} + \left(p_a \cdot \frac{\partial a}{\partial q}\right) = \frac{\partial P_h}{\partial q} \tag{11-42}$$

The marginal willingness to pay for q can be separated into its own amenity- and health-related components. The marginal implicit price of q measures the sum of the components, but it provides no information on the relative importance of the amenity and the health components of marginal value. In an earlier analysis, Portney (1981) examined the relationship between property values and the value of reduced mortality risks due to air pollution. He inferred a value of mortality risk reduction by assuming that the amenity component of $\partial P_h / \partial q$ was very small.

Manipulations similar to those described in Chapter 10 can be used to derive an alternative expression for w_q:

$$w_q = \left[\frac{(\partial u / \partial s)}{\lambda} - p_w\right]\frac{ds}{dq} - p_a \cdot \frac{\partial a^*}{\partial q} + \frac{\left[\frac{\partial u}{\partial v} \cdot \frac{\partial v}{\partial q}\right]}{\lambda} \tag{11-43}$$

where a^* is the optimal quantity of a, given the constraints. As before, the first two terms represent the disutility and the lost wages associated with the observed change in number of sick days and the observed change in averting expenditures after optimal adjustment to the change in q. The marginal implicit price of q captures both the health-related and amenity values associated with the improvement in q. The amenity value could be found by subtraction if the marginal implicit price of q and the health benefit were estimated separately.

This model can be extended to the case where health is affected by air quality both at home and at some other location, say work. Suppose the number of days of illness depended on air quality at home, q_H, air quality at work, q_W, and averting activities. The individual can choose q_H through the selection of a place to live. As long as q_H and q_W vary independently, they can be treated separately for purposes of valuation. The two MWTPs take the by-now familiar form

$$w_{q_H} = \frac{\left[\frac{\partial u}{\partial v} \cdot \frac{\partial v}{\partial q_H}\right]}{\lambda} + p_a \cdot \frac{(\partial s/\partial q_H)}{(\partial s/\partial a)} = \frac{\partial P_H}{\partial q_H} \tag{11-44}$$

and

$$w_{q_W} = p_a \cdot \frac{(\partial s/\partial q_W)}{(\partial s/\partial a)} \tag{11-45}$$

The marginal implicit price of q_H captures the welfare effect of changes in air quality at home but does not say anything about the welfare effects of changes in q elsewhere. To implement either measure, it is necessary to know the separate contributions of pollution at home and at work to inducing illness.

Applying the Model to Agricultural and Commercial Land

Up to this point I have presented the hedonic model in the context of the market for housing. However, the model can be applied to the market for any differentiated product. In this section I show how the hedonic price model can be reformulated to deal with the land as an input into production processes. The individual's utility function is replaced by the profit function of the firm. Bid and offer curves for land as a function of its productive characteristics can be derived. The model can be used for measuring the values of improvements in these characteristics. Possible applications include the markets for commercial land (for example, for shopping and offices) and for agricultural land. Palmquist (1989) developed the analytical model for agricultural land. Miranowski and Hammes (1984), Palmquist and Danielson (1989), and Mendelsohn and others (1994) present estimates of hedonic price functions and marginal implicit prices for characteristics of agricultural land.

Let $\boldsymbol{Q}$ be a vector of exogenously determined productive characteristics of land and let $\mathbf{Z}$ be a vector of characteristics that can be produced by landowners at some cost, $C(\boldsymbol{Q}, \boldsymbol{Z}, \boldsymbol{P}_f)$ where $\boldsymbol{P}_f$ is a vector of factor input prices. Following Rosen's original development (1974) as described in Chapter 4, the cost function can be used to derive a family of offer curves for each characteristic.

On the demand side of the property market, there are producers who use land as an input. For ease of exposition, assume that each producer uses one parcel of land. Suppose that each of the n producers has a multi-product production function:

$$\boldsymbol{X}_i \, (\boldsymbol{X}_i, \boldsymbol{F}_i, \boldsymbol{Q}_i, \boldsymbol{Z}_i) \geq 0 \qquad (11\text{-}46)$$

where $\boldsymbol{X}_i$ is the vector of outputs, and $\boldsymbol{F}_i$ is the vector of variable factor inputs. Each producer selects output and input levels and characteristics of the parcel of land he or she rents to maximize variable profits, $\boldsymbol{P} \cdot \boldsymbol{X}_i - \boldsymbol{P}_f \cdot \boldsymbol{F}_i$ subject to the constraints of the production function and the exogenous $\boldsymbol{Q}_i$. The solution to this problem can be expressed as the variable profit function $N(\boldsymbol{P}, \boldsymbol{P}_f, \boldsymbol{Q}_i, \boldsymbol{Z}_i)$. Bid curves for each characteristic can be derived from the variable profit function. An equilibrium for the hedonic rental market is given by the double envelope of the bid and offer curves. The equilibrium is represented by the hedonic rental function, $R(\boldsymbol{Q}_i, \boldsymbol{Z}_i)$. This function can be differentiated to give the marginal implicit prices in equilibrium for all the productive characteristics of land. Inverse demand functions and marginal bid functions (holding profits constant) can be derived for each characteristic. The estimation and identification of these functions are subject to the same kinds of problems discussed in the earlier section on the identification problem, but for present purposes, assume that these problems have been resolved.

Now suppose that there has been an increase in one exogenous characteristic, say q_j. If we assume that input and output prices are unaffected by this improvement, the approaches to welfare measurement discussed in an earlier section can be extended to deal with this problem. In the case where the improvement in the environmental characteristic leads to changes in output prices, there will be further changes in producers' surplus and in consumers' surplus. These must be taken into account in deriving a welfare measure. In such a case it may be more feasible to pursue welfare measurement through the output market, as described in Chapter 9. The derivative of the variable profit function with respect to the price of each output gives the supply function for that output. Changes in any of the characteristics of land shift the supply functions of each of the outputs. The framework developed in Chapter 9 provides a consistent way of deriving welfare measures as areas between shifting supply curves for each of the multiple outputs.

The Repeat-Sales Model

The standard hedonic price model is based on the analysis of a cross section of housing prices at a point in time. The repeat-sales model is a variant of the hedonic price model that exploits a time series of sale prices of houses whose structural characteristics have not changed over time. The repeat-sales model was developed by Palmquist (1982). It substantially simplifies the task of model specification and estimation because characteristics that do not change over time can be omitted from the regression equation.

Suppose for simplicity that houses are not bought and sold but are rented for one-year terms, and suppose that rents, R, just clear the market every year. Suppose for the moment that there is no depreciation of the quality of the housing services provided by each unit of housing and that the overall price level for housing is constant over time. Then, in any particular year, we could estimate a hedonic price function of the form

$$R_i = R(\boldsymbol{S}_i, \boldsymbol{N}_i, \boldsymbol{Q}_i) \tag{11-47}$$

where i indexes dwellings. Assume that there is one environmental attribute, q, which varies across space and also over time at all or most locations. If all the other characteristics in equation 11-47 are constant over time and if the hedonic price function itself is unchanged, then for any two years, t and t', we have

$$R_i' - R_i = R\left(\boldsymbol{S}_i, \boldsymbol{N}_i, \boldsymbol{Q}_i^*, q_i'\right) - R\left(\boldsymbol{S}_i, \boldsymbol{N}_i, \boldsymbol{Q}_i^*, q_i\right) = f\left(q_i' - q_i\right) \tag{11-48}$$

where $\boldsymbol{Q}_i^*$ is all elements of $\boldsymbol{Q}_i$ other than q, and the prime indicates the variable for the second year. All terms involving characteristics other than q cancel out. This is obvious for the linear specification. For the semi-log form,

$$\ln R = a + c \cdot q + \sum_{i=1}^{n} c_i \cdot \boldsymbol{Z}_i \tag{11-49}$$

where $\boldsymbol{Z}_i$ is all other characteristics ($\boldsymbol{S}_i$, $\boldsymbol{N}_i$, and $\boldsymbol{Q}_i^*$), the change in rents is given by

$$\ln R' - \ln R = \ln\left(\frac{R'}{R}\right) = a + c \cdot q \tag{11-50}$$

Even the quadratic Box–Cox specification with all its interaction terms substantially simplifies in the repeat-sales formulation.

The assumption that the hedonic price function is unchanged is problematic. There is nothing in the theory of hedonic price functions that ensures that this assumption will hold. Changes in the level of any characteristic could change the whole hedonic price function. Palmquist (1982, 345) presents evidence that supports the assumption of stability of the hedonic price function over time. However, Edmonds (1985) found that the hedonic price function for housing in Tokyo changed significantly over time.

Unfortunately, application of the repeat-sales model in the real world is not quite as simple as portrayed here because other things change over time. Specifically, houses depreciate with age, and there may be fluctuations in the general price level for houses. Now let us consider a more realistic situation in which the prices of houses, P_h, are observed, houses as assets depreciate over time at a constant percentage rate, d, and the price level for housing in general fluctuates over time.

Let $I_t(t = 0, \ldots, t, \ldots, T)$ be the price index for housing in this housing market, with $T_0 = 1$. For houses having no changes in characteristics (and still assuming that the basic hedonic price function is stable over time), sales prices for houses at different points in time are governed by

$$P_{h_t} = I_t \cdot e^{-d \cdot t} \cdot P_{h_0}(\cdot) \tag{11-51}$$

Because price fluctuations and depreciation cause proportional rather than absolute changes in housing prices, this relationship places restrictions on the way changes in other housing characteristics can be specified, at least if the advantages of the repeat-sales model are to be retained. Specifically, the price changes associated with changes in characteristics over time must also be proportionate, that is, the basic hedonic price function must take the semi-log form:

$$P_{h_t} = e^{\left(a + c \cdot q_{jt} + \Sigma c_k \cdot \mathbf{Z}_k\right)} \tag{11-52}$$

Then for a house that sells twice, at t and t', we have

$$\ln\left(\frac{P_{h_{t'}}}{P_{h_t}}\right) = \ln\left(\frac{I_{t'}}{I_t}\right) e^{-d(t'-t)} \cdot e^{c(q_{t'} - q_t)} \tag{11-53}$$

So, although this model does not allow for flexibility in the functional form of the hedonic equation, it does substantially reduce data costs and chances for misspecification of other housing characteristics.

One potential econometric problem must be noted. Suppose a property is sold n times during the period being studied. This results in $n - 1$ independent pairs of sales prices that can be used as observations in the regression equation. If n is greater than 2, the error terms for different pairs of sales of the same house are correlated because they share a common sale price term. Palmquist (1982) shows that estimation of the repeat-sales model by ordinary least squares results in biased coefficient estimates and shows how this problem can be remedied through use of Aitken's estimator.

In conclusion, the repeat-sales model has the advantages of reducing the data costs and the possibility of specification error by eliminating from the regression equation those structural and other characteristics that do not change over time. However, to use the repeat-sales model it is necessary to assume that the coefficients of the hedonic price function are constant over time. The unavoidable facts of depreciation and fluctuating price levels place restrictions on the way the environmental characteristic of concern can be specified.

Discrete Choice Models

The hedonic price model is based on the assumption that each attribute of the housing bundle is a continuous variable and that an individual can choose any point on the continuous and differentiable hedonic price function in the n-dimensional attribute space. As noted above, this is clearly not a completely realistic assumption, and in some respects it may seriously misrepresent the problem of choosing a bundle of housing attributes. For example, the number of bedrooms in a house is not a continuous variable. There may be no one-bedroom houses on one-acre lots or four-bedroom houses with swimming pool and attached garage on quarter-acre lots. Discrete choice models provide an alternative way of looking at housing choice and inferring values for housing attributes.

Some of the discrete choice models that have been used in the literature focus on the individual's bid function for housing bundles. Such models are based on the probability that an individual will be the highest bidder for a specified bundle of housing attributes. These are known as bid rent models or random bidding models. An alternative approach is to focus on the individual's utility function defined on housing attributes. These models investigate the probability that a specified bundle of housing attributes (including the price of the bundle) will be chosen by the individual, that is, will be revealed to convey the highest level of utility. These are known as random utility models.

Both types of models can be used to derive the marginal bid or MWTP function for individual attributes from an estimate of the bid function or indirect utility function. Thus, in principle, both types of models allow for the calculation of the benefits of changes in an environmental attribute, at least assuming no relocation and no changes in the hedonic price function. Both types of models start with the assumption of utility maximization subject to the standard budget constraint defined by income, prices of market goods, and the hedonic price function.

In the bid rent model, the utility maximization problem is solved to obtain the individual's bid function, that is, the bid as a function of the housing attributes and income, holding utility constant. Then, for an individual of any given income level, the bid can be written as a function of the observable housing attributes plus a random error term reflecting unobserved attributes of either the individual or the housing bundle:

$$B_i = B_i(\mathbf{Z}) + \varepsilon_i \tag{11-54}$$

where, for simplicity, **Z** represents the vector of all housing attributes and location-specific amenities. Because this bid function is derived for a given income level, and all households face the same market prices, these terms can be omitted from that function. For examples of this type of model, see Ellickson 1981; Lerman and Kern 1983; and Gross 1988.

The probability that individual i will choose housing bundle k rather than any of the other available bundles, indicated by k', is

$$P(k|\mathbf{Z}) = P(B_{ik} > B_{ik'}) = P[B_i(\mathbf{Z}_k) - B_i(\mathbf{Z}_{k'}) + \varepsilon_{ik} - \varepsilon_{ik'} > 0] \tag{11-55}$$

If the random error terms are independently and identically distributed with a Type I extreme value distribution, this probability can be written in the logit form:

$$P(k \mid \mathbf{Z}) = \frac{\exp B_{ik}}{\Sigma_k \exp B_{ik}} \tag{11-56}$$

Finally, if the bid function is specified to be linear in the parameters, maximum likelihood methods can be used to estimate these parameters. As Lerman and Kern (1983) and Gross (1988) point out, estimation of equation 11-56 fixes only the slope of the bid function, not its level. However, information on the bids actually paid can be used in the estimation process to fix these values and to make it possible to calculate the marginal bid functions for individual attributes.

In the random utility model, given the multinomial logit specification, the probability that individual i will choose housing bundle k over bundle k' can be written as (following Cropper et al. 1993)

$$P[V_i(\mathbf{Z}_k) - V_i(\mathbf{Z}_{k'}) + (\varepsilon_{ik} - \varepsilon_{ik'})] = \exp\left\{\frac{V_i(\mathbf{Z}_k)}{\Sigma_k \exp\left[V_i(\mathbf{Z}_{k'})\right]}\right\} \tag{11-57}$$

where $V_i(\cdot)$ is the indirect utility function. If $V_i(\cdot)$ is specified to be linear in the parameters, this, too, can be estimated with maximum likelihood methods. Knowledge of the parameters of the indirect utility function makes it possible to compute welfare measures for changes in any of the housing attributes, including an environmental amenity.

In an interesting simulation study, Cropper and her coauthors (1993) have compared estimates of welfare measures derived from a hedonic price model with those from the random utility model for given known household preferences. After simulating an equilibrium in an urban housing market, they used the resulting hedonic price and individual choice data to estimate both a hedonic price model with its marginal bid functions and a random utility model. They then calculated welfare measures for 25% and 100% changes in each of 10 attributes, including both neighborhood attributes and attributes of individual houses (such as number of bathrooms, lot size, and age). They found that the random utility model provided more accurate estimates of the known welfare measure than the hedonic price model. This was true even when they assumed that the researcher did not know the true form of individuals' utility functions. They suggest that the reason for this is the difficulty in identifying and obtaining accurate estimates of the marginal bid functions with the hedonic price model when data are generated by only a single market.

Other researchers (see for example, Bartik and Smith 1987, 1224–1225; Palmquist 1991, 119) have suggested that although one strength of the discrete choice model is its ability to generate welfare measures for nonmarginal changes relatively easily, the model does so because it forces the researcher to make strong assumptions about the functional form of the utility function or bid function. If similar strong assumptions are made about the functional form of the inverse demand functions for attributes in the hedonic model, these functions can be identified too. Cropper and her coauthors (1993) suggest that even when the functional form of preferences is known, the discrete choice model outperforms the hedonic model as a way of measuring welfare change. For recent examples of empirically based efforts to compare welfare measures from standard hedonic models with random utility and random bidding models, see Chattopadhyay 1998, 2000, and Palmquist and Israngkura 1999.

Summary

It is now time to summarize the answers to the two questions raised at the beginning of this chapter. The first question concerned the specification and estimation of a model of housing prices for purposes of valuing environmental amenities. Hedonic price theory provides a coherent basis for explaining the prices of houses in an urban market as a function of the levels of characteristics embedded in each house. The dynamic version of this model relates changes in the prices of houses to changes in the levels of one or more characteristics, other things being equal. The major limitation of the hedonic model is its assumption that consumers of housing can select their most preferred bundle of characteristics from a complete range of levels of all characteristics. However, the discrete choice model can be adapted to deal with this problem. Both the random bidding model and the random utility model can be developed to explain housing market behavior and choices as functions of, among other things, the characteristics of houses.

The second question concerned the derivation of measures of economic value and welfare change for changes in the levels of environmental amenities. Measures of value for marginal and nonmarginal changes can be derived from a properly specified hedonic price model. Values for marginal changes in amenity levels are found simply by adding up the observed or computed MWTPs for all affected individuals. However, for nonmarginal amenity changes, welfare measurement requires knowledge of the inverse demand function or the income-compensated bid function for the amenity. And these, in turn, require a solution to the daunting identification problem.

The discrete choice models provide means for direct estimation of either bid functions or indirect utility functions, provided that data on the socioeconomic characteristics of housing buyers are available. However, they are based on assumptions about the functional form of buyers' utility functions that are not testable.

There are some limitations to the property value models for estimating welfare effects. First, because the property value models are based on the consequences of individuals' choices of residence, they do not capture willingness to pay for improvements in environmental amenities at other points in the urban area, for example, in the work place, shopping areas, or parks and recreational areas. Second, because the property value models are based on observing behavioral responses to differences in amenity levels across houses, they only capture willingness to pay for perceived differences in amenities and their consequences. For example, if there are subtle, long-term health effects associated with reduced environmental quality at some housing sites but people are unaware of the causal link of these effects to the housing site, their willingness to pay to avoid the effects will not be reflected in housing price differences.

On the other hand, property value measures can capture the value of all the possible effects of changes in environmental quality at a housing site in a single number. If air pollution, for example, causes increases in the incidence of respiratory disease, damages ornamental vegetation, and reduces the quality of the view, the property value model summarizes these effects in a single number.

References

Anderson, Robert J., and Thomas D. Crocker. 1972. Air Pollution and Property Values: A Reply. *Review of Economics and Statistics* 54(4): 470–473.

Atkinson, Scott E., and Thomas D. Crocker. 1987. A Bayesian Approach to Assessing the Robustness of Hedonic Property Value Studies. *Journal of Applied Econometrics* 2(1): 27–45.

Bartik, Timothy J. 1987. The Estimation of Demand Parameters in Hedonic Price Models. *Journal of Political Economy* 95(1): 81–88.

———. 1988. Measuring the Benefits of Amenity Improvements in Hedonic Price Models. *Land Economics* 64(2): 172–183.

Bartik, Timothy J., and V. Kerry Smith. 1987. Urban Amenities and Public Policy. In *Handbook of Regional and Urban Economics,* edited by Edwin S. Mills. Amsterdam: Elsevier.

Brown, James N., and Harvey S. Rosen. 1982. On the Estimation of Structural Hedonic Price Models. *Econometrica* 50(3): 765–768.

Cassel, Eric, and Robert Mendelsohn. 1985. The Choice of Functional Forms for Hedonic Price Equations: Comment. *Journal of Urban Economics* 18(2): 135–142.

Chattopadhyay, Sudip. 1998. An Empirical Investigation into the Performance of Ellickson's Random Bidding Model, with an Application to Air Quality Valuation. *Journal of Urban Economics* 43(2): 292–314.

———. 1999. Estimating the Demand for Air Quality: New Evidence Based on the Chicago Housing Market. *Land Economics* 75(1): 22–38.

———. 2000. The Effectiveness of McFadden's Nested Logit Model in Valuing Amenity Improvement. *Regional Science and Urban Economics* 30(1): 23–43.

Colwell, Peter F., and Gene Dilmore. 1999. Who Was First? An Examination of an Early Hedonic Study. *Land Economics* 75(4): 620–626.

Cropper, Maureen L., Leland B. Deck, Nalin Kishor, and Kenneth E. McConnell. 1993. Valuing Product Attributes Using Single Market Data: A Comparison of Hedonic and Discrete Choice Approaches. *Review of Economics and Statistics* 75(2): 225–232.

Cropper, Maureen L., Leland B. Deck, and Kenneth E. McConnell. 1988. On the Choice of Functional Form for Hedonic Price Functions. *Review of Economics and Statistics* 70(4): 668–675.

Edmonds, Radcliffe G., Jr. 1985. Some Evidence on the Intertemporal Stability of Hedonic Price Functions. *Land Economics* 61(4): 445–451.

Ellickson, Bryan. 1981. An Alternative Test of the Hedonic Theory of Housing Markets. *Journal of Urban Economics* 9(1): 56–79.

Epple, Dennis. 1987. Hedonic Prices and Implicit Markets: Estimating Demand and Supply Functions for Differentiated Products. *Journal of Political Economy* 87(1): 59–80.

Freeman, A. Myrick, III. 1971. Air Pollution and Property Values: A Methodological Comment. *Review of Economics and Statistics* 53(4): 415–416.

———. 1974a. Air Pollution and Property Values: A Further Comment. *Review of Economics and Statistics* 56(4): 454–456.

———. 1974b. On Estimating Air Pollution Control Benefits from Land Value Studies. *Journal of Environmental Economics and Management* 1(1): 74–83.

Goodman, Allen C. 1978. Hedonic Prices, Price Indices and Housing Markets. *Journal of Urban Economics* 5(4): 471–484.

Graves, Phil, James C. Murdoch, Mark A. Thayer, and Don Waldman. 1988. The Robustness of Hedonic Price Estimation: Urban Air Quality. *Land Economics* 64(3): 220–233.

Gross, David J. 1988. Estimating Willingness To Pay for Housing Characteristics: An Application of the Ellickson Bid–Rent Model. *Journal of Urban Economics* 24(1): 95–112.

Halvorsen, Robert, and Henry O. Pollakowski. 1981. Choice of Functional Form for Hedonic Price Equations. *Journal of Urban Economics* 10(1): 37–49.

Horowitz, Joel L. 1984. Estimating Compensating and Equivalent Income Variations from Hedonic Price Models. *Economics Letters* 14(4): 303–308.

———. 1986. Bidding Models of Housing Markets. *Journal of Urban Economics* 20(2): 168–190.

———. 1987. Identification and Stochastic Specification in Rosen's Hedonic Price Model. *Journal of Urban Economics* 22(2): 165–173.

Kanemoto, Yoshitsugu. 1988. Hedonic Prices and the Benefits of Public Projects. *Econometrica* 56(4): 981–989.

Kiel, Katherine A., and Jeffrey E. Zabel. 1999. The Accuracy of Owner-Provided House Values: The 1978–1991 American Housing Survey. *Real Estate Economics* 27(2): 263–298.

Lerman, Steven R., and Clifford R. Kern. 1983. Hedonic Theory, Bid Rents, and Willingness-To-Pay: Some Extensions of Ellickson's Results. *Journal of Urban Economics* 13(3): 358–363.

Lind, Robert C. 1973. Spatial Equilibrium, the Theory of Rents, and the Measurement of Benefits from Public Programs. *Quarterly Journal of Economics* 87(2): 188–207.

Mäler, Karl-Göran. 1977. A Note on the Use of Property Values in Estimating Marginal Willingness To Pay for Environmental Quality. *Journal of Environmental Economics and Management* 4(4): 355–369.

McConnell, Kenneth E., and T.T. Phipps. 1987. Identification of Preference Parameters in Hedonic Models: Consumer Demands with Nonlinear Budgets. *Journal of Urban Economics* 22(1): 35–52.

Mendelsohn, Robert. 1984. Estimating the Structural Equations of Implicit Markets and Household Production Functions. *Review of Economics and Statistics* 66(4): 673–677.

———. 1985. Identifying Structural Equations with Single Market Data. *Review of Economics and Statistics* 67(3): 525–529.

———. 1987. A Review of Identification of Hedonic Supply and Demand Functions. *Growth and Change* 18(1): 82–92.

Mendelsohn, Robert, William D. Nordhaus, and Daigee Shaw. 1994. The Impact of Global Warming on Agriculture: A Ricardian Analysis. *American Economic Review* 84(4): 753–771.

Miranowski, John A., and Brian D. Hammes. 1984. Implicit Prices of Soil Characteristics for Farmland in Iowa. *American Journal of Agricultural Economics* 66(5): 745–749.

Murdoch, James C., and Mark A. Thayer. 1988. Hedonic Price Estimation of Variable Urban Air Quality. *Journal of Environmental Economics and Management* 15(2): 143–146.

Niskanen, William A., and Steve H. Hanke. 1977. Land Prices Substantially Underestimate the Value of Environmental Quality. *Review of Economics and Statistics* 59(3): 375–377.

Palmquist, Raymond B.1982. Measuring Environmental Effects on Property Values without Hedonic Regressions. *Journal of Urban Economics* 11(3): 333–347.

———. 1984. Estimating the Demand for Characteristics of Housing. *Review of Economics and Statistics* 64(3): 394–404.

———. 1989. Land as a Differentiated Factor of Production: A Hedonic Model and Its Implications for Welfare Measurement. *Land Economics* 65(1): 23–28.

———. 1991. Hedonic Methods. In *Measuring the Demand for Environmental Improvement*, edited by John B. Braden and Charles D. Kolstad. Amsterdam: North-Holland.

———. 1992a. Valuing Localized Externalities. *Journal of Urban Economics* 31(1): 59–68.

———. 1992b. A Note on Transactions Costs, Moving Costs, and Benefit Measurement. *Journal of Urban Economics* 32(1): 40–44.

———. Forthcoming. Property Value Models. In *Handbook of Environmental Economics*, edited by Karl-Göran Mäler and Jeffery R. Vincent. Amsterdam: North-Holland.

Palmquist, Raymond B., and Leon E. Danielson. 1989. A Hedonic Study of the Effects of Erosion Control and Drainage on Farmland Values. *American Journal of Agricultural Economics* 71(1): 55–62.

Palmquist, Raymond B., and Adis Israngkura. 1999. Valuing Air Quality with Hedonic and Discrete Choice Models. *American Journal of Agricultural Economics* 81(5): 1128–1133.

Parsons, George R. 1990. Hedonic Prices and Public Goods: An Argument for Weighting Locational Attributes in Hedonic Regressions by Lot Size. *Journal of Urban Economics* 27(3): 308–321.

Pines, David, and Yoram Weiss. 1976. Land Improvement Projects and Land Values. *Journal of Urban Economics* 3(1): 1–13.

Polinsky, A. Mitchell, and Steven Shavell. 1976. Amenities and Property Values in a Model of an Urban Area. *Journal of Public Economics* 5(1–2): 119–129.

Portney, Paul R. 1981. Housing Prices, Health Effects, and Valuing Reductions in Risk of Death. *Journal of Environmental Economics and Management* 8(1): 72–78.

Quigley, John M. 1982. Nonlinear Budget Constraints and Consumer Demand: An Application to Public Programs for Residential Housing. *Journal of Urban Economics* 12(2): 177–201.

Ridker, Ronald G. 1967. *Economic Costs of Air Pollution: Studies in Measurement.* New York: Praeger.

Ridker, Ronald G., and John A. Henning. 1967. The Determinants of Residential Property Values with Special Reference to Air Pollution. *Review of Economics and Statistics* 49(2): 246–257.

Rosen, Sherwin. 1974. Hedonic Prices and Implicit Markets: Product Differentiation in Perfect Competition. *Journal of Political Economy* 82(1): 34–55.

Smith, V. Kerry, and Ju Chin Huang. 1995. Can Markets Value Air Quality? A Meta-Analysis of Hedonic Property Value Models. *Journal of Political Economy* 103(1): 209–227.

Sonstelie, Jon C., and Paul R. Portney. 1980. Gross Rents and Market Values: Testing the Implications of the Tiebout Hypothesis. *Journal of Urban Economics* 7(1): 102–118.

Straszheim, Mahlon. 1974. Hedonic Estimation of Housing Market Prices: A Further Comment. *Review of Economics and Statistics* 56(3): 404–406.

Strotz, Robert H. 1968. The Use of Land Value Changes To Measure the Welfare Benefits of Land Improvements. In *The New Economics of Regulated Industries*, edited by Joseph E. Haring. Los Angeles: Occidental College.

Taylor, Laura O. 2003. The Hedonic Method. In *A Primer on Nonmarket Valuation*, edited by Patricia A. Champ, Kevin J. Boyle, and Thomas C. Brown. Dordrecht, Netherlands: Kluwer Academic Press.

Zabel, Jeffrey E., and Katherine A. Kiel. 2000. Estimating the Demand for Air Quality in Four U.S. Cities. *Land Economics* 76(2): 174–194.

CHAPTER

12

Hedonic Wage Models

The hedonic wage model is a formalization of the concept of compensating wage differentials, which can be traced back to Adam Smith. The basic idea is that, other things being equal, workers will prefer jobs with more pleasant working conditions over those that are less pleasant. The greater supply of workers for pleasant jobs will depress the wage levels of such jobs. In equilibrium, the difference in wages between two jobs with different working conditions, will reflect the workers' monetary valuations of the differences in working conditions.

The basic hedonic wage model has been refined and applied empirically to two important questions of particular interest to environmental and resource economists and policymakers. One question concerns the value of reducing the risk of death, injury, or illness. The hedonic wage model has been used to estimate the wage–risk trade-off as a revealed preference measure of this value. The other question concerns the values of the environmental and social amenities that vary across regions. Wage differences across regions have been used as indicators of the values of region-specific environmental, cultural, and social amenities.

From a worker's perspective, a job can be viewed as a differentiated product, that is, a good with a bundle of characteristics such as working conditions, prestige, training and enhancement of skills, and levels of risk of accidental injury and exposure to toxic substances. If workers are free to move from one urban area to another, then jobs are also differentiated, in part, by the environmental and other characteristics of the urban areas in which the jobs are located. If workers are free to choose from a menu of differentiated jobs, the hedonic price technique can be applied to the data on wages, job characteristics (including their locations), and worker characteristics to estimate the marginal implicit prices of these job characteristics.

Employers, from their perspective, can be viewed as choosing from among a set of workers of different characteristics. This is a distinguishing feature of labor markets. In the typical application of the hedonic theory to differentiated goods, producers are viewed as selling a good embodying a package of characteristics and as being indifferent to the characteristics of the purchaser of the good. In hedonic wage studies, the employer is viewed as selling a package of job characteristics (including the quality of the work environment), but at the same time the employer is purchasing work effort and cannot be indifferent to the productive characteristics of the firm's employees. Thus the hedonic wage equation must be interpreted as an equilibrium relationship that reflects not only the interaction of supply and demand for job characteristics but also the interaction of supply and demand of worker characteristics (see Lucas 1977; Rosen 1979). This means that both worker and job characteristics must be included as arguments in the estimated hedonic wage equation.

As in the case of hedonic property values, the derivative of the hedonic wage function with respect to any job characteristic can be interpreted as the marginal implicit price of that characteristic. If the worker is maximizing utility, the marginal implicit price can be taken as an estimate of the worker's marginal willingness to pay for the characteristic. It gives the change in income necessary to just compensate for a small change in the characteristic. Because the hedonic wage function need not be linear, these marginal values may be different for different workers. Similarly, the derivative of the hedonic wage function with respect to any worker characteristic gives its marginal implicit price and, assuming profit maximization, the marginal value of that characteristic to the employer.

To estimate the value of a nonmarginal change in a characteristic, it is necessary to know the compensated inverse demand function for it. As in the case of hedonic property values, the inverse demand function cannot be estimated from data from a single labor market unless additional restrictions are imposed. Some examples of efforts to identify these functions are described in the next section.

The interpretation of the hedonic wage function as revealing marginal implicit prices and marginal values requires that all of the transactions that make up the data be undertaken in the same market. In other words, each buyer (seller) in the market must have had the opportunity to match up with any of the other sellers (buyers) and to choose the most preferred given prices, and so forth. In the terminology of Chapter 11, the market must be in equilibrium and must not be segmented into submarkets with incomplete mobility among segments. When hedonic wage equations are estimated using data from several urban areas, it is necessary to assume that these areas are part of a single market. In practice, labor markets can be segmented on the basis of geography, with moving costs and lack of information on job alternatives imposing barriers between labor markets in different parts of the country. Markets can also be segmented on the basis of education and skill requirements and between blue-collar and professional or managerial workers. Geographic segmentation can lead to different marginal implicit price schedules in different regions. Segmentation on the basis of occupation or education level can lead to different marginal implicit price functions across occupational categories. One approach to the problem of geographic segmentation is to estimate the hedonic wage function only for occupational groups that are believed on a priori grounds to be part of a national labor market. In general, the extent of market segmentation and its significance for empirical estimation of hedonic wage functions are not known.

In the next section of this chapter, I discuss the application of the hedonic wage model to measuring the value of risk reduction and briefly review the results of some of the empirical applications of this model. I then outline and describe some of the models that have been developed to explain and interpret interregional wage differences as reflections of the values people place on regional amenities.

Wage Differences and the Value of Reducing Risks

In Chapter 10, I discussed the concept of the statistical value of life and its role in estimating the value of policies that reduce the risk of death (by reducing, for example, exposures to harmful chemicals). I also presented models that show how to draw inferences about the value of reductions in risk and the statistical value of life from revealed preferences (through the wage–risk trade-off, for example). Here I discuss the use of the hedonic wage model as one approach to estimating the individual's willingness to pay for reductions in risk. Most of the applications of the hedonic model to risk valuation have dealt with risks of death due to accidents on the job.

As in the simple static model in Chapter 10, suppose that each individual chooses a job to maximize expected utility from consumption of the numeraire, X, and from the vector of job characteristics, $\boldsymbol{J}$. In addition to $\boldsymbol{J}$, each job is characterized by its risk of accidental death, δ_i, where i indexes jobs. Individuals face a hedonic wage function that is the locus of points at which firms' marginal wage offers (as functions of job characteristics) equal workers' marginal acceptance wages (see Chapter 4 for a description of equilibrium in hedonic markets). This function is

$$p_w = p_w(\delta, \boldsymbol{J}) \tag{12-1}$$

where p_w is the weekly or monthly wage and where hours of work per period could be one of the characteristics in J. The individual chooses job i to maximize expected utility subject to the wage constraint, that is

$$\max E[u] = \pi \cdot u(X, J) + \lambda\left[p_w(\delta, J) - X\right] \tag{12-2}$$

where π is the probability of surviving the period and being able to consume X. In wage–risk studies risk of death, δ, rather than survival probability, π, is observed. The relationship between the two is given by $\pi = (1 - \delta)(1 - \phi)$, where ϕ is the probability of dying from nonwork-related causes. Because ϕ is usually small for the working-age population, π is approximately equal to $1 - \delta$.

The first-order conditions governing the choices of X and job risk are

$$\pi_i \cdot \left(\frac{\partial u}{\partial X}\right) = \lambda \tag{12-3}$$

$$\frac{u(\cdot)}{\lambda} = \frac{\partial p_w}{\partial \delta_i} \tag{12-4}$$

and

$$\frac{\pi_i \cdot (\partial u / \partial J_i)}{\lambda} = -\frac{\partial p_w}{\partial J_i} \tag{12-5}$$

for all job characteristics, J_i.

From equation 12-3, λ is the expected marginal utility of consumption, which by assumption is positive. According to equation 12-4, the marginal willingness to pay for an increase in the probability of surviving the job risk must equal its marginal implicit price. Equation 12-4 also implies that wages must be lower for jobs that are safer, that is, the marginal implicit price of an increase in π_i is a decrease in the wage rate. Equation 12-5 requires that the marginal willingness to pay for each job characteristic equal its marginal implicit price.

If workers know the relationship between market wages and job attributes and risks, then each worker selects the collection of job attributes and risks that equates the marginal benefit of each attribute to its marginal cost. In the case of risk of death, the marginal cost of working in a less risky job is the lower wage received, $\partial p_w / \partial \delta_i$; this must equal the marginal willingness to pay for lower job risk. In other words, the risk premium associated with a higher risk job must be equal to the individual's marginal willingness to accept compensation for risk.

Estimating Marginal Values for Risk

Data on wages, job attributes, and worker attributes are used to estimate the hedonic wage function—an equilibrium relationship between the wage, on one side, and the job characteristics and variables affecting worker productivity, on the other. If the hedonic wage function can be estimated satisfactorily, we can calculate the risk premium for a marginal change in risk by evaluating the partial derivative of the function at a given risk level and set of job attributes. However, if the wage–risk trade-off locus is nonlinear, this marginal willingness to pay (MWTP) will vary with the baseline risk of each worker. For nonmarginal changes in risk, the value of the change to the individual cannot be calculated from the wage–risk trade-off curve alone because of the convexity of the individual's indifference curve or bid curve. For example, see Figures 4-5 and 4-6.

One of the major questions in interpreting estimates of willingness to pay for risk reduction is whether individuals perceive differences in risks across jobs and if so, whether these perceptions are accurate. Compensating wage differentials for risk can exist in the labor market only if workers perceive differences in risks across jobs. The absence of compensating differentials need not mean that workers do not value reducing the risk of accidental death, only that they are unaware of the differences in risks. If individuals have inaccurate estimates of job risks, then risk premiums can exist. However, these risk premiums will yield biased estimates of individuals' marginal willingness to pay for risk reduction unless the researcher can identify what individuals thought they were buying when they accepted a particular job with its bundle of characteristics. Because most hedonic wage studies use objective measures of job risk, it is important to find out if individuals' perceptions correspond well with these objective measures.

The only evidence pertaining directly to individuals' perceptions of job-related risks is found in research that compares workers' risk perceptions with data on frequency of job-related death and injury. This evidence suggests that workers' perceptions are positively correlated with objective risks but may overstate them. Viscusi (1979) reported a positive correlation between a dichotomous risk variable (posed as a question, "Is your job dangerous?") and accident rate data from the Bureau of Labor Statistics (BLS). In a subsequent study, Viscusi and O'Connor (1984) reported that workers in the chemical industry perceived risk of injury on their job to be about 50% higher than BLS estimates. Gerking and others (1988) also found that workers' perceptions of risk of death on the job overstated BLS accidental death rates.

Other studies providing evidence on the accuracy of risk perceptions compare relative frequency of deaths, by cause, with individuals' perceptions of these frequencies. Slovic and others (1979) found that, on average, people overestimate the likelihood of infrequent causes of death (deaths due to botulism, floods, and tornadoes) but underestimate the probability of deaths with higher frequencies (deaths due to heart disease or cancer). However, Fischhoff and coauthors (1981) note that one must distinguish between an individual's perception of the relative frequency of death in some population and the individual's estimate of his or her *own* risk of death. There is evidence that the latter is often underestimated (see Fischhoff et al. 1981; Hamermesh 1985). What matters in assessing the evidence provided by the hedonic wage model is the accuracy of people's perceptions of the *differences* in risks among different jobs, not the absolute level of these risks. I am not aware of any evidence on this point, and the lack of evidence makes it hard to reach firm conclusions about the validity of studies that use objective rather than subjective risk estimates. In future research it will be important to measure each respondent's perceptions of his or her own risk of death.

In practice, the biggest difficulties in estimating hedonic wage functions have come from lack of sufficiently accurate and detailed data on job-related risk of death and injury and from lack of adequate data on other job and worker characteristics, especially those that may be correlated with risk of accidental death. See Viscusi 1993 for a discussion of this and other data- and model-specification issues.

Identifying Marginal Willingness-To-Pay Functions

We know that the partial derivative of the hedonic wage function with respect to each characteristic is its marginal implicit price and that in equilibrium we can take the marginal implicit price to be a point estimate of each worker's marginal willingness to pay for that characteristic. However, as I explained in Chapter 11, a second stage of analysis is required to identify the MWTP function for each characteristic. In contrast to the empirical analysis of property values, there have been very few efforts to identify the willingness-to-pay (WTP) function for job characteristics.

Two recent studies have dealt with the identification problem in somewhat different ways. In both cases, the authors imposed structure on the problem by making explicit assumptions about the form of the underlying utility functions. In one study (Biddle and Zarkin 1988), identification of the MWTP function for reducing risks of an accidental injury on the job was achieved by an instrumental variables approach. The second study (Viscusi and Moore 1989) is noteworthy for two reasons. The first is that the authors identified the MWTP function for reduced risks of death by making use of interregional variation in the hedonic wage function. To identify the parameters of the marginal bid function, the marginal price of job risk was estimated for different regions of the United States, thus ensuring variation in marginal price that is independent of the variables entering the marginal bid function. The second reason is that the authors explicitly took into account the age of each worker and the number of expected life years at risk.

Viscusi and Moore specified a lifetime utility maximization problem, which was a simplification of equation 10-18 (in Chapter 10) in this book in that the exogenous probability of death was constant across all time periods. This assumption made it possible to derive a simplified expression for the first-order conditions and the MWTP for any given explicit utility function. The model also made it possible to derive estimates of the implied discount rate on life years. Viscusi and Moore used data from the 1982 wave of the University of Michigan Panel Study of Income Dynamics, supplemented by National Traumatic Occupational Fatality data published by NIOSH. Depending upon the specification of the utility function in the estimation technique, the implied discount rate ranged from 11% to 17% for the sample as a whole. See also Moore and Viscusi 1990.

Conclusions

The finding of a positive wage–risk premium is quite robust across data sets, time periods, theoretical models, and even countries (Kniesner and Leeth 1991; Liu et al. 1997; Shanmugam 2000). It appears that (1) workers perceive differences in risks across jobs; (2) these perceptions are correlated with objective measures of on-the-job risks; and (3) workers prefer jobs with lower risks, other things being equal, and are willing to pay for safety in the form of reduced wages. In assessing the accuracy of empirical estimates of the value of risk reduction, there are still questions about the accuracy of workers' perceptions of risks, the accuracy of the available risk data that are aggregated over occupational groupings or industry categories, and the ability to control for other individual and job-related determinants of wage differentials. It has also become clear that there is no one single value for risk reduction; rather, individuals' values depend upon such things as age and income. Understanding these individual determinants of behavior toward risk and the preferences and values that lie behind them is a rich field for further theoretical and empirical research.

All these studies share the common characteristic of assuming that the location of the job is unimportant; that is, they do not control for differences in urban amenities and the potential effects of amenities on wage levels across cities. However, in a series of papers Smith and Gilbert have included location-specific characteristics such as air pollution levels in their hedonic wage equations; see Smith 1983 and Smith and Gilbert 1984, 1985. The modeling of determinants of interurban wage differentials is discussed in the next section.

Interurban Wage Differences and the Value of Amenities

Those cities that are more desirable places to live and work will attract workers from less desirable cities and regions. The in-migration of labor will exert downward pressure on wage rates in the desirable city. An equilibrium occurs when wages have fallen to the point where the marginal worker is indifferent between moving to this city and staying in his or her next best alternative location. The difference in wages between this city and the next-best alternative is a compensating wage differential.

The possibility that such compensating differentials could be used as estimates of the monetary value of amenity differences spawned a series of empirical studies during the 1970s (Nordhaus and Tobin 1972; Hoch and Drake 1974; Hoch 1977; Meyer and Leone 1977). In these studies, measures of average wage rates, earnings, or income were regressed on variables reflecting such things as climate (e.g., temperature, humidity, frequency of rain), environmental quality (say, measures of air pollution, water pollution, and access to recreational resources such as beaches), cultural amenities (for example, numbers of museums, newspapers, and universities), the disamenities of urban life (e.g., crime rates, population density), and city size itself. Some of these studies controlled for differences in the occupational structure of urban labor forces by estimating separate equations for individual occupations. Others used more aggregated measures such as average earnings. In none of these early studies was the estimating equation derived from a formal model of individual choice, interurban migration, or supplies and demands of labor in a system of interconnected urban labor markets.

Rosen (1979) was apparently the first to attempt to provide a formal model for deriving the structural equations relating wages to urban amenities and disamenities. Such a formal model is necessary to provide a welfare theoretic interpretation of regressions that explains wage differences across cities. Rosen pointed out that there are really two hedonic markets in which individuals are making choices—one for labor and one for land or housing. A decision to work in one city is also a decision to purchase housing in that city. As individuals are drawn toward the more desirable cities and push wages down in those cities, they also push out the demand for land and housing, thus increasing the prices of land and housing. Not only are there compensating wage differentials, then, but there are also compensating land rent and housing price differentials across cities. The labor market model must also provide a coherent explanation for why firms in some cities can pay higher wages and still compete in markets for goods traded among cities. A formal model of both sets of markets is required to draw inferences about amenity values and WTPs from data on wage differentials and housing prices.

A number of such formal models have now been presented in the literature. See, for example, Rosen (1979), Cropper and Arriaga-Salinas (1980), Cropper (1981), Henderson (1982), Hoehn and others (1987), Blomquist and others (1988), and Roback (1982, 1988). Bartik and Smith (1987) review most of these models, discuss similarities and differences in their theoretical structures, and describe some of the major empirical results. See also Palmquist 1991. These models share the common feature of explicitly dealing with the interaction between markets for labor and for land across urban areas. However, they differ in how they model specific features of the determination of the interurban equilibrium. For example, some models treat the size of each city as fixed, whereas others allow the city boundary to expand to accommodate higher populations. Some models treat the costs of firms as exogenous, whereas other models allow costs to be affected by environmental amenities and the population of the city itself. Some models treat amenity levels as uniform within each city, whereas others allow for variation in amenity levels within as well as among urban areas.

Models that allow all the relevant variables (such as city size, firm location, and cost) to be determined endogenously are analytically intractable. As a consequence, the results of these models depend upon the specific features of the model. None of these models fully captures all the complexity of the general interurban equilibrium. The models produce what Bartik and Smith have termed "partial descriptions" of the more complex reality (1987, 1232).

In this section I do not try to explain all the models, with their different results. Instead, I first employ a simple model of individual choice to show how the interaction between wages and rent or housing price affects our ability to interpret regression coefficients as welfare measures. Then I sketch out one version of a simple model of the equilibrium of the land and labor markets that permits the estimation of a marginal amenity value. I end this discussion by examining in a more qualitative manner the implications of constructing richer models with more endogenous variables and more interactions.

The Welfare Measure with Two Hedonic Markets

Most of the early studies of interurban wage differences were based either implicitly or explicitly on the assumption that wage differences measure the values of amenity differences. For example, Nordhaus and Tobin (1972) and Meyer and Leone (1977) used their results to make adjustments to the national income accounts for nonpriced positive and negative welfare effects of urbanization. More recently Clark and Kahn (1989) interpreted their coefficient on an interurban wage equation as a marginal implicit price and used it in a second-stage estimation of a marginal bid function for an amenity. However, a very simple model of individual choice of a city to live and work in can be used to show that the assumption that wage differences measure amenity values is not valid. This is because the wage differential is also affected by an interaction between the markets for labor and land.

To see this, consider the simplest case of an individual who derives utility from the consumption of a numeraire good, X, the quantity of housing consumed, h, and the level of an urban-specific amenity, q. By selecting a city to live in, the individual determines his or her annual wage, p_w, and the level of the urban amenity. Assume that q_i is the same at all locations within city i. Let X be normalized with a price of 1 that is constant across cities because X is a nationally traded good. The individual selects q_i along with X and h to maximize

$$u = u(X, h, q_i) + \lambda[p_w(q_i) - p_h(q_i) \cdot h - X] \tag{12-6}$$

Both wages and the price of housing vary across cities according to the level of the amenity in each city. The intuition for this follows: If one city is relatively more desirable as a place to live and work, other things being equal, workers will move to that city; the increase in the supply of labor will push down wages in that city; and, because people must live in the same city in which they work, the increased demand for housing will bid up its price. Of course, a complete model must aggregate across individuals as well as specify the supply sides of these markets to solve for the wage and housing price equations. I return to this point below.

The first-order conditions for the individual choice problem are

$$\frac{\partial u}{\partial X} = \lambda \tag{12-7}$$

$$\frac{\partial u}{\partial h} = \lambda \cdot p_h \tag{12-8}$$

$$\frac{\partial u}{\partial q} = \lambda\left(h \cdot \frac{\partial p_h}{\partial q} - \frac{\partial p_w}{\partial q}\right) \tag{12-9}$$

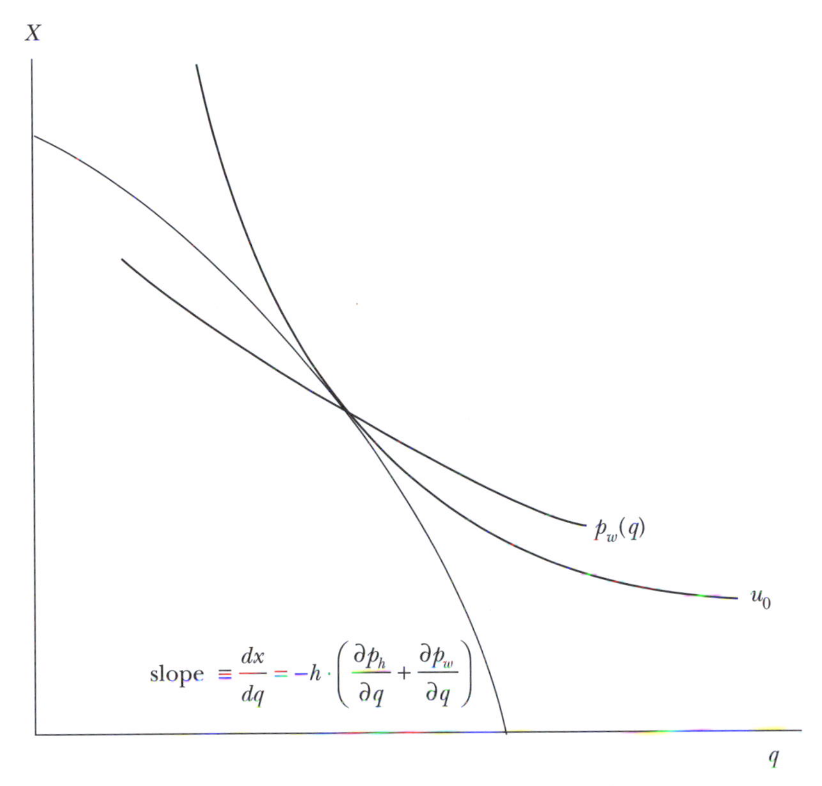

Figure 12-1. The individual's choice of an urban amenity

Using the first-order conditions, the individual's optimum choice of X and q can be found. Graphically, the optimum combination is at the tangency between the individual's indifference curve for X and q and the budget line giving the terms at which X and q can be exchanged, both holding the quantity of housing h constant. See Figure 12-1. The marginal rate of substitution (MRS) between X and q is

$$\text{MRS}_{Xq} = \frac{(\partial u / \partial q)}{(\partial u / \partial X)} = -\frac{dX}{dq} \Big| \, u^0 \tag{12-10}$$

The slope of the budget line is found by taking the total differential of the budget constraint and treating h as a constant. The slope is

$$\frac{dX}{dq} = \frac{\partial p_w}{\partial q} - h \cdot \frac{\partial p_h}{\partial q} \tag{12-11}$$

The marginal rate of substitution can also be interpreted as the marginal willingness to pay for q or w_q. In equilibrium, w_q can be inferred from knowledge of the slope of the budget line:

$$w_q = h \cdot \frac{\partial p_h}{\partial q} - \frac{\partial p_w}{\partial q} \tag{12-12}$$

This MWTP has two components. The first component is the willingness to spend more on housing in a higher quality city with a higher price for housing; the second is the willingness to accept a lower wage in the higher quality city.

This analysis makes it clear that knowledge of the hedonic wage gradient across cities is not sufficient for inferring marginal amenity values. The hedonic wage gradient is also shown in Figure 12-1. The budget line is more steeply sloped than the wage gradient because it includes the negative term for the effect of q on the price of housing. This means that the marginal implicit price of q in the labor market is an underestimate of the individual's marginal valuation of q when q is purchased simultaneously through two hedonic markets.

A General Model of Interurban Equilibrium

To estimate a welfare measure such as equation 12-12, it is necessary to develop a formal model of the interurban equilibrium so that the hedonic wage and housing price equations can be properly specified. The model I present here does not capture all the complexities of the interurban location equilibrium problem, but it does show the general features of most of the models that have been presented in the literature. This model is patterned most closely after those of Hoehn and others (1987) and Roback (1982). It incorporates the following features:

- It takes account of the variation in housing prices within cities caused by the spatial character of a city. Specifically, because land rents vary inversely with commuting costs, distance from the city center is an argument in the rent function.
- It treats the size of a city as endogenous. As more desirable cities draw workers to them and the price of housing is bid up, it becomes profitable to convert some of the surrounding agricultural land to housing. This helps to dampen the upward pressure on housing prices.
- It treats the number of firms in each city as endogenous. Firms' costs depend on the level of one or more amenities and disamenities and, perhaps, on city size to reflect agglomeration effects.

In this model I first assume that all individuals have identical preferences and identical endowments of wealth, which in turn are assumed to be zero for simplicity. From my initial assumption it follows that an equilibrium must be characterized by equal levels of utility for all individuals. Each individual's preferences are represented by the utility function

$$u = u(X, s, q) \tag{12-13}$$

where X is the numeraire good traded in a "world" market at a price of 1, s is the quantity of land occupied by the individual for housing, and q is the amenity that varies across cities but is uniform within each city. Let there be n cities in this economy. Each city offers a wage of p_{w_i} and an amenity level q_i, $i = 1, \ldots, i, \ldots, n$.

Each city also has the standard circular form in which all jobs are located at the city center. Individuals choose a residential location at distance d and incur commuting costs of $t \cdot d$ per period, where t is the unit per period transportation cost. To compensate for higher commuting costs, the price of land p_s will decline with increasing distance from the city center. The boundary of the city will be at distance d^* where in equilibrium $p_s(d^*)$ is equal to p_{s^*}, the rental price of undeveloped or agricultural land.

Each individual chooses a wage and amenity package by selecting a city in which to live and work, a location for his or her residence, a quantity of residential land, and a quantity of the numeraire. Formally, each individual's choice problem is to maximize

$$u = u(X, s, q) + \lambda[p_w(q_i) - t \cdot d - p_s(d) \cdot s - X] \tag{12-14}$$

The solution to this problem gives the indirect utility function:

$$u = v[p_w(q_i) - t \cdot d, p_{s_i}(d), q_i] \tag{12-15}$$

An equilibrium is achieved when wages and land prices have adjusted so that all individuals are indifferent as to the choice of a city and their location within the city chosen. From equation 12-15, this equilibrium must satisfy

$$v\left[p_w(q_i) - t \cdot d,\ p_{s_i}(d),\ q_i\right] = v^* = v\left[p_w(q_i) - t \cdot d_i^*,\ p_{s^*},\ q_i\right] \tag{12-16}$$

The left-hand side of equation 12-16 can be solved for each city's land price schedule, $p_{s_i}(d, q_i, t, v^*)$. The right-hand side can be solved for the city's physical size, $d_i^*(t, p_{s^*}, q_i, v^*)$. The application of Roy's Identity to equation 12-16 gives the individual's demand function for land,

$$s = -\frac{(\partial v / \partial p_s)}{(\partial v / \partial p_w)} = a(p_w - t \cdot d,\ p_{s_i},\ q_i) \tag{12-17}$$

Because the supply of land at any distance is given by $2 \cdot \pi \cdot d$, the total number of people that any city can accommodate is

$$P = \int_0^{d^*} \frac{2 \cdot \pi \cdot d}{d(\cdot)} \cdot dd \tag{12-18}$$

The model must be closed by specifying the production sector of the economy, which also determines the demand for labor. Assume that the industry producing good X is competitive and subject to constant returns to scale. Then price, which has been normalized at 1, equals average cost as well as marginal cost. If firms purchase materials and capital at fixed world prices, then some economic characteristic of each city must be an argument in the firms' cost functions; otherwise, all firms would be forced by competition to pay the same wage. Hoehn and others (1987) assume that q and total population are arguments in the cost function, the latter through an agglomeration effect. Alternatively, Roback (1982) assumes that firms use land in production, so that p_s is an argument in the cost function, along with q.

This system of equations can be solved to determine p_w and P for each city and p_s and d for each individual within the city. These variables depend upon the exogenous q_i, t, and p_{s^*}, which are the variables included in hedonic wage and land price equations. Because of all the interdependencies, the comparative statics of this type of model can be fairly complex. Hoehn and others (1987) analyzed the comparative statics of their model with respect to changes in q. The results depend on how q and P affect the unit cost function.

To derive the marginal willingness to pay for q, we return to equation 12-15, take its total differential, and set it equal to zero, obtaining

$$du = \frac{\partial v}{\partial p_w} \cdot dp_w + \frac{\partial v}{\partial p_s} \cdot dp_s + \frac{\partial v}{\partial q} \cdot dq = 0 \tag{12-19}$$

and

$$w_q \equiv \frac{(\partial v / \partial q)}{(\partial v / \partial p_w)} = -\frac{(\partial v / \partial p_s)}{(\partial v / \partial p_w)} \cdot \frac{dp_s}{dq} - \frac{dp_w}{dq} \tag{12-20}$$

Again, employing Roy's Identity, this becomes

$$w_q = s \cdot \frac{dp_s}{dq} - \frac{dp_w}{dq} \qquad (12\text{-}21)$$

This expression is similar to the one derived above from the model of individual choice; see equation 12-12. Here, each individual's marginal willingness to pay for q consists of two components, the change in the expenditure on land associated with an increase in q and the willingness to accept a lower wage rate for an improvement in q. The comparative static analysis of Hoehn and others (1987) shows that it is possible for dp_s/dq to be negative or for dp_w/dq to be positive. Although the level of q is uniform within each city, individuals can have different MWTPs for q, depending upon the other elements in their consumption bundle, in particular, the quantity of land.

Other models of the interurban equilibrium have been developed to examine other forms of interaction. The additional features that can be captured in these models include the following:

- Intraurban variation in amenity levels. Some amenities and disamenities that vary on average across cities also vary systematically within each city. For example, crime rates and air pollution tend to be higher in the center of each city than at the city boundary. This variation in amenity levels within cities will affect the spatial pattern of land rents and housing prices within each city; see Cropper 1981.
- The existence of both traded and nontraded goods. With zero transportation costs, traded goods sell at the same prices in all cities. However, the prices of nontraded goods can vary across cities, leading to differences in the cost of living. The explicit treatment of nontraded goods is necessary to determine how cost of living differences should be treated in the specification of the hedonic wage function. Cropper (1981) showed that housing prices should not be included in the index of prices used as an argument in hedonic wage functions.
- Variation in the cost of supplying housing across cities. Although the simpler urban models are formulated in terms of land rent, housing prices are easier to observe for purposes of hedonic price estimation. Because the concept of interurban equilibrium is inherently long-run, it is necessary to model the supply side of the housing market. The cost of producing housing will depend on, among other things, the price of land (which varies within and across cities) and the wage rate (which also varies across cities); see, for example, Roback 1982 and Hoehn et al. 1987.

Summary

In this chapter we have examined two types of application of hedonic price theory to labor markets. In one application, the focus is on the intrinsic characteristics of jobs. These applications seek to infer willingnesses to pay for changes in characteristics such as risk of accidental death or injury on the job. In these applications, it is usually assumed that the location of the job is unimportant. An exception is the work of Smith and Gilbert (1984, 1985) described above. In some wage–risk studies, regional dummy variables may be included to control for differences in wage levels across broad regions of the country, but the location of the job and the amenities or disamenities that go with living near the job are not explicitly considered.

In the other application, the focus is on the choice of a city in which to live and work and, in some cases, where in the city to live. Jobs are treated as homogeneous, except for their location. These applications seek to infer willingness to pay for urban amenities. In some of these applications, job characteristics are controlled to some extent by examining interurban differences within broad occupational categories. A key feature of applications of this kind is the need to model explicitly the interaction between the two hedonic markets—those for labor and for land or housing. Introducing heterogeneous jobs into this interurban labor market does not require adding another hedonic market into the analysis. However, it does require a more detailed treatment of the determinants of wage differences within a multicity, multicharacteristic market for differentiated labor. This is an important subject for further theoretical and empirical research.

References

Bartik, Timothy J., and V. Kerry Smith. 1987. Urban Amenities and Public Policy. In *Handbook of Regional and Urban Economics,* edited by Edwin S. Mills. Amsterdam: Elsevier.

Biddle, Jeff E., and Gary A. Zarkin. 1988. Worker Preferences and Market Compensation for Job Risk. *Review of Economics and Statistics* 70(4): 660–667.

Blomquist, Glenn C., Mark C. Berger, and John P. Hoehn. 1988. New Estimates of Quality of Life in Urban Areas. *American Economic Review* 78(1): 89–107.

Clark, David E., and James R. Kahn. 1989. The Two-Stage Hedonic Wage Approach: A Methodology for the Evaluation of Environmental Amenities. *Journal of Environmental Economics and Management* 16(2): 106–120.

Cropper, Maureen L. 1981. The Value of Urban Amenities. *Journal of Regional Science* 21(3): 359–373.

Cropper, Maureen L., and A.S. Arriaga-Salinas. 1980. Inter-City Wage Differentials and the Value of Air Quality. *Journal of Urban Economics* 8(3): 236–254.

Fischhoff, Baruch, Sarah Lichtenstein, Paul Slovic, Stephen L. Derby, and Ralph L. Keeney. 1981. *Acceptable Risk.* New York: Cambridge University Press.

Gerking, Shelby, Menno de Haan, and William Schulze. 1988. The Marginal Value of Job Safety: A Contingent Valuation Study. *Journal of Risk and Uncertainty* 1(2): 185–199.

Hamermesh, Daniel S. 1985. Expectations, Life Expectancy, and Economic Behavior. *Quarterly Journal of Economics* 100(2): 389–408.

Henderson, J. Vernon. 1982. Evaluating Consumer Welfare Amenities and Interregional Welfare Differences. *Journal of Urban Economics* 11(1): 32–59.

Hoch, Irving. 1977. Variations in the Quality of Urban Life among Cities and Regions. In *Public Economics and the Quality of Life*, edited by Lowdon Wingo and Alan Evans. Baltimore, MD: The Johns Hopkins University Press for Resources for the Future and the Centre for Environmental Studies.

Hoch, Irving, and Judith Drake. 1974. Wages, Climate, and the Quality of Life. *Journal of Environmental Economics and Management* 1(3): 268–295.

Hoehn, John P., Mark C. Berger, and Glenn C. Blomquist. 1987. A Hedonic Model of Interregional Wages, Rents, and Amenity Values. *Journal of Regional Science* 27(4): 605–620.

Kniesner, Thomas J., and John D. Leeth. 1991. Compensating Wage Differentials for Fatal Injury Risk in Australia, Japan, and the United States. *Journal of Risk and Uncertainty* 4(1): 75–90.

Liu, Jin-Tan, James K. Hammitt, and Jin-Long Liu. 1997. Estimated Hedonic Wage Function and the Value of Life in a Developing Country. *Economics Letters* 57(3): 353–358.

Lucas, Robert E.B. 1977. Hedonic Wage Equations and Psychic Wages in the Returns to Schooling. *American Economic Review* 67(4): 549–558.

Meyer, John R., and Robert A. Leone. 1977. The Urban Disamenity Revisited. In *Public Economics and the Quality of Life*, edited by Lowdon Wingo and Alan Evans. Baltimore, MD: The Johns Hopkins University Press for Resources for the Future and the Centre for Environmental Studies.

Moore, Michael J., and W. Kip Viscusi. 1990. Models for Estimating Discount Rates for Long-Term Health Risks Using Labor Market Data. *Journal of Risk and Uncertainty* 3(4): 381–401.

Nordhaus, William D., and James Tobin. 1972. Is Growth Obsolete? In *Economic Research: Retrospect and Prospect, Economic Growth*, Fiftieth Anniversary Colloquium series, vol. 5. New York: National Bureau of Economic Research.

Palmquist, Raymond B. 1991. Hedonic Methods. In *Measuring the Demand for Environmental Quality*, edited by John B. Braden and Charles D. Kolstad. Amsterdam: North-Holland.

Roback, Jennifer. 1982. Wages, Rents, and the Quality of Life. *Journal of Political Economy* 90(6): 1257–1278.

———. 1988. Wages, Rents, and Amenities: Differences among Workers and Regions. *Economic Inquiry* 26(1): 23–41.

Rosen, Sherwin. 1979. Wage-Based Indices of Urban Quality of Life. In *Current Issues in Urban Economics*, edited by Peter Mieszkowski and Mahlon Straszheim. Baltimore, MD: The Johns Hopkins University Press.

Shanmugam, K.R. 2000. Valuation of Life and Injury Risks. *Environmental and Resource Economics* 16(4): 379–389.

Slovic, Paul, Baruch Fischhoff, and Sarah Lichtenstein. 1979. Rating the Risks. *Environment* 21(3): 14–20, 36–39.

Smith, V. Kerry. 1983. The Role of Site and Job Characteristics in Hedonic Wage Models. *Journal of Urban Economics* 13(4): 296–321.

Smith, V. Kerry, and Carol C.S. Gilbert. 1984. The Implicit Valuation of Risks to Life: A Comparative Analysis. *Economics Letters* 16(3–4): 393–399.

Smith, V. Kerry, and Carol C.S. Gilbert. 1985. The Valuation of Environmental Risks Using Hedonic Wage Models. In *Horizontal Equity, Uncertainty, and Economic Well-Being*, edited by Martin David and Timothy Smeeding. Chicago: University of Chicago Press.

Viscusi, W. Kip. 1979. *Employment Hazards: An Investigation of Market Performance.* Cambridge, MA: Harvard University Press.

———. 1993. The Value of Risks to Life and Health. *Journal of Economic Literature* 31(4): 1912–1946.

Viscusi, W. Kip, and Michael J. Moore. 1989. Rates of Time Preference and Valuation of the Duration of Life. *Journal of Public Economics* 38(3): 297–317.

Viscusi, W. Kip, and Charles J. O'Connor. 1984. Adaptive Responses to Chemical Labeling: Are Workers Bayesian Decision Makers? *American Economic Review* 74(5): 942–956.

CHAPTER

13

Recreational Uses of Natural Resource Systems

Many natural resource systems such as lakes, rivers and streams, estuaries, and forests are used extensively by people for various kinds of recreation activities, including fishing, hunting, boating, hiking, and camping. As places to conduct such activities, natural resource systems provide valuable services to people. From an economic perspective, these services have two important features. The first is that the economic value of these services depends upon the characteristics of the natural resource system. The characteristics determining value can be affected by air and water pollution and by resource management decisions about such things as the rates of harvest of timber and fish, the extraction of minerals and petroleum, and the allocation of water flows between diversionary uses and various in-stream uses. Knowledge of the values of these services may be important for a variety of resource management decisions.

The second important feature is that access to the resource for recreation is typically not allocated through markets. Rather, access is typically open to all comers at a zero price or a nominal entrance fee that bears no relationship to the cost of providing access. There is little or no variation in these access prices over time or across sites to provide data for econometric estimation of demand functions.

In this chapter I address three questions. First, how is the value of the flow of recreation services from an existing natural resource system defined and measured? Second, how is the value of introducing a new recreation site estimated ex ante? And third, how can we estimate the value of changes in the quality of a recreation site or changes in the quality of the flow of recreation services from the system?

The value of the services from a recreation site is the sum of the willingnesses to pay (WTPs) for these services of all the affected people. On the assumption that their WTP is for use values (see Chapter 5), the value of the site is the area under the aggregate compensated demand curve for visits to that site. Similarly, the value of adding a site to the existing supply of sites is the value of reducing the price of a visit to the new site from one that is so high that no one visits the site to whatever the actual price of visits to that site will be. Following the argument of weak complementarity, the value of a change in the quality of the site (positive or negative) is the area between the old and new compensated demand curves for visits to the site (see Chapter 4). So, for all three questions, we need information on the demand for visits to the site in question.

All the models for estimating demands for visits take advantage of the fact that each visit to a recreation site involves an implicit transaction in which the cost of traveling to the site is incurred in return for access to the site. Different individuals will face different travel costs to any site, and one individual will face different travel costs for the different sites he or she might visit. The responses of people to this variation in the implicit prices of visits are the basis for estimating the values of recreation sites and changes in site quality.

The choices that people make in response to differences in travel costs have usually been modeled from one of two perspectives regarding time. In one perspective, there is a substantial period such as a season or a year during which people can make several trips, perhaps to more than one site. People choose the number of trips to make to each site. In the other perspective, there is a time at which people decide whether to visit any site, and if so, which site to visit.

In models using the first perspective, variation in the cost of traveling to the site (the implicit price of access) can be exploited to estimate a travel cost demand function. The value of the flow of services from the site is the area under the compensated demand curve for these services or for access to the site, aggregated over all who visit the site. The value of a change in the site's services can be found by invoking the concept of weak complementarity and estimating the area between two compensated demand curves that are conditioned on different levels or qualities of services at the site.

Models using the second perspective are a form of discrete choice or random utility model (RUM). Welfare values for individuals are calculated from the parameters of the indirect utility function, which is estimated from observations of individual choices. Because the characteristics of sites are arguments in the indirect utility function, it is a straightforward matter to calculate welfare values for changes in site characteristics or qualities.

In the first section of this chapter I outline the basic travel cost model of the demand for a single site, and I explain how this model can be used to value recreation sites and changes in the characteristics of sites. In the second section I discuss the random utility model of recreation choice and show how it can be used to estimate the value of adding or subtracting sites from the set of available sites and to value changes in site attributes. In the third section I discuss a major issue that must be addressed in implementing and estimating either model, that is, the measurement of travel cost and especially the cost of time. In the final section, I examine the hybrid hedonic travel cost model of recreation demand and other approaches to estimating recreation values.

Parsons (2003) is another introduction and overview of recreation demand modeling. This work emphasizes the econometric specification and estimation of recreation demand models. Phaneuf and Smith (forthcoming) provide a more rigorous treatment of modeling issues and the estimation of welfare measures. Finally, the papers collected in Herriges and Kling 1999 deal in more detail with a number of specific topics in recreation demand analysis, especially with regard to RUMs.

The Travel Cost Model of Recreation Demand

Although the demand for a site can be modeled as an aggregate or market demand, the usual practice is to estimate demand functions at the level of the individual and to calculate site values by adding up individuals' values for the site. As noted, the value of a recreation site to an individual is defined as the area under the individual's compensated demand curve for that site. However, the compensated demand may not be observable, so the usual practice is to begin with the ordinary demand curve for the site. The individual's ordinary demand curve gives the number of visits an individual of given socioeconomic characteristics makes to the site as a function of the price or cost of a visit to the site. This price includes the admission fee to the site. At any time, all individuals will face the same admission fee or price of access. This fee typically does not change over time. Thus estimation of individuals' demand functions may not be feasible from the available data on admission fee and number of visits. The travel cost model is based on the recognition that the cost of traveling to a site is an important component of the full cost of a visit and that, for any given site, there will usually be wide variation in travel cost across any sample of visitors to that site.

The Basic Model

To formalize this insight, let us look at a model of an individual's choice of the number of visits to make to a recreation site. This model is adapted from McConnell (1985). Assume for now that there is only one site available and that all visits have the same duration. Modeling of the choice of the length of a visit to a site and of the choice of a site or sites to visit when there are alternatives will be taken up in later sections of this chapter. We assume that the individual's utility depends on the total time spent at the site, the quality of the site, and the quantity of a numeraire. With the duration of a visit fixed for simplicity, the time on site can be represented by the number of visits. The individual solves the following utility maximization problem:

$$\max: u(X, r, q) \tag{13-1}$$

subject to the twin constraints of monetary and time budgets:

$$M + P_w \cdot t_w = X + c \cdot r \tag{13-2}$$

and

$$t^* = t_w + (t_1 + t_2)r \tag{13-3}$$

where

X = the quantity of the numeraire whose price is 1,
r = number of visits to the recreation site,
q = environmental quality at the site,
M = exogenous income,
p_w = wage rate,
c = monetary cost of a trip,
t^* = total discretionary time,
t_w = hours worked,
t_1 = round-trip travel time, and
t_2 = time spent on site.

Assume that r and q are complements in the utility function. As discussed in Chapter 4, this means that the number of visits will be an increasing function of the site's environmental quality. The time constraint reflects the fact that both travel to the site and time spent on the site take time away from other activities. Thus there is an opportunity cost to the time spent in the recreation activity. Assume also that the individual is free to choose the amount of time spent at work and that work does not convey utility (or disutility) directly. Thus the opportunity cost of time is the wage rate. I return to the nature of the time constraint and the opportunity cost of time in a later section. Finally, assume that the monetary cost of a trip to the site has two components: the admission fee, f, which could be zero, and the monetary cost of travel. This cost is $p_d \cdot d$, where p_d is the per-mile cost of travel and d is the round-trip distance to the site.

Substituting the time constraint (equation 13-3) into the monetary budget constraint (equation 13-2) yields:

$$M + p_w \cdot t^* = X + p_r \cdot r \tag{13-4}$$

where p_r is the full price of a visit given by

$$p_r = c + p_w \cdot (t_1 + t_2) = f + p_d \cdot d + p_w \cdot (t_1 + t_2) \tag{13-5}$$

As equation 13-5 makes clear, the full price of a visit consists of four components: the admission fee, the monetary cost of travel to the site, the time cost of travel to the site, and the cost of time spent at the site. On the assumption that individuals are free to choose the number of hours worked at a given wage rate, the two time costs are valued at the wage rate. In a more realistic model with income and payroll taxes, time would be valued at the after-tax wage rate.

Maximizing equation 13-1 subject to the constraint of equation 13-4 will yield the individual's demand function for visits:

$$r = r(p_r, M, q) \tag{13-6}$$

If all individuals spend the same amount of time at the site and have the same wage, then this component of the price of a visit is the same for all individuals. Given these assumptions, the data on rates of visitation, travel costs, and variation in entry fees (if any) can be used to estimate the coefficient on p_r in a travel cost–visitation function. Because of the additivity of the components of price reflected in equation 13-5, the coefficient on p_r can be used to derive the individual's demand for visits to a site as a function of the admission fee.

Let us review the assumptions that make this basic version of the travel cost model work.

- First, it is assumed that the wage rate is the relevant opportunity cost of time. In a provocative article, Alan Randall argues that for several reasons "travel cost is inherently unobservable" (1994, 88). In my view the most compelling part of his argument concerns the difficulties in defining and measuring the opportunity cost of time spent in travel, which is characterized by Randall as "an empirical mystery" (1994, 90). I will return to this issue in a later section.
- Second, it is assumed that all visits entail the same amount of time spent on the site. This assumption plays two important roles in the simple form of the model: it makes it possible to measure site usage by the scalar r, the number of visits; and it makes the full price of a visit, p_r, a

parameter to the individual. If the individual chooses the amount of time of each visit, then p_r is an endogenous variable. Modeling the choice of time spent on site is discussed in the section on time below.

- Third, it is assumed that there is no utility or disutility derived from the time spent traveling to the site. If part of the trip involves the pleasures of driving through a scenic countryside, then travel cost is overestimated by equation 13-5. The way various uses of time affect utility and the shadow value of time is discussed in the section on time below.
- Fourth, it is assumed that each trip to the site is for the sole purpose of visiting the site. If the purpose of the trip is to visit two or more sites or to visit a relative en route, then at least part of the travel cost would be a joint cost that cannot be uniquely allocated among different purposes. If an alternative destination can be identified as the primary purpose of the trip, then the relevant cost of the visit to the recreation site is the incremental cost of adding the site visit to the trip given the trip to the primary destination. Parsons (2003) discusses this and other approaches to dealing with multiple-destination trips.
- Fifth, it is assumed that there are no alternative recreation sites available to these individuals. The role of substitute sites and the possibility of multisite models are discussed later in this section.
- Sixth, it is assumed that the individual's choice of where to live (which is one determinant of the cost of a trip to a recreation site) is independent of preferences for recreation visits. If people choose residential locations to be near preferred recreation sites, then the price of a visit is endogenous. Parsons (1991) has suggested an instrumental variables approach that may avoid the bias that such choices would otherwise impart to the estimation of trip demand functions.

Valuing Recreation Sites

The value of a recreation resource is the discounted present value of the flow of recreation services to those who visit the site, as measured by the area behind the aggregate compensated demand curve for visits to the site as a function of the admission fee, that is, the amount that individuals would be willing to pay to avoid having the admission fee increased to the point that aggregate visits would fall to zero. The fact that this can be estimated from equation 13-6 depends on the assumption that individuals perceive and respond to changes in the travel-related component of the cost of a visit in the same way they would respond to a change in the admission price.

Because the travel cost model yields a Marshallian demand curve, the usual practice is to rely on Willig's (1976) analysis to argue that the ordinary consumer's surplus measure is a close approximation to the true value. However, Willig's bounds may not be appropriate for valuing recre-

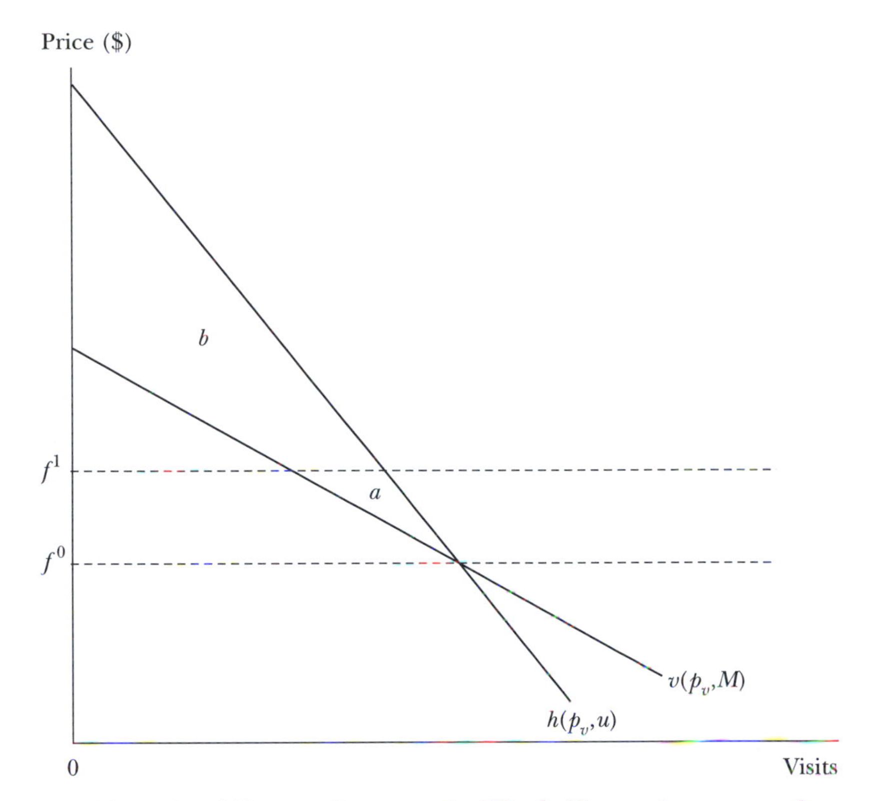

Figure 13-1. The difference between the Marshallian and compensating measures of the value of a recreation site

ation sites. His analysis is based on price changes for which positive quantities would be demanded at both the initial and second prices. However, the value of a site is found by postulating a price change sufficient to drive the quantity demanded to zero. For a relatively small price increase, such as from f^0 to f^1 as shown in Figure 13-1, the error in using consumer's surplus is the area *a*. However, the error in using the ordinary consumer's surplus as an approximation to the compensating variation measure of the site value is the area $a + b$. Because this error could be substantial, some investigators have begun to apply the methods of exact welfare measurement for determining the values of recreation sites. The functional form for the travel cost demand equation can be derived from an explicit specification of the utility function. Then, as discussed in Bockstael et al. 1986, the methods of Hausman (1981) or Vartia (1983) can be used to obtain the indirect utility function. For an application, see Bockstael et al. 1987.

The statement that the area under the demand curve is the measure of the value of the flow of services is based implicitly on the assumption that the demand curve is known with certainty. This implies that the welfare measure is deterministic. However, because the demand curve is estimated from a statistical model, there are random errors in the estimates of the parameters. This fact makes the welfare measure itself a random variable, and it has implications for the way in which welfare measures are calculated. Bockstael and Strand (1987) have shown how different assumptions about the sources of error in the regression estimates lead to different methods for calculating the consumer's surplus measure. (See also Graham-Tomasi et al. 1990.) Smith (1990) has argued that when the purpose of the statistical estimation of demand functions is the derivation of welfare measures, estimators should be selected on the basis of the statistical properties of the welfare estimates rather than the statistical properties of the demand function parameters.

Sometimes policymakers wish to know the value of adding a new site to an existing recreation resource system. Conceptually, the value is the area under the compensated demand curve for the new site, but there are no data for estimating this demand curve directly before the site is built. However, knowledge of the pattern of demands for the existing sites can be exploited to obtain an estimate of the value of adding a site to the system, at least if the new site is qualitatively similar to one of the existing sites. For applications of this approach, see Burt and Brewer 1971 and Cicchetti et al. 1976.

The first step is to estimate a system of demand functions for the set of existing sites using a multisite model of the type described below. The second step is to identify one of the existing sites as the closest substitute to the new site on the basis of nearby location or similar qualitative characteristics, or both. Then the travel costs to the new site are used to generate a set of price reductions for those for whom the new site is closer than the existing site. These price changes are inserted in the demand function for the existing site to calculate a welfare measure. The principal limitation of this approach arises from the inability of the multisite travel cost model to explicitly incorporate quality differences across sites. Also, the approach is implicitly based on the assumption that the only source of value associated with the new site is the reduction in travel costs for some visitors.

Substitute Sites

In the basic model described here, it was assumed that there was only one site to visit. However, suppose that several sites are available to the population whose recreation behavior is being studied. Then it is plausible that the number of visits that any individual makes to the site in question will depend not only on its implicit price but also on the implicit prices of any substitute sites in the region. If the travel costs of the relevant substitute sites are omitted from the estimated equation, its parameters will be biased.

Omitting the price of a substitute site will bias the estimates of both the intercept term and the coefficient on the price of the site (Caulkins et al. 1985; McKean and Revier 1990). The sign of the bias on the own price coefficient depends on the correlation between it and the omitted price variable. If the correlation between the two travel cost variables is positive, then omitting the substitute price biases the own price elasticity toward zero. However, if the two travel costs are inversely correlated, the estimated own price coefficient is subject to a negative bias and the price elasticity of demand for visits is biased upward. The sign of the correlation between the two travel costs depends upon the spatial distribution of the population relative to the two sites.

When the recreation site being investigated is part of a regional recreation resource system, there is the difficult question of deciding which other sites are substitutes whose prices should be included in the site demand equation. The comprehensive approach of including all alternative sites is cumbersome and may require more data than are available. An alternative is to ask each individual what other site that person visits most frequently and to include only that site's price as the relevant substitute price. However, different respondents could name different sites, and this approach may miss important substitution effects. For example, if an individual alternates between fishing at the site in question and swimming at another site, the swimming site would be selected as the alternative. However, the more relevant alternative would be the next-best fishing site. There is no single right answer to the question of how to select substitute sites. This is a problem that requires good judgement on the part of the researcher.

Multisite Models

Analysis of the demand for a single site would be appropriate if the researcher is interested in valuing the availability of that single site or changes in the characteristics of only that site. However, many policy-relevant questions involve changes in the value of a set of sites due to changes in the number and availability of sites or to changes in the qualities of these sites. In such cases, the interactions and the substitution effects among sites must be modeled explicitly. This calls for some form of multisite model.

Multisite models are estimated as systems of demand equations. For each site j $(j = 1, \ldots, j, \ldots, m)$, a demand equation of the following form is specified:

$$r_{ji} = r_j\left(p_{r_{ji}}, p_{r_{ki}}, M_i, q_j\right)$$
$$(i = 1, \ldots, i, \ldots, n), (k = 1, \ldots, k, \ldots, m), \text{and } k \neq j \qquad (13\text{-}7)$$

where r_{ji} is the number of visits individual i makes to the site, $p_{r_{ji}}$ is the full price of a visit by i to j, and $p_{r_{ki}}$ is the set of substitute prices for visits to other sites. This type of model can be estimated from data on individual observations (see, for example, Burt and Brewer 1971) or from data aggregated into zones of origin (see, for example, Cicchetti et al. 1976).

As the number of sites and demand equations increases, multisite models can become cumbersome and difficult to estimate. And, as will be discussed in more detail in the section called Valuing Changes in Site Quality, it is not possible to take into account differences in site qualities and characteristics in this form of model. These difficulties have led researchers to experiment with several forms of alternative multisite models.

One modeling simplification is to aggregate all the visits made by an individual to all the available sites into a number of visits to a "typical site," where the typical site is the site most frequently visited by that individual. Specifically, a single demand equation is estimated from individual data, where that equation takes the form

$$r_i = r\left(p_{r_{ti}}, q_t, M_i\right) \tag{13-8}$$

where $p_{r_{ti}}$ and q_t are the price and quality of the site most frequently visited by the individual. In one application of this approach (Caulkins et al. 1986), an effort was made to control for substitutes by including proxy variables for substitute prices and qualities. The proxies were weighted averages of the prices and qualities of other sites actually visited. The weights were the relative frequencies of visits to the alternative sites.

The typical-site model does not fully capture the complexity of recreation behavior in a multisite setting. It provides no basis for explaining the allocation of visits across sites in the recreation system. Also, if people visit different sites to engage in different types of activities, the proxy variables for substitute sites are not likely to capture the true substitution behavior of individuals faced with changes in the availability or quality of their most frequently visited site. The example of the individual who visits one site for fishing and another site for swimming applies here as well.

Another type of simplification of the multisite model is to treat all observations of r_{ji} as belonging to a single travel cost demand equation. In this "pooled" model, a single equation of the following form would be estimated:

$$r_{ji} = r\left(p_{r_{ij}}, q_j, M_i\right) \tag{13-9}$$

For one example of this type of model, see Smith et al. 1986.

In another approach, Morey, in a series of papers (1981, 1984, 1985), has developed a model of how individuals allocate a given recreation time and money budget among alternative sites on the basis of site prices and qualities. However, this ability to model site substitution behavior formally comes at the expense of losing the ability to explain the total level of recreation activity for each individual. In Morey's model, utility is a function of the number of visits to each site and the attributes of these sites. Utility is maximized subject to a fixed, exogenous time budget, where time is allocated to the on-site activity, travel to the site, and working to earn the money costs of travel and admission to the sites visited. A specific functional form for the utility function is chosen so that share equations can be derived as solutions to the constrained maximization problem. The share of visits made to site j is a function of site prices and characteristics. Because the share equations are derived from an explicit utility function, it is possible to calculate exact measures of welfare for changes in the availability or qualities of sites (Morey 1985). In its focus on the allocation of activities among sites, the Morey share model has some similarity to the RUMs described later in this chapter.

Valuing Changes in Site Attributes

An improvement in an attribute of environmental quality at a site shifts the demand curve for visits to the site out and to the right. Given the assumptions of weak complementarity (that there is a choke price and that at the choke price, the marginal utility of quality is zero), the area between the two compensated demand curves for the site is an exact measure of the welfare change. However, the travel cost model provides only Marshallian demand curves. As pointed out in Chapter 4, the Willig error bounds do not apply to areas between ordinary demand curves. Therefore, value measures based on the travel cost model must be used with caution. An alternative approach, as discussed above, is to use one of the methods of exact welfare measurement.

In using an exact welfare measure of value, there is no need to take into account changes in recreation use at other sites or savings in travel cost as people substitute visits to the improved site for visits to other sites. These components of benefits are fully captured in the area between the demand curves for the improved site. For example, if recreation is switched from an alternative site, the demand curve for that site shifts to the left. However, it would be incorrect to measure the area between those demand curves to adjust the measure of benefits. There has been no change in any quality measure at the alternative site, so the integral of the demand function for that site over the range of quality change at that site is zero.

If there are simultaneous changes in quality at several recreation sites, the theory of welfare change under multiple price or quantity changes applies. Consider, for example, two sites where there is a simultaneous improvement in quality. In principle, benefits are measured by evaluating the welfare change for site *A*, holding (by assumption) the quality at site *B* at its original level, then evaluating the welfare change for site *B* given the improved level of quality at site *A*. The welfare measure is independent of the order in which the quality changes are evaluated.

Two questions must still be resolved in implementing this measure of the value of site-quality changes. First, what attributes of the site matter to people, and how are they measured empirically? Second, how are the relevant quality measures to be incorporated into models of site demand so that shifts in demand curves can be predicted? In this section, I take up each of these questions in turn. I then consider the special problems raised by on-site congestion as one relevant measure of the quality of an individual's experience at a site.

Measuring Site Quality. The word "quality" is a proxy for a variety of characteristics of a site, some of which are affected negatively by pollution or by competing economic activities, such as logging and mineral extraction, or positively by investments in enhancing the flow of recreation services from a resource. Quality measures can be based on scientific data, such as the dissolved oxygen concentration or pH of a water body, or quality can be represented by some measure of perceived quality derived from questions asked of recreation visitors. Also to be considered is whether a single quality measure should be employed, or whether multiple measures are appropriate, either entered separately or aggregated into some kind of quality index.

Objective measures of quality are reproducible in the sense that two observers should obtain the same measure, within the bounds of measurement error. However, people might make choices about recreation on the basis of their perceptions of quality rather than the objective measure. If individuals' perceptions are functions of objective measures and personal characteristics, then it may be possible to estimate a "perception function" and to use this function to model choices and measure welfare values. For some kinds of recreation activity, individuals' choices may depend on an interaction between an objective measure of quality and an individual characteristic such as skill or ownership of specialized equipment. An increase in the abundance of a particular species of fish, for example, will have a bigger effect on the behavior of an angler who has acquired skill in fishing for that species than on anglers who have not. See Phaneuf and Smith (forthcoming) for further discussion of these issues.

Incorporating Site Quality in the Recreation Demand Model. To estimate the value of a quality change at a site, we need to know how the demand curve for that site shifts with the change in the relevant quality characteristic. Consider first the case in which recreation demand is modeled for only a single site. If the demand function of equation 13-6 were estimated from data collected during one season, there would be no variation in site quality attributes across the sample. Thus there would be no way of identifying the influence of objective measures of site quality on recreation demand. If the data were collected over a longer period of time, during which site quality changed, it might still be difficult to separate the effects of quality change from other factors that were also changing over time. However, if behavior were based on perceptions of quality, and if perceptions varied independently of travel cost, then the influence of perceived quality on behavior could be measured.

One way to obtain an estimate of the impact of quality changes at a single site on the demand for the site is to ask users contingent activity questions. For example, McConnell (1986) asked residents of New Bedford, Massachusetts, how many times they actually visited beaches around New Bedford Harbor and how many times they would expect to visit those same beaches if the pollution due to polychlorinated biphenyls (PCBs) were eliminated. Responses to these questions, along with estimates of travel costs for respondents, provided the data to estimate a demand function for visits to these beaches that incorporated the effect of PCB pollution on demand.

When several sites are studied using a multisite recreation demand model, there may be substantial variation in qualities across sites. However, it may be difficult to disentangle the effects of differences in site qualities from the effects of differences in travel costs. All the empirical attempts to incorporate site quality in multisite demand models have involved some kind of simplification; thus they also have various limitations in their ability to characterize recreation demand accurately. These approaches include the pooled visitation model (e.g., Smith et al. 1986), the "typical trip" model (e.g., Caulkins et al. 1986), and the varying parameter model (e.g., Vaughan and Russell 1982a; Smith et al. 1983a; Smith and Desvousges 1985; Smith et al. 1986). Because of the superiority of the RUMs in incorporating site attributes in multisite choice models, these approaches have fallen out of favor in the past decade or so. Readers interested in more details about these models should see Freeman 1993.

Congestion as a Quality Attribute. Congestion of a recreation site occurs when the number of users is so large that it diminishes the utility and therefore the WTP of those users. The degree of crowding at a site can be considered one of the site attributes that influence the quality of the recreation services each individual experiences and the utility each individual obtains from visiting the site. The presence of congestion at a recreation site has implications for the estimation of recreation demand and the measurement of the value of environmental quality change.

Congestion, like pollution, is a negative externality. It affects all users of a site. Users may have different marginal willingnesses to pay to avoid congestion. Also like pollution, the congestion case satisfies the conditions of weak complementarity. Thus the benefits of avoiding congestion can be evaluated by measuring areas between demand curves. In this subsection, a simple model of congestion is developed to show the effects of congestion on demand for a site and on the prediction of recreation use. The model is then used to show how the value of improving quality at a site is reduced if congestion is present, and how improving quality at one site can lead to additional benefits by reducing congestion at other sites. This model is developed more fully in Freeman and Haveman 1977. That paper developed rules for optimal pricing and showed how these are affected by assumptions about the incidence of congestion costs. The earliest analyses of the problem of congested recreation facilities include Fisher and Krutilla 1972 and Cicchetti and Smith 1973. Other theoretical analyses can be found in McConnell and Duff 1976; McConnell 1980; and Smith 1981. Dorfman (1984) provides an elegant treatment of the optimal congestion and optimal pricing problems.

Suppose that we have recorded the level of recreation use at a particular site that we know to be congested; the case is illustrated in Figure 13-2. Let this level of use be V^1. Suppose also that the admission fee at the site is zero. Assume that by application of the travel cost model, we have estimated individuals' demand curves for the site and aggregated them to obtain D^1V^1. This will be called a constant-congestion demand curve because it reflects the behavior of a set of users, all of whom are experiencing the same level of congestion when they visit this site. The triangular area, $0D^1V^1$ is the consumer's surplus associated with having this recreation site available, given its present level of use, congestion, and site quality.

Suppose now that it is known how much each individual would be willing to pay to use this site if it were uncongested. Aggregating the individuals' willingnesses to pay for an uncongested site gives an uncongested demand curve, D^2V^2. According to Figure 13-2, congestion deters some users ($V^2 - V^1$) from using the site; their loss is the area CV^2V^1. We can also identify the costs of congestion to existing users. At use level V^1, those users incur congestion costs of $D^1D^2CV^1$, the area between their uncongested demand curve and the observed constant-congestion demand curve out to V^1. This is their total willingness to pay to avoid the present level of congestion at the site.

Now suppose a gate fee of p^1 is imposed. Can the new level of use be predicted from the information at hand? The answer is no. The constant-congestion demand curve D^1V^1 would predict a reduction in recreation use to

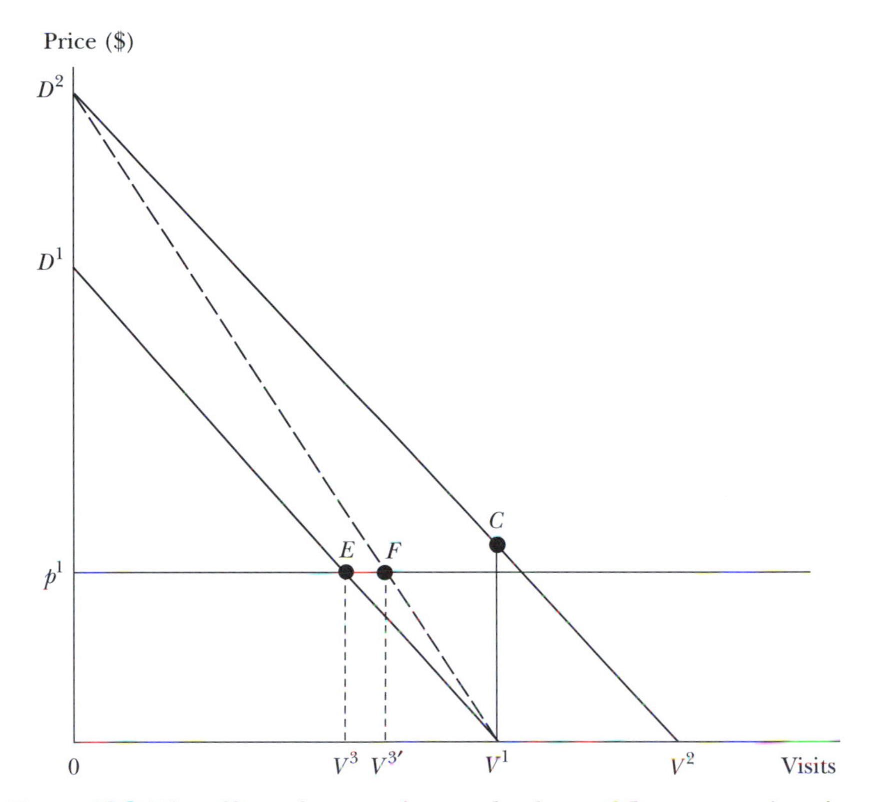

Figure 13-2. The effect of congestion on the demand for a recreation site

V^3. However, with lower use, congestion is reduced. With less congestion, WTP is higher, and more people are willing to use the site at a given price. Thus we would expect use to be greater than that predicted by the constant-congestion demand curve. Suppose that actual use is $V^{3\prime}$. There is another constant-congestion demand curve through point *F*. This curve is conditioned on the level of congestion associated with the observed level of use, $V^{3\prime}$. This curve is not shown in Figure 13-2. In fact, there is a whole family of constant-congestion demand curves. Each one represents the aggregation of individual willingnesses to pay conditional on a level of total use and its associated congestion. Only one point on a particular constant congestion demand curve can ever be observed; that is the point that corresponds to the level of use on which the curve is conditioned. However, any one of these constant-congestion demand curves could be estimated by the travel cost technique if that level of use were actually observed because the travel cost technique predicts individuals' responses to price changes conditioned on a given observed level of total use.

The locus of these observable points traces out another kind of demand curve, a congestion-adjusted demand curve. This is represented as D^2V^1 in

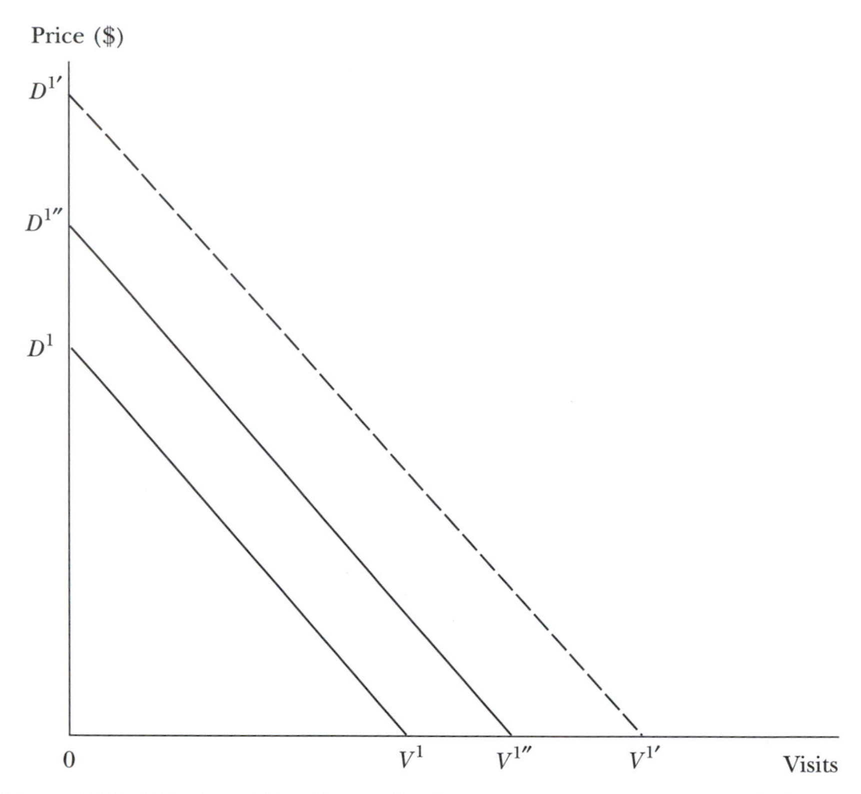

Figure 13-3. The benefits of a quality improvement at a congested site

Figure 13-2. The congestion-adjusted demand curve can be used for predicting the effect of changes in entry fees on use levels. However, the area under this curve has no significance for benefit evaluation. The curve could not be derived from a travel cost study because for the latter all recreationists experience the same congestion level.

Now let us examine the benefits of improving some quality characteristic at a site when congestion exists; the case is illustrated in Figure 13-3. The constant-congestion demand curve D^1V^1 in Figure 13-3 has been computed from travel cost data based on a level of use of V^1. Now let there be a quality improvement at the site. Suppose the constant-congestion demand curve were shifted out to the right to $D^{1'}V^{1'}$. This would indicate the potential increase in use and potential benefits if increases in congestion could be avoided; it is the area $D^1D^{1'}V^{1'}V^1$. The additional users impose congestion costs on one another, though, which partially offset the effects of improved quality on use and benefits.

Suppose now that the actual increase in use is to $V^{1''}$. There is then a new constant-congestion demand curve through this point, $D^{1''}V^{1''}$. The net benefit of improving site quality, taking into account the increased congestion, is the area $D^1D^{1''}V^{1''}V^1$. The increase in congestion has dissipated some of the potential benefits from improving site quality.

To empirically account for the separate effect of congestion, it is necessary to include terms for congestion as a site characteristic in the system of visitation equations for a set of sites in a given region. The estimation of the coefficients on the congestion term poses the same set of problems as those found in the estimation of the effects of site quality on visitation, discussed above.

Consider two sites, A and B, that are substitutes for each other where A is currently uncongested, perhaps because of poor quality, and B is congested. Suppose that there is an improvement in quality at site A. We expect visits to A to increase and visits to B to decrease. The reduction in visits would be reflected in a leftward shift of the constant-congestion demand curve for site B. If B were uncongested, the shift in the demand curve for B would be ignored in the calculation of the increase in value for site A. However, given that B is congested, the shift in recreation use from B toward A reduces that congestion. Because this reduction in congestion is a consequence of the site quality improvement at A, the value of this reduction should be added to the value of the quality improvement at site A in computing the total welfare value of the change in quality at site A.

Random Utility Models

It is difficult to adapt the standard travel cost model of recreation site demand to fully capture the values of site characteristics or the effects of variations in site quality on the demand for a site. The discrete choice or random utility model (RUM), with its emphasis on explaining the choice among sites as a function of the characteristics of the available sites, is well suited to this task. However, the gain in the ability to explain choices among sites comes at the expense of an inability to explain the total demand for a recreation activity, at least with the simple form of the model. In this section I briefly sketch the logic of the simple RUM, its use in estimating the value of a site, and the benefits of changes in characteristics. I then describe alternative approaches to modeling overall recreation activity within the RUM framework.

For more detailed treatments of the theory and estimation of RUMs, readers should consult Bockstael et al. 1986; Bockstael et al. 1991; and Herriges and Kling 1999. Interesting examples of applications include early studies by Hanemann (1978) (see also Bockstael et al. 1987), Feenberg and Mills (1980), and Morey and others (1993).

Consider a set of n individuals ($i = 1, \ldots, i, \ldots, n$). For each recreation occasion, these individuals can choose one site from a set of m alternatives

($j = 1, \ldots, j, \ldots, m$). Each individual has an income M_i and vector of socioeconomic characteristics $\boldsymbol{S}_i$. Each site has a vector of environmental quality attributes $\boldsymbol{Q}_j$. The travel cost for individual i to site j is C_{ij}. For individual i choosing site j, utility is given by the conditional indirect utility function:

$$u_{ij} = v_i(M_i - C_{ij}, \boldsymbol{Q}_j, \boldsymbol{S}_i) + \varepsilon_{ij} \quad (13\text{-}10)$$

where ε_{ij} is the unobservable component of utility and is assumed to be random. The prices of market goods are omitted for notational simplicity. Although many RUM applications are based on an assumed linear form for equation 13-10, utility functions that are nonlinear in attributes can be specified.

The probability of individual i choosing to visit site j is the probability that u_{ij} is greater than u_{ik} for all $k \neq j$. If all the ε_{ij} are independently and identically distributed with a Type I extreme value distribution with zero mean, the probability of individual i visiting site j^* is

$$Pr_{j^*}(i) = \frac{e^{v_i\left(M_i - C_{ij^*},\, \boldsymbol{Q}_{j^*},\, \boldsymbol{S}_i\right)}}{\sum_{j=1}^{m} e^{v_i\left(M_i - C_{ij},\, \boldsymbol{Q}_j,\, \boldsymbol{S}_i\right)}} \quad (13\text{-}11)$$

For two sites with identical attributes and travel costs, the probability of choosing either is 0.5. Once a functional form for $v_i(\cdot)$ has been specified, its parameters can be estimated using maximum likelihood methods.

Care must be taken in specifying the choice set to avoid possible biased estimates of coefficients by omitting relevant sites. The analyst should include all sites with a nonzero probability of being visited. However, this could result in a very large choice set. In practice, the choice set can be reduced by aggregating sites of similar attributes and distance into composite sites, by random drawing of sites from the complete set, or by truncating the choice set at some maximum distance from the recreationist. See Parsons and Kealy 1992 and Pendleton 1999 for further discussion of this issue.

Another issue is avoiding violation of the property of independence of irrelevant alternatives (IIA). Such violations are likely to occur if two or more different types of recreation resource or recreation activity are included in the choice set. For example, suppose that there are three alternatives from which to choose: Lake A, Beach B, and Beach C. An improvement in the quality of Beach C is likely to reduce the probability of visiting Beach B relative to the probability of visiting Lake A, a violation of IIA. One way to avoid this problem is to specify a nested model of choice in which the individual is assumed, first, to choose a type of recreation resource or activity and second, to choose a site within that category. For an example, see Bockstael et al. 1986.

Once the parameters of $v_i(\cdot)$ have been estimated, the monetary value of a change in $\boldsymbol{Q}_j$ from $\boldsymbol{Q}_j^0$ to $\boldsymbol{Q}_j^1$ can be calculated. For individual i, this value is defined implicitly by

$$v_i\left(M_i - C_{ij} - CS_{ij}, \boldsymbol{Q}_j^1, \boldsymbol{S}_i\right) + \varepsilon_{ij} = v_i\left(M_i - C_{ij}, \boldsymbol{Q}_j^0, \boldsymbol{S}_i\right) + \varepsilon_{ij} \tag{13-12}$$

where CS_{ij} is the compensating surplus measure of the welfare change associated with the change in $\boldsymbol{Q}_j$. Before CS_{ij} can be calculated from equation 13-12, two hurdles must be overcome. The first hurdle arises because for each individual there is no variation in income across alternatives in the choice set, so no independent estimate of the parameter on income gives the marginal utility of income. This hurdle can be gotten over by recognizing that the relevant income measure is total income less the cost of the activity, C_{ij}. Thus the coefficient on income is the negative of the estimated coefficient on the cost of the activity.

The second hurdle arises because of the unobserved component of utility. Because of that, it is not possible to know whether any specific individual will visit the site in question. If individual i does not visit the site, CS_{ij} is zero. Bockstael and others (1991, 261–262) describe one approach to dealing with the researcher's uncertainty. They define the welfare measure as the payment that equates the researcher's expected value of realized utility with and without the change in $\boldsymbol{Q}_j$. Specifically, CS_{ij} is defined implicitly by

$$E\left[v*\left(M_i - C_{ij} - CS_{ij}, \boldsymbol{Q}_j^1, \boldsymbol{S}_i\right)\right] = E\left[v*\left(M_i - C_{ij}, \boldsymbol{Q}_j^0, \boldsymbol{S}_i\right)\right] \tag{13-13}$$

where $v*(\cdot)$ is the maximum of

$$v_i(M_i - C_{ij}, \boldsymbol{Q}_j, \boldsymbol{S}_i) \quad \text{all } j. \tag{13-14}$$

Given the assumptions about the distribution of ε_{ij} and assuming that the marginal utility of income is constant,

$$CS_{ji} = \frac{\ln\left[\sum_{j=1}^{m} e^{v\left(\boldsymbol{Q}_j^1\right)}\right] - \ln\left[\sum_{j=1}^{m} e^{v\left(\boldsymbol{Q}_j^0\right)}\right]}{b} \tag{13-15}$$

where b is the coefficient on income and the other arguments in $v_i(\cdot)$ are omitted for simplicity.

Similar calculations can be used to obtain the value of adding or deleting a site with a specified set of characteristics from the individual's choice set. For the addition of site $m + 1$, the expression is

$$CS_{m+1,i} = \frac{\ln\left[\sum_{j=1}^{m+1} e^{v(Q_j)}\right] - \ln\left[\sum_{j=1}^{m} e^{v(Q_j)}\right]}{b} \tag{13-16}$$

and for deleting site m, the expression is

$$CS_{m-1,i} = \frac{\ln\left[\sum_{j=1}^{m} e^{v(Q_j)}\right] - \ln\left[\sum_{j=1}^{m-1} e^{v(Q_j)}\right]}{b} \tag{13-17}$$

which is equal to or less than zero. (Again, see Bockstael et al. 1991, especially 262–263.) There are other approaches to defining a welfare measure using a random utility framework. Hanemann describes two of them and examines the relationships among them (Hanemann 1999, 43–48).

Because the expressions derived above give the welfare measure associated with one choice occasion, it is necessary to find some way of predicting each individual's total number of visits per year or per recreation season. One approach is to model the determination of the number of choice occasions or activity days as a separate problem. See, for example, Bockstael et al. 1986. This approach is discussed in a later section on the estimation of participation rates. A second approach is to extend the discrete choice framework in various ways to include choices among activities as well as among sites and to make undertaking no recreation activity an option for each choice occasion.

There are several ways of doing this. One is to specify that one of the m alternatives is no activity and to estimate a single logit equation (see, for example, Feenberg and Mills 1980). Another way is to specify a nested form of logit model in which the first equation predicts whether the individual will undertake any activity on a given choice occasion, while the second equation gives the probability that the individual will choose a particular site or activity. The first equation would include as explanatory variables some measures of the average availability, quality, or both, of recreation sites accessible to the individual. The second equation would include measures of the costs and qualities of each of the available alternatives. This latter model specification is more consistent with the property of IIA. The expected number of visits to a site is given by the product of the number of choice occasions and the joint probability of undertaking the activity and choosing site *j*, given that some activity is chosen. See Morey (1999) for a careful analysis of the properties of the nested RUMs. Parsons (2003) also discusses nesting strategies for RUMs.

Time and the Implicit Price of Visits

The travel cost model of recreation demand and RUM model of site choice described above were kept quite simple to focus attention on the underlying intuition. In this section I discuss the role of time as a cost and as an element of choice in these models. In this section, I first show why it is important to include the correct measure of time cost in the travel cost demand equation and what bias results from omitting the time cost of travel. I then discuss alternative approaches to estimating the shadow price of time. Finally I take up the problem of modeling the length of the visit. This is important because time on site influences both the utility of the visit and the time cost of the visit.

The Importance of Time Costs

Both travel to a site and the recreation activity itself take time, and time is scarce. Also, the time spent in any activity could be a separate argument in individuals' utility functions. For example, time at work could yield either positive or negative utility, and almost certainly time spent at a recreation site has a positive marginal utility, at least up to some point. The opportunity cost of time must be included in a properly specified model of the demand for recreation visits. The appropriate shadow price of time will depend on the alternative uses to which the time could be put and on the nature of the constraints on individual choice.

It has long been recognized that the time spent in travel to a site should be included as a component of travel cost for purposes of estimating the demand for visits in a travel cost model (for example, Cesario and Knetsch 1970) or the probability of selecting a site in a RUM model. From equation 13-5, travel cost is defined as the sum of the monetary and time costs of travel to the site. If the time component of costs is omitted in the estimation of the demand function for visits, the cost variable is biased downward. The result will be that the estimated parameter on cost or price will be biased upward. As a consequence, the estimated demand curve will be more elastic than the true demand curve, and the benefits of the recreation site (the area under the demand curve) will be underestimated.

The choice of the shadow price of time is critical to the estimation of the elasticity of demand for the site and the calculation of the value of the site. The choice of a high shadow price of time raises the importance of time cost in explaining visits as a function of distance. With a higher shadow price of time, the predicted reduction in the number of visits because of an increase in the entry fee is smaller, and the estimated demand curve is less elastic. Clearly, though, using a shadow price of time that is too high can have the opposite effect on estimates of the elasticity of site demand and value, that is, underestimates of site demand and overestimates of site value.

Measuring the Time Cost of Travel

One of the key assumptions in the simple travel cost model described above is that individuals are free to choose the hours they work at a given wage and that this wage governs the trade-off between work and leisure. Given this assumption, the individual maximizes utility by allocating time among alternative activities to equate the marginal values of time in these activities with the wage rate. Thus the wage can be taken as an indicator of the shadow value or marginal opportunity cost of time.

One practical difficulty in using the wage rate in recreation demand studies is that surveys often provide data on family income rather than the hourly wage rate. If the wage is inferred by dividing family income by some estimate of hours worked (typically, 2,000 hours per year), measurement error is introduced. One way to avoid this problem is to use a separately estimated hedonic wage equation to predict the wage rate for each individual in the sample (see Smith et al. 1983b).

There is also empirical evidence from other types of behavior that is inconsistent with the standard model of the labor–leisure trade-off at the market wage. Cesario (1976) was the first to point this out in the context of the analysis of recreation demand. He cited evidence from choices of transportation mode by commuters that the revealed opportunity cost of time was perhaps only one-third of the market wage. For more recent evidence from stated preference analysis of transportation mode choice, see Calfee et al. 2001. Following Cesario, the practice of using a shadow price of time of one-third of the wage became common in recreation demand analysis (Shaw and Feather 1999). More recently Englin and Shonkwiler (1995) used a latent-variable approach in an effort to estimate the unobserved true cost of travel time (including time cost) as a function of observable indicator variables. In their application of the model to data on recreation visits to Lake Champlain, they found that travel time has an estimated value that is about 40% of the average wage of the sample.

The evidence of a lower shadow price of time could be explained either by institutional constraints on the choice of hours of work, such as the standard 40-hour workweek and monthly or annual salaries, or by a richer theory of choice that recognized that time spent in various activities including recreation, traveling, and at work might convey utility (positive or negative) directly. We consider each possibility in turn.

Suppose that all (or most) jobs offer a 40-hour workweek at a fixed weekly salary. Then the individual cannot equate values and costs of leisure time at the margin. Models in which there is no trade-off between work and recreation time have been analyzed by several authors. Smith and others (1983b) assumed that individuals could allocate a fixed nonrecreational time budget between work and other activities. However, the recreation time budget was fixed and exogenous. Individuals could only allocate time between two or more sites. They found that the relevant price of a visit to a site was the sum of the money travel cost and a proportion of the total time per visit ($t_1 + t_2$ in our notation). This proportionality factor depended upon the wage rate and the shadow prices attached to the constraints on money and time. Thus the proportionality factor was not observable, but it could be estimated from the data.

Bockstael and others (1987) presented a model that was flexible in that it could accommodate either of two different cases regarding the individual's labor-market equilibrium. These authors assumed that the duration of a visit was fixed. If the individual could alter hours of work at the margin, the relevant price was the sum of the money travel cost and the total time of the visit ($t_1 + t_2$) valued at the marginal wage rate. If the individual could not make marginal adjustments in the number of hours worked, the money travel cost and time cost per visit entered the demand function for visits separately. The parameter for the time cost component was a function of parameters of the preference function. Bockstael and others (1987) assumed a specific functional form for preferences, derived the alternative demand functions, and estimated the parameters from a data set that distinguished between those who could and those who could not alter their work time at the margin. See also McKean et al. 1995 for a similar modeling approach.

Using a different model, Feather and Shaw (1999) used individuals' responses to survey questions to classify them according to whether they were able to choose freely the number of hours they worked at a given wage, were required to work more hours than they wished at the given wage (were "overemployed"), or would prefer to work more hours at the given wage (were "underemployed"). They were then able to estimate a shadow wage function for their sample and to compute shadow wages for the individuals in each group. As the standard model predicts, the shadow wage for those able to freely choose their hours of work was not significantly different from the mean observed wage, whereas for those who were overemployed (underemployed), the shadow wage was greater (less) than the observed wage. Because the three groups each represented about one-third of the working sample, this result does not support the practice of using a shadow price of time of one-third the wage for all recreationists.

Another line of research questions whether the wage rate, however measured, is the appropriate shadow price of time even for those people who can freely trade off time and work at the going wage. One of the key assumptions in the standard model, in which the wage rate is made the appropriate shadow price, is that there are no time variables appearing directly or implicitly as arguments in the utility function. Suppose that the time spent at work has either a positive or a negative utility at the margin. Then there will be a divergence between the shadow price or scarcity value of time in other activities and the wage rate as a measure of the opportunity cost of time at work. This basic idea can be traced back to Johnson (1966). Cesario (1976) first analyzed its implications in the context of the travel cost model of recreation demand. To illustrate this point, we will modify the simple travel cost model outlined above to include time at work as an argument in the utility function.

The constrained maximization problem becomes

$$\max: u(X, r, t_w) + \lambda(M - p_w \cdot t_w - X - c \cdot r) + \mu[t^* - t_w - (t_1 + t_2) \cdot r] \quad (13\text{-}18)$$

where site quality q has been omitted for simplicity. Because μ is the marginal utility of time, μ/λ is the scarcity value or marginal willingness to pay for time. The relevant first-order conditions include

$$\frac{(\partial u / \partial r)}{\lambda} = c + \frac{\mu}{\lambda}(t_1 + t_2) \quad (13\text{-}19)$$

and

$$\frac{(\partial u / \partial t_w)}{\lambda} + p_w = \frac{\mu}{\lambda} \quad (13\text{-}20)$$

Equation 13-19 says that the marginal willingness to pay for a visit must equal the full cost of a visit. The full cost is the sum of the monetary cost c and the time cost of the visit, where the time cost $t_1 + t_2$ should be valued at the scarcity value of time, μ/λ. However, equation 13-20 shows that the wage rate may not be a good measure of the scarcity value of time. If the marginal utility of time spent working is negative, then the wage rate is an overestimate of the scarcity value of time because the wage is also compensating for the disutility of work. In other words, the opportunity cost of diverting an hour of time away from work is less than the wage rate.

In contrast to this suggestion, McConnell and Strand (1981) derived an alternative model that allowed for the estimation of the relationship between the scarcity value of time and the wage rate from the sample data.

Suppose the demand for visits can be specified as a linear function of travel cost:

$$r = a + b_1[c + (\mu/\lambda)(t_1 + t_2)] \tag{13-21}$$

First, let s be the ratio of the shadow price of time to the wage rate. Substitute $s \cdot p_w$ for μ/λ in equation 13-21. Then estimate

$$r = a + b_1 \cdot c + b_2 \cdot p_w \cdot (t_1 + t_2) \tag{13-22}$$

Because $b_2 \cdot p_w$ must equal $b_1 \cdot s \cdot p_w$, then $s = b_2/b_1$. For McConnell and Strand's sample, s was about 0.6.

In the McConnell and Strand model, although the scarcity value of time depends in part on the wage rate, it depends also on the alternative use of time, specifically on the marginal utility of work. A richer model of time and choice would allow for other utility-yielding uses of time. For example, there could be other leisure activities, or time could be an input into the household production of utility-yielding goods.

Time on Site

As the models described above have shown, time spent on the recreation site is part of the cost of the recreation visit. Therefore, in principle, time on site should be included in the estimated demand function for recreation visits. However, there are some assumptions under which this is not necessary. As McConnell (1985) has shown, if all individuals in a sample choose visits of the same duration and if they all have the same opportunity cost of time, time cost on site becomes part of the constant term in the estimated equation. The assumption of identical duration of visits may be reasonable in some cases, for example, where day-trip activities predominate, or if separate equations are estimated for visits of different duration. However, the assumption of identical opportunity costs of time is probably not reasonable, especially given that most travel cost studies now are based on observations of individuals rather than aggregates of people grouped by distance zone.

A related issue is how to model the choice of the duration of a visit. McConnell (1992) has investigated the implications of incorporating the choice of time on site for the proper specification of the recreation demand function. He specifies the following utility function.

$$u = u(X, r, t_2) \tag{13-23}$$

and assumes that it displays what he calls joint weak complementarity between r and t_2. By this he means that if $r = 0$, the marginal utility of on-site time is zero, and if on-site time is zero, the marginal utility of a trip is zero. The individual must choose r, t_2, and X to maximize equation 13-23, subject to the constraint of equation 13-4, or

$$M + p_w \cdot t^* = X + p_r \cdot r \tag{13-24}$$

where $p_r = f + p_d \cdot d + p_w \cdot (t_1 + t_2)$. Because p_r is not exogenous to the choice problem, it cannot be an argument in a Marshallian demand function. McConnell shows that it is still possible to define a Marshallian demand function based on the exogenous components of trip cost. First, define a new term:

$$p_r^* = f + p_d \cdot d + p_w \cdot t_1 \tag{13-25}$$

After maximizing equation 13-23 and solving for the demand functions, the indirect utility function can be obtained:

$$u = v(p_r^*, p_w, M) \tag{13-26}$$

McConnell then shows that Roy's Identity can be employed to derive a Marshallian demand function that contains only exogenous variables:

$$r(p_r^*, p_w, M) \tag{13-27}$$

This expression can be estimated. For other efforts at modeling the determinants of time on site, see Smith et al. 1983b; Kealy and Bishop 1986; and Wilman 1987.

Conclusions

The role of time in explaining recreation demand and in valuing recreation visits and sites raises some thorny issues for both the standard travel cost and RUM approaches of analysis. Clearly, time is an important variable in the analysis of recreation demand and value. However, numerical estimates of demand and value require either that the numerical value of the shadow price of time be known or that it can be estimated from a model of the choices made regarding the uses of time. A variety of models of choice and time are available in the literature, but as yet, different model structures yield quite different estimates of the shadow price of time. There is no clear basis for preferring one model and its value over other models. Until these issues can be resolved, estimates of recreation values should be presented as conditional upon a specific value of the shadow price of time or a specific modeling approach regarding the role of time. The uncertainty in the estimates that this implies should be acknowledged.

Other Approaches to Valuation

The Hedonic Travel Cost Model

Up to this point, the modeling strategies we have discussed have focused on the demand for visits to a site. Where site demand depends in part on the characteristics of the site, information on site demand can be used to value site characteristics. An alternative strategy for estimating the value of a site characteristic is to examine the implicit prices of characteristics themselves. When the issue is formulated in this manner, hedonic methods immediately come to mind. Brown and Mendelsohn (1984) developed a formal hedonic model of the relationship between the travel cost to a site as a form of price and the characteristics of the site. They implemented the model using data on visits to sites of different characteristics by steelhead anglers. Other empirical applications of the model include Mendelsohn 1983; Bockstael et al. 1987; Smith and Kaoru 1987; and Englin and Mendelsohn 1991.

The key feature of the hedonic travel cost model is the assumption that the cost of visiting any site j from origin i is functionally related to the characteristics of site j. Specifically,

$$C_{ij} = C_i(\boldsymbol{Q}_j) \tag{13-28}$$

where C_{ij} is the per-trip travel cost from origin i to site j and $\boldsymbol{Q}_j$ is a vector of site characteristics q_k, $k = 1, \ldots, k, \ldots, p$, at site j. Equation 13-28 is estimated separately for each origin. So in effect, each origin and the sites visited by people from that origin are treated as belonging to a single hedonic market. If an individual visits more than one site, each site is treated as a separate observation for purposes of estimating equation 13-28. It should be noted that estimating equation 13-28 from data on all visitors from the origin requires assuming that they all have the same opportunity cost of time used for travel.

Smith and Kaoru (1987) are critical of the hedonic travel cost model for requiring this assumption, among other reasons. However, in principle, it should be possible to estimate separate hedonic price functions for subsets of the origin population that are homogeneous on the basis of the opportunity cost of time. This approach would be equivalent to recognizing that the hedonic market is segmented or stratified on the basis of the cost of time. In practice, this approach to the problem would require far more observations and detailed individual data than are likely to be available.

For the representative individual at origin i, utility depends upon the characteristics of the sites visited, the number of visits to each site, and the quantity of the numeraire good. The individual's choice problem can be represented by

$$\max u(r_j, \boldsymbol{Q}_j, X) \tag{13-29}$$

subject to

$$M = X + \sum_{j=1}^{m} r_j \cdot C_i\left(\boldsymbol{Q}_j\right) \tag{13-30}$$

where r_j is the number of visits to site j. The first-order conditions for a maximum include:

$$\frac{\left(\partial u / \partial q_k\right)}{\lambda} = r_j \cdot \left(\frac{\partial C_i}{\partial q_k}\right) \quad \text{for all } k \tag{13-31}$$

and

$$\frac{\left(\partial u / \partial r_j\right)}{\lambda} = C_i\left(\boldsymbol{Q}_j\right) \quad \text{for all } j \tag{13-32}$$

Given equation 13-31, the marginal implicit price of q_k can be used in a second-stage regression to estimate marginal WTP functions for characteristics. Implementation of the second stage must deal with the problems of identification and endogenous prices discussed in Chapter 11 in the context of the hedonic property value model.

All the hedonic travel cost studies cited above report negative implicit prices for at least some characteristics. This means that some individuals from the given origin are actually visiting more distant sites with higher travel costs but lower levels of the attribute in question. This behavior is not consistent with the model of choice laid out in equations 13-28 through 13-32. It is difficult to explain except as the result of misclassification of attributes that are actually bads at the margin or because of statistical problems such as measurement error or multi-collinearity (Pendleton 1999). Smith and others (1991) have proposed an approach that avoids negative prices. They calculate marginal implicit prices by estimating a "best-practice" or least-cost frontier of the availability of characteristics. This approach constrains the marginal implicit prices of desirable characteristics to be non-negative.

The hedonic travel cost model as described here has come under significant criticism, especially by Bockstael and others (1991, 233–234), Smith and Kaoru (1987), and Bockstael and McConnell (1999). Much of this criticism has focused on the following problem. Unlike the hedonic property value model, in the hedonic travel cost model there is no marketlike process that brings the marginal implicit travel costs of each characteristic into a market-clearing relationship with the quantities of those characteristics. Whereas travel cost is an implicit price affecting individuals' behavior, there is no feedback from this behavior to adjust the implicit price of the characteristic. In the property value model, an increase in the quantity of the characteristic at one property makes that property relatively more attractive. The market bids its price up until a new equilibrium is reached. In the hedonic travel cost model, there is no similar mechanism operating to clear the market should there be an increase in the level of some characteristic at a recreation site. In the absence of appropriate adjustment mechanisms, observations of marginal implicit prices will not necessarily reveal individuals' marginal values. However, see Pendleton 1999 and Pendleton and Mendelsohn 2000 for vigorous defenses of the hedonic travel cost model.

Estimating Participation Rates

So far, our discussion has been about site-specific analyses that have as their objective the measurement of economic demand and value. Another approach to the analysis of recreation demand and value involves less stringent data requirements, assumptions, and estimation techniques. Such less stringent requirements can be considered advantages; but they entail loss of the ability to infer values from the empirical analysis. This approach is to estimate equations relating the participation in specified recreation activities by a given population to the socioeconomic characteristics of that population and to the supply and quality of recreation opportunities available to that population. If population-specific participation equations can be estimated, it would be possible to predict the increase in participation to be expected with an increase in the supply of recreation opportunities or with an improvement in site amenities.

Examples of this approach include Miller and Hay's analysis of the determinants of participation in duck hunting (Miller and Hay 1981) and Russell and Vaughan's analysis of the effects of improved water quality on recreational fishing activity (Russell and Vaughan 1982; Vaughan and Russell 1982b). In these applications, levels of recreation activity are usually predicted with a two-equation model. The first equation predicts an individual's probability of undertaking the recreation activity in question. The second equation predicts the number of days of activity conditional on being a participant in the activity. Explanatory variables would include one or more measures of the quantity, quality, or both, of the recreation resource available, along with socioeconomic variables. Theory gives no guidance on how to define the areas within which supply or quality are measured or on how different measures might be aggregated into an index of availability.

Once the participation equations have been estimated, changes in participation can be calculated by plugging in alternative values for the relevant quantity and/or quality variables that explain the probability of participation and the number of days of the activity. Although this approach bears a superficial resemblance to a nested RUM, the key difference is that this approach focuses on the determinants of activity levels or participation rates rather than on the determinants of the choices of sites actually visited. Also, this approach has not been derived explicitly from a utility-maximizing framework and does not purport to estimate the parameters of the indirect utility function. Finally, it produces only estimates of changes in the quantity of recreation, not monetary estimates of welfare changes.

Valuing Increases in Activities

In valuing changes in recreation activities as measured by the number of recreation days, the typical practice is to obtain a prediction of the change in recreation days by some means and to multiply this by an assumed constant unit-day value derived independently. See, for example, Miller and Hay 1981 and Vaughan and Russell 1982a. Such unit values could be derived from the results of either revealed preference travel cost and RUM studies or from stated preference studies of recreation values. For example, both Walsh and others (1990) and Smith and Kaoru (1990) have provided lists of unit values or consumer's surpluses per trip derived from reviews of the available literature. When travel cost studies are the source of unit values, the total value or consumer's surplus of the site is divided by the number of activity days or visits to obtain an average consumer's surplus or consumer's surplus per visit or per day.

This approach to valuing changes in activity levels can be criticized on the grounds that it is based upon average values when marginal values are relevant. Moreover, if the average quality of the recreation resource influences activity rates, applying a unit value to the change in activity levels does not capture the value of the change in quality to the existing users.

As Morey (1994) has shown, the validity of using a consumer's surplus per day to value changes in recreation activity depends on what has caused the change. If the change arises from a change in the price of the activity, there is a constant consumer's surplus per activity day and it is equal to the change in price. The welfare value of the change in price is bounded from below (above) by the product of the constant unit value and the old (new) levels of recreation activity. If the increase in activity is the result of an increase in the supply of resources suitable for recreation, for example from the creation of new sites or the opening of sites previously closed because of pollution, it could be argued that the increase in supply was equivalent to a reduction in the effective price of recreation.

We would then need to compute this change in effective price of the site in question.

However, if the change in activity is due to a change in the quality of the site, a constant consumer's surplus per day of activity implies that the consumer's surplus per day is independent of the number of days at or visits to the site. Although this assumption is implied by the structure of the simple RUM model of value with its emphasis on a single choice occasion, it is not consistent with the standard travel cost model and the general property of diminishing marginal rates of substitution between recreation visits and all other goods.

Summary

Both of the behavioral models presented in this chapter, the travel cost model and the random utility model, draw inferences about individuals' willingness to pay for the recreation services of a site by treating the cost of travel to the site as an implicit price. However, the two models are based on quite different views of how individuals make recreation choices. The standard travel cost model is a direct offshoot of conventional consumer demand theory. It views individuals as selecting quantities of visits to different sites as a function of the vectors of prices and quantities. It models behavior over a time period long enough so that individuals can choose many visits, perhaps to several different sites. The strength of the standard model is that it captures individuals' choices of total visits over a recreation season, but the weakness of the standard model is its difficulty in capturing the role of site-specific characteristics or qualities in influencing the distribution of total visits across different sites. One might also question whether the implicit assumption that people choose a total quantity of visits and their distribution across sites at the beginning of the recreation season reflects a realistic picture of recreation behavior.

In contrast, the random utility model focuses specifically on each separate choice occasion when individuals are deciding which site to visit. The selection of a site is explained by the price and characteristics of that site relative to the alternatives in the choice set, so site characteristics play a central role in explaining recreation decisions. This is the strength of the random utility model. The weakness of the model is its difficulty in explaining total visits over a season.

The travel cost model is better suited to answering the first of the three questions posed at the beginning of this chapter, that is, how the value of an existing site is to be measured. This is because the travel cost model is better at predicting the total number of trips to a site. The random utility model does a better job of answering the third question, how the value of a change in site quality is to be measured. This is because the random utility model does a better job of capturing the role of site characteristics in influencing the decisions about whether to participate in recreation and the choice of a site to visit. The random utility model also probably does a better job of answering the second question, how the value of adding a site to a system is to be estimated. This is because the random utility model provides an estimate of the indirect utility function itself. This function can be used to calculate the welfare values of changes in the vectors of available characteristics and their costs or prices.

Both models require that there be an observable cost or price of a visit. A major difficulty in implementing either model is observing the monetary value of the time component of cost of a visit. At least one author (Randall 1994) has argued that this is a fundamental limitation of the travel cost and RUM models. However, we have shown that there are modeling approaches that make it possible to draw inferences about the monetary value of the opportunity cost of time.

References

Bockstael, Nancy E., W. Michael Hanemann, and Catherine L. Kling. 1987. Estimating the Value of Water Quality Improvements in a Recreational Demand Framework. *Water Resources Research* 23(5): 951–960.

Bockstael, Nancy E., W. Michael Hanemann, and Ivar E. Strand, Jr. 1986. *Measuring the Benefits of Water Quality Improvements Using Recreation Demand Models.* Report to the U.S. Environmental Protection Agency. College Park, MD: University of Maryland.

Bockstael, Nancy E., and Kenneth E. McConnell. 1999. The Behavioral Basis of Non-Market Valuation. In *Valuing Recreation and the Environment: Revealed Preference Methods in Theory and Practice,* edited by Joseph A. Herriges and Catherine L. Kling. Cheltenham, U.K.: Edward Elgar.

Bockstael, Nancy E., Kenneth E. McConnell, and Ivar [E.] Strand. 1991. Recreation. In *Measuring the Demand for Environmental Quality,* edited by John B. Braden and Charles D. Kolstad. Amsterdam: North-Holland.

Bockstael, Nancy E., and Ivar E. Strand. 1987. The Effect of Common Sources of Regression Error on Benefit Estimates. *Land Economics* 63(1): 11–20.

Bockstael, Nancy E., Ivar E. Strand, and W. Michael Hanemann. 1987. Time and the Recreational Demand Model. *American Journal of Agricultural Economics* 69(2): 293–302.

Brown, Gardner, Jr., and Robert Mendelsohn. 1984. The Hedonic Travel Cost Method. *Review of Economics and Statistics* 66(3): 427–433.

Burt, Oscar R., and Durward Brewer. 1971. Estimation of Net Social Benefits from Outdoor Recreation. *Econometrica* 39(5): 813–827.

Calfee, John, Clifford Winston, and Randolph Stempki. 2001. Econometric Issues in Estimating Consumer Preferences from Stated Preference Data: A Case Study of the Value of Automobile Travel Time. *Review of Economics and Statistics* 83(4): 699–707.

Caulkins, Peter P., Richard C. Bishop, and Nicolaas W. Bouwes [Sr.]. 1985. Omitted Cross-Price Variables in the Linear Travel Cost Model: Correcting Common Misperceptions. *Land Economics* 61(2): 182–187.

———. 1986. The Travel Cost Model for Lake Recreation: A Comparison of Two Methods for Incorporating Site Quality and Substitution Effects. *American Journal of Agricultural Economics* 68(2): 291–297.

Cesario, Frank J. 1976. Value of Time and Recreation Benefit Studies. *Land Economics* 52(1): 32–41.

Cesario, Frank J., and Jack L. Knetsch. 1970. Time Bias in Recreation Benefit Estimates. *Water Resources Research* 6(3): 700–704.

Cicchetti, Charles J., Anthony C. Fisher, and V. Kerry Smith. 1976. An Econometric Valuation of a Generalized Consumer Surplus Measure: The Mineral King Controversy. *Econometrica* 44(6): 1259–1276.

Cicchetti, Charles J., and V. Kerry Smith. 1973. Congestion, Quality Deterioration and Optimal Use: Wilderness Recreation in the Spanish Peaks Primitive Area. *Social Science Research* 2(1): 15–30.

Dorfman, Robert. 1984. On Optimal Congestion. *Journal of Environmental Economics and Management* 11(2): 91–106.

Englin, Jeffrey, and Robert Mendelsohn. 1991. A Hedonic Travel Cost Analysis for Valuation of Multiple Components of Site Quality: The Recreation Value of Forest Management. *Journal of Environmental Economics and Management* 21(3): 275–290.

Englin, Jeffrey, and J.S. Shonkwiler. 1995. Modeling Recreation Demand in the Presence of Unobservable Travel Costs: Toward a Travel Price Model. *Journal of Environmental Economics and Management* 29(3): 368–377.

Feather, Peter, and W. Douglass Shaw. 1999. Estimating the Cost of Leisure Time for Recreation Demand Models. *Journal of Environmental Economics and Management* 38(1): 49–65.

Feenberg, Daniel, and Edwin S. Mills. 1980. *Measuring the Benefits of Water Pollution Abatement.* New York: Academic Press.

Fisher, Anthony C., and John V. Krutilla. 1972. Determination of Optimal Capacity of Resource-based Recreation Facilities. *Natural Resources Journal* 12(3): 417–444.

Freeman, A. Myrick, III. 1993. *The Measurement of Environmental and Resource Values: Theory and Methods.* Washington, DC: Resources for the Future.

Freeman, A. Myrick, III, and Robert H. Haveman. 1977. Congestion, Quality Deterioration, and Heterogeneous Tastes. *Journal of Public Economics* 8(2): 225–232.

Graham-Tomasi, Theodore, Wiktor L. Adamowicz, and Jerald J. Fletcher. 1990. Errors of Truncation in Approximations to Expected Consumer Surplus. *Land Economics* 66(1): 50–55.

Hanemann, W. Michael. 1978. "A Methodological and Empirical Study of the Recreation Benefits from Water Quality Improvement." Ph.D. dissertation, Harvard University, Cambridge, MA.

———. 1999. Welfare Analysis with Discrete Choice Models. In *Valuing Recreation and the Environment: Revealed Preference Methods in Theory and Practice,* edited by Joseph A. Herriges and Catherine L. Kling. Cheltenham, U.K.: Edward Elgar.

Hausman, Jerry A. 1981. Exact Consumer's Surplus and Dead Weight Loss. *American Economic Review* 71(4): 662–676.

Herriges, Joseph A., and Catherine L. Kling (eds.). 1999. *Valuing Recreation and the Environment: Revealed Preference Methods in Theory and Practice.* Cheltenham, U.K.: Edward Elgar.

Johnson, M. Bruce. 1966. Travel Time and the Price of Leisure. *Western Economics Journal* 4(1): 135–145.

Kealy, Mary Jo, and Richard C. Bishop. 1986. Theoretical and Empirical Specification Issues and Travel Cost Demand Studies. *American Journal of Agricultural Economics* 68(3): 660–667.

McConnell, Kenneth E. 1980. Valuing Congested Recreation Sites. *Journal of Environmental Economics and Management* 7(4): 289–294.

———. 1985. The Economics of Outdoor Recreation. In *Handbook of Natural Resource and Energy Economics.* Vol. 1, edited by Allen V. Kneese and James L. Sweeney. Amsterdam: North-Holland.

———. 1986. *The Damages to Recreational Activities from PCBs in New Bedford Harbor.* Cambridge, MA: Industrial Economics.

———. 1992. On-Site Time in the Demand for Recreation. *American Journal of Agricultural Economics* 74(4): 918–925.

McConnell, Kenneth E., and Virginia A. Duff. 1976. Estimating Net Benefits of Outdoor Recreation under Conditions of Excess Demand. *Journal of Environmental Economics and Management* 2(1): 224–230.

McConnell, Kenneth E., and Ivar [E.] Strand. 1981. Measuring the Cost of Time in Recreation Demand Analysis: An Application to Sport Fishing. *American Journal of Agricultural Economics* 63(1): 153–156.

McKean, John R., Donn M. Johnson, and Richard G. Walsh. 1995. Valuing Time in Travel Cost Demand Analysis: An Empirical Investigation. *Land Economics* 71(1): 96–105.

McKean, John R., and Charles F. Revier. 1990. An Extension of "Omitted Cross-Price Variable Biases in the Linear Travel Cost Model: Correcting Common Misperceptions." *Land Economics* 66(4): 431–436.

Mendelsohn, Robert. 1983. An Application of the Hedonic Travel Cost Framework for Recreation Modeling to the Valuation of Deer. In *Advances in Applied Microeconomics,* edited by V. Kerry Smith and Anne D. Witte. Greenwich, CT: JAI Press.

Miller, Jon R., and Michael J. Hay. 1981. Determinants of Hunter Participation: Duck Hunting in the Mississippi Flyway. *American Journal of Agricultural Economics* 63(4): 677–684.

Morey, Edward R. 1981. The Demand for Site-Specific Recreational Activities: A Characteristics Approach. *Journal of Environmental Economics and Management* 8(4): 345–371.

———. 1984. The Choice of Ski Areas: Estimation of a Generalized CES Preference Ordering with Characteristics. *Review of Economics and Statistics* 66(4): 584–590.

———. 1985. Characteristics, Consumer Surplus, and New Activities: A Proposed Ski Area. *Journal of Public Economics* 26(2): 221–236.

———. 1994. What Is Consumer Surplus Per Day of Use, When Is It a Constant Independent of the Number of Days of Use, and What Does It Tell Us about Consumer Surplus? *Journal of Environmental Economics and Management* 26(3): 257–270.

———. 1999. TWO RUMS unCLOAKED: Nested-Logit Models of Site Choice and Nested-Logit Models of Participation and Site Choice. In *Valuing Recreation and the Environment: Revealed Preference Methods in Theory and Practice,* edited by Joseph A. Herriges and Catherine L. Kling. Cheltenham, U.K.: Edward Elgar.

Morey, Edward R., Robert D. Rowe, and Michael Watson. 1993. A Repeated Nested-Logit Model of Atlantic Salmon Fishing. *American Journal of Agricultural Economics* 75(3): 578–592.

Parsons, George R. 1991. A Note on Choice of Residential Location in Travel Cost Demand Models. *Land Economics* 67(3): 360–364.

———. 2003. The Travel Cost Model. In *A Primer on Nonmarket Valuation,* edited by Patricia A. Champ, Kevin J. Boyle, and Thomas C. Brown. Dordrecht, Netherlands: Kluwer Academic Press.

Parsons, George R., and Mary Jo Kealy. 1992. Randomly Drawn Opportunity Sets in a Random Utility Model of Lake Recreation. *Land Economics* 68(1): 93–106.

Pendleton, Linwood. 1999. Reconsidering the Hedonic *vs.* RUM Debate in the Valuation of Recreational Environmental Amenities. *Resource and Energy Economics* 21(2): 167–189.

Pendleton, Linwood, and Robert Mendelsohn. 2000. Estimating Recreation Preferences Using Hedonic Travel Cost and Random Utility Models. *Environmental and Resource Economics* 17(1): 89–108.

Phaneuf, Daniel J., and V. Kerry Smith. Forthcoming. Recreation Demand Models. In *Handbook of Environmental Economics,* edited by Karl-Göran Mäler and Jeffery R. Vincent. Amsterdam: North-Holland.

Randall, Alan. 1994. A Difficulty with the Travel Cost Method. *Land Economics* 70(1): 88–96.

Russell, Clifford S., and William J. Vaughan. 1982. The National Recreational Fishing Benefits of Water Pollution Control. *Journal of Environmental Economics and Management* 9(4): 328–354.

Shaw, W. Douglass, and Peter Feather. 1999. Possibilities for Including the Opportunity Cost of Time in Recreation Demand Models. *Land Economics* 75(4): 592–602.

Smith, V. Kerry. 1981. Congestion, Travel Cost Recreational Demand Models, and Benefit Evaluation. *Journal of Environmental Economics and Management* 8(1): 92–96.

———. 1990. Estimating Recreation Demand Using the Properties of the Implied Consumer Surplus. *Land Economics* 66(2): 111–120.

Smith, V. Kerry, and William H. Desvousges. 1985. The Generalized Travel Cost Model and Water Quality Benefits: A Reconsideration. *Southern Economics Journal* 52(2): 371–381.

Smith, V. Kerry, William H. Desvousges, and Ann Fisher. 1986. A Comparison of Direct and Indirect Methods for Estimating Environmental Benefits. *American Journal of Agricultural Economics* 68(2): 280–290.

Smith, V. Kerry, William H. Desvousges, and Matthew P. McGivney. 1983a. Estimating Water Quality Benefits: An Econometric Analysis. *Southern Economic Journal* 50(2): 422–437.

———. 1983b. The Opportunity Cost of Travel Time in Recreation Demand Models. *Land Economics* 59(3): 259–278.

Smith, V. Kerry, and Yoshiaki Kaoru. 1987. The Hedonic Travel Cost Model: A View from the Trenches. *Land Economics* 63(2): 179–192.

———. 1990. Signals or Noise? Explaining the Variation in Recreation Benefit Estimates. *American Journal of Agricultural Economics* 72(2): 419–433.

Smith, V. Kerry, Raymond B. Palmquist, and Paul Jakus. 1991. Combining Farrell Frontier and Hedonic Travel Cost Models for Valuing Estuarine Quality. *Review of Economics and Statistics* 73(4): 694–699.

Vartia, Yrjo O. 1983. Efficient Methods of Measuring Welfare Change and Compensated Income in Terms of Ordinary Demand Functions. *Econometrica* 51(1): 79–98.

Vaughan, William J., and Clifford S. Russell. 1982a. *Freshwater Recreational Fishing: The National Benefits of Water Pollution Control.* Washington, DC: Resources for the Future.

———. 1982b. Valuing a Fishing Day: An Application of a Systematic Varying Parameter Model. *Land Economics* 58(4): 451–463.

Walsh, Richard G., Donn M. Johnson, and John R. McKean. 1990. Nonmarket Values from Two Decades of Research on Recreation Demand. In *Advances in Applied Microeconomics,* vol. 5, edited by Arthur N. Link and V. Kerry Smith. Greenwich, CT: JAI Press.

Willig, Robert D. 1976. Consumer's Surplus without Apology. *American Economic Review* 66(4): 589–597.

Wilman, Elizabeth A. 1987. A Simple Repackaging Model of Recreational Choices. *American Journal of Agricultural Economics* 69(3): 604–612.

CHAPTER
14

Conclusions

In this chapter, I first take up three additional topics that do not fit easily into any of the previous chapters. The first is the use of values borrowed from studies of values in other settings to estimate the value of the environmental change of interest. This practice has come to be known as "benefits transfer." The second is the use of data from both revealed preference and stated preference analyses to estimate a single valuation model. The third is the estimation of values for the services provided by ecosystems. I then conclude with an assessment of the current state of the art of nonmarket valuation.

Benefits Transfer

Benefits transfer refers to the practice of applying nonmarket values obtained from primary studies of resource or environmental changes undertaken elsewhere to the evaluation of a proposed or observed change that is of interest to the analyst. Examples include

- using values per day for recreational angling at one lake obtained from a travel cost demand study to value an increase in the same activity at another lake;
- using the willingness to pay to avoid an asthma attack obtained from an averting behavior study to value the prevention of a day of respiratory symptoms; and
- using the willingness to pay to preserve a square mile of tropical rain forest in one country estimated from a stated preference study to value the preservation of a square mile of rain forest in another country.

The practice of benefits transfer became common in the economic analysis of environmental regulations in the United States in the mid-1980s even before any systematic development of either terminology or procedures and protocols and certainly before any rigorous testing of the validity of the practice. All of this changed in 1992 with the publication of a set of nine papers on benefits transfer in a special issue of *Water Resources Research*; see Brookshire and Neill 1992 for an introduction and overview of these papers.

It is common to refer to the environmental policy being evaluated as the "policy site" and the source of the values being used as the "study site," even though many of the policies for which benefits transfer are used are not site specific. Desvousges and others (1998) suggested using "transfer context" and "original context" as alternative terminology, but their suggestion has not caught on as of this writing.

In principle, the values at the policy site can be different from those of the study site for two sets of reasons: differences in the characteristics of the two environmental features being valued ("supply side" factors) and differences between the populations making use of, or at least valuing, the environmental or resource change ("demand side" factors). The latter can include differences in income, tastes and preferences, and other relevant socioeconomic characteristics. Thus we must consider the question of how values can be adjusted in the transfer process to reflect these two types of differences. To examine this question, we look at the types of procedures available for benefits transfer from the simplest to the most sophisticated. There are two useful sources for more detailed information on how to design and implement a benefits transfer. One (Rosenberger and Loomis 2003) is a fairly basic "how to do it" that walks the reader through the steps involved, using three case studies. The other (Desvousges et al. 1998) is a more comprehensive and rigorous treatment that includes discussions of several econometric issues in analyzing and summarizing study site data.

The simplest procedure is to select the study site that in the analyst's judgement is most similar to the policy site in terms of both demand-side and supply-side characteristics. If there is more than one suitable study site (or more than one estimated value for a given study site), one could use a range, calculate a mean or median value for the distribution, or perhaps subjectively weight the values based on judgements of their quality before taking the mean. If the study site and policy site values are for different years, the study site value should be adjusted for changes in prices using a suitable price index. However, adjusting for inflation in this way involves the assumption that the budget shares of expenditures of both the study site and policy site populations are similar to the weights used in constructing the price index (Eiswerth and Shaw 1997). If there is information available on the income levels of the two populations, it would also be appropriate to adjust the study site WTP using an estimate of the income elasticity of WTP. The relevant elasticity is not the income elasticity of demand for the good. As Flores and Carson (1997) have shown, the relationship between the two elasticity concepts is complex.

This simple procedure does not allow for more systematic adjustments of study site values to account for differences in site characteristics and populations. To make such adjustments, it is necessary to have information on how values are related to the relevant characteristics. Some information of this sort is available if the study site data includes a valuation function or WTP function, that is, a statistical relationship between the WTPs of individuals in the sample and their socioeconomic characteristics. The adjusted policy site value can be calculated by plugging mean values for the policy site population into the valuation function.

A more sophisticated approach is to apply meta-analytical techniques to a larger sample of study site data that represents a variety of population and site characteristics. The meta-analysis equation can also include variables that reflect differences in methods, for example, revealed preference versus various forms of stated preference methods. Examples include Smith and Kaoru 1990 for recreation and Mrozeck and Taylor 2002 for the value of statistical life. Desvousges et al. 1998 and Rosenberger and Loomis (forthcoming) contain references to additional examples.

Finally, Smith and others (2002) have developed and applied a method, based on data from several studies, using different methods to "calibrate" the parameters of an assumed underlying preference function over site characteristics. The calibrated preference function can then be used to calculate values for a variety of policy site proposals.

An important question concerns the validity of using benefit transfer values as a substitute for doing original valuation research for the policy being evaluated. Any assessment of the validity of benefits transfer requires a yardstick. The yardstick used by those who have investigated this question is an estimate of value obtained by stated preference or revealed preference methods at the policy site, so these are tests of convergent validity, not criterion validity (see Chapter 6).

Rosenberger and Loomis 2003 and Shrestha and Loomis 2001 both review recent studies of validity; see these papers for references. Also Shrestha and Loomis (2001) performed their own validity test of benefits transfer of outdoor recreation activities across countries based on a meta-analysis of more than 600 recreation values from U.S. studies. Overall the results are mixed, but one finding stands out. Convergent validity is generally higher when the transfer is done using either a valuation function or meta-analysis. This confirms the importance of using systematic methods to adjust study site values for differences in population and site characteristics. Nevertheless, significant differences between transfer values and original values often occur. For example, Shrestha and Loomis (2001) conclude that the absolute average percentage difference between transfer values and the original values was 28%. However, they observe that this may still be acceptable for many transfer applications because this degree of imprecision would not obscure large differences between the benefits and costs of policy actions being evaluated.

In recognition of the popularity of using benefits transfer and the potential savings in analytical costs that are possible, several organizations have created and made available on the Internet large databases of valuation studies that can be used in benefits transfer. They provide not only the estimated values from each study but other data on study methodology in a standardized format so that studies can be evaluated for their suitability for the transfer exercise. The largest and most detailed of these is the Environmental Valuation Reference Inventory (EVRI), which was created and is maintained by the Canadian federal environmental agency, Environment Canada. It is a searchable storehouse of empirical studies on the economic value of environmental benefits and human health effects. Access to EVRI is free for Canadian residents, and subscription information is available at the EVRI website (http://www.evri.ec.gc.ca/evri/). The Environmental Protection Agency of New South Wales has also created its own environmental valuation database, called ENVAL, which focuses on Australian and southeast Asian studies (http://www2.epa.nsw.gov.au/envalue/).

There are also more specialized databases. For example the University of California has created a database of economic values for beneficial uses of water (http://buvd.ucdavis.edu/), and the U.S. Fish and Wildlife Service maintains a sports fishing value database (http://www.indecon.com/fish/).

Combining Revealed Preference and Stated Preference Data

It seems natural to think of revealed preference (RP) and stated preference (SP) methods as substitutes or alternative ways of estimating values for a given change in environmental or resource conditions. However, several authors have shown that it is possible to use RP and SP methods as complements by combining RP and SP data to estimate a single valuation model. These combined models embody the hypothesis that the RP and SP data come from the same structure of preferences or utility function. Testing this hypothesis can be viewed as a kind of convergent validity test.

In the first published study using this approach, Cameron (1992) specified a travel cost demand model to explain the numbers of fishing trips taken in a year by individuals in her sample of Texas anglers. She also specified a stochastic utility difference model to explain the same individuals' responses to a contingent behavior question of the form "If the travel cost of your trips were $\$X$ more, would you stop visiting these sites altogether?" The functional forms of the two equations were derived from an assumed underlying quadratic utility function. She constrained the common parameters of the two equations to be equal, estimating them using a simultaneous equation technique. She then used the estimated utility parameters to generate welfare measures for various changes in the conditions of access to the sites to illustrate the method.

Adamowicz and others (1994) designed a survey explicitly to generate both RP and SP data of recreation choices. Their RP data were used to estimate a random utility model of recreation site choices. This was then combined with responses to a set of SP questions about site choice based on the same set of site attributes. The SP questions were designed to eliminate collinearity among attributes. They estimated the random utility model and SP models separately and then combined the two data sets for a joint estimation. Their results supported the hypothesis that both sets of data came from the same underlying preference structures.

McConnell and others (1999) have also combined RP and SP data on recreation activities to estimate willingness to pay for a recreation trip. Their method allowed differences in the preferences generating the RP and SP data because the SP responses came after individuals had visited the sites in question. However, they could not reject the hypothesis that the parameters in the two models were equal. More recently Earnhart (2001, 2002) has combined SP data with hedonic housing discrete choice data to examine housing choices and to estimate values for environmental amenities linked to residential locations.

As Kling (1997) has shown through simulation studies, combining SP and RP data can reduce bias and improve the precision of welfare estimates. Also the design of the choice sets for the SP part of the study presents an opportunity for examining portions of the individuals' response surfaces or preference structures for which there is no observed behavior.

Valuing the Services of Ecosystems

An ecosystem can be defined as "a spatially explicit unit of the earth that includes all of the organisms, along with all of the components of the abiotic environment within its boundaries" (Likens 1992, as cited in Christensen and Franklin 1997). It is a human construct whose size and scope depend on how an investigator draws the boundaries around the objects of interest. The unit of analysis can be as small as the gut of an insect and as large as the delta area of a major river system. Each ecosystem is characterized by its spatial dimensions, its species composition, the functions or processes that it carries out, and the services that it provides to people. By ecosystem functions or processes, I mean the normal characteristic actions or activities of the system that are necessary for its self-maintenance (Whigham 1997, 231). By ecosystem services, I mean the results of these functions and processes that accrue to people and "sustain and fulfill human life" (Daily 1997, 3; Daily et al. 1997, 1), that is, that enhance human well-being.

A number of authors have provided lists of ecosystem services. Gretchen Daily's (1997, 3–4) list is representative and includes the following:

- purification of air and water,
- mitigation of floods and droughts by regulation of hydrological cycles,
- detoxification and decomposition of wastes,
- generation and renewal of soils,
- pollination of crops and natural vegetation,
- control of agricultural and other pests,
- nutrient recycling,
- partial climate stabilization, and
- providing aesthetic beauty and intellectual stimulation.

Other authors (for example, Costanza et al. 1997) have added such things as erosion control, decomposition of wastes, habitat or refugia for species and preservation of biodiversity, and production of food and raw materials. Some of these services affect humans directly, as in the cases of food and raw material production. These direct services can be either market or nonmarket services. Other services affect people only indirectly, as in the cases of nutrient recycling and the regeneration of soil. In these cases, it is necessary to establish the link between the ecosystem service being valued and the channel through which it affects people. I have more on this point below.

Given this anthropocentric definition of ecosystem services, the economic concepts of willingness to pay and willingness to accept compensation provide the conceptual basis for defining the economic values of these services. The methods and models described in this book provide the means for estimating values, at least in principle. Estimating the economic value of an ecosystem service involves three steps. The first is determining the nature and size of the environmental change affecting ecosystem structure and function. The change could be in the spatial area of a particular type of habitat, for example, freshwater wetlands, in the populations of species present, or in the fluxes of energy or nutrients through the system. The second step involves determining how these changes affect the quantities and qualities of ecosystem service flows to people. The third step involves using existing economic methods where available to assess the changes in people's well-being, as measured in dollars.

When an ecosystem service supports the production of a marketed commodity, the value of a change in that service is the sum of the changes in consumers' and producers' surpluses in that market. For example, an increase in the population of a pollinating insect could increase the output of agricultural crops, resulting in lower prices to consumers, greater quasi-rents to producers, or both. Tidal wetlands shelter the young of commercially valuable fish species. Changes in the area of wetlands have been related to changes in commercial harvest of fish and blue crabs (Lynne et al. 1981; Bell 1989). The methods described in Chapter 9 are available for this type of ecosystem service.

When ecosystem functions support nonmarket environmental services, we may be able to draw on the tool kit of nonmarket valuation methods to determine the economic values of changes in these service flows. For example, when a change in an ecosystem service results in an improvement in the quality of outdoor recreational experiences, travel cost and related models of recreational demand can be used to estimate the value of the service flow, and SP methods might be used to value aesthetic services.

To estimate the economic value of a basic ecosystem function, we need to know the link between that function and the ecosystem service flows that it supports. This will not always be easy to uncover. One approach to establishing this link is to think of the relevant components of the ecosystem as being involved in a production process. For discussions of the production function approach, see Barbier 1994, 2000. Under this approach, the ecosystem is assumed to be an equilibrium system that can be subjected to comparative static analysis to determine changes in service flows in response to changes in ecosystem conditions. For an example, see Crocker and Tschirhart 1992.

One complication, however, is that in terms of production theory, ecosystems are multiproduct production systems in which jointness in production is likely to be a dominant feature. For example, a species of bird might be valued both for its pollination of a commercial fruit species and for its control of insects that damage some other commercially valuable plant. The value of the bird species is the sum of the values of all its services. However, the jointness in production must be taken into account when estimating the values of these individual service flows (Chapter 9). Another complication is that the responses of ecosystems to perturbations might display nonlinearities, discontinuities, multiple endpoints, and even chaotic behavior, especially for changes in the populations of species and fluxes of energy or nutrients (Levin and Pacala forthcoming). In fact some aspects of ecosystem behavior might be fundamentally unpredictable (Huisman and Weissing 2001). For these reasons, economists may have more success in estimating the values of changes in the spatial extent of an ecosystem than changes in other characteristics of the system.

It is sometimes suggested that the cost of replacing a function of an ecological system with a human-engineered system can be used as a measure of the economic value of the function itself. In a classic example, Gosselink and others (1974) used an estimate of the cost of a tertiary sewage treatment as the economic value of the nutrient removal function of a tidal wetland. More recently Chichilnisky and Heal (1998) estimated the water purification function of a large protected watershed area in the Catskill Mountains of New York to be $6–8 billion. Replacement cost can be a valid measure of economic value only if three conditions are met: the human-engineered system must provide services of equivalent quality and magnitude, the human-engineered system must be the least costly alternative, and individuals in aggregate must be willing to incur these costs if the natural service were not available (Shabman and Batie 1978). When these conditions are not met, there is no presumption that replacement cost is either an overestimate or an underestimate of true economic value; the two numbers are measures of different things. In the Catskill Mountain case, the above conditions arguably were met because the city of New York would have been forced to spend $6–8 billion building a water filtration plant for the city water supply if it did not invest in watershed protection.

The State of the Art

In the concluding chapter of *The Benefits of Environmental Improvement: Theory and Practice* (Freeman 1979), I wrote:

> Suppose the administrator of EPA wished to know the magnitude of the benefits accruing from a given pollution control policy. If asked, I believe an economist could specify the economic theory and models he would use, the data he would like to have, and the empirical techniques he would apply to the data to obtain measures of benefits. (Freeman 1979, 248)

I added five qualifications to this optimistic assessment of the state of the art. Briefly, they were:

1. Where RP methods were not available (for example, for measuring nonuse values), the economist would have to resort to SP measures. I was concerned about "the accuracy of responses of individuals in the necessarily hypothetical situations they pose" (Freeman 1979, 249).
2. In the valuation of reductions in mortality risks, there did not appear to be broad acceptance of the idea that individuals' behavior could or should be the basis of welfare measures.

3. Because all welfare measures are conditioned upon the existing distribution of income, to accept these welfare measures for making public policy choices involved accepting the existing distribution as satisfactory.
4. Economic measures of welfare change "must be built upon noneconomic data such as dose–response functions" (Freeman 1979, 250), which often were not available.
5. The data required to implement many of the models and methods would often be difficult and costly to obtain.

In looking back over this list of qualifications today, I see less reason to be concerned about the first two of them. Regarding the first qualification, I think that substantial progress has been made in the development of SP methods. I am now more optimistic about the ability of properly framed questions to generate meaningful data on preferences and values. This does not mean that all the investigator has to do is go out and ask people some questions. However, well-designed SP studies have a place in economists' toolboxes. As to the second qualification, although there is still some debate about it, I think that there is now fairly wide acceptance of the use of RP measures of the value of risk reduction for evaluating public policies that affect health and safety. Such measures are sanctioned, for example, by the U.S. Environmental Protection Agency for use in performing regulatory impact analyses (U.S. EPA 2000). As noted in Chapters 10 and 12, we have a substantial body of evidence on the magnitude of risk premiums revealed in various markets. Rather, the terms of the debate have shifted; it now revolves around whether, in valuing risk reduction policies, differences in age, income, health status, cause of death, and the level of baseline risk should be taken into account, and if so, how.

Having said this, however, I also want to add a new qualification to the list. It has to do with the process of fitting data to a model to generate a welfare measure and the effects of the choice of a model and functional form on the welfare measure. There has been a virtual explosion in the number of models available to be exploited, and a number of rich data sets have been created. Several studies show that welfare measures can be sensitive to the choice of a model or the choice of a functional form for a specific model. Some of these studies involve simulations in which the "true" welfare measure is known and can be compared with estimates derived by fitting the simulated data to alternative models; for examples, see Cropper et al. 1988, 1993. Other studies involve examining the sensitivity of welfare estimates to changes in some aspect of the specification of the model being applied to real data; for examples, see Kaoru and Smith (1990) and Cooper and Loomis (1992). Because the true model cannot be known, we must add model uncertainty to the list of sources of uncertainty in welfare measures. An important area for future research is to obtain a better understanding of the sources and properties of model uncertainty.

In the first chapter of this book, I discussed several ways of classifying the types of environmental and resource service flows for which value measures might be desired. I conclude now by briefly outlining the types of models that can be applied to value each type of service flow. This information is summarized in Table 14-1. The first thing to note is that SP models apply to the estimation of all types of service flows except those that work through changes in market prices (producers' and consumers' surpluses). SP methods are all that is available for estimating nonuse values.

For valuing mortality risks, some version of the hedonic wage model is probably the method of choice, although averting behavior and SP studies have been conducted in this area with some success. Some form of averting behavior model would be most appropriate for valuing the nonfatal health effects as well. Even the health production function model can be considered a special case of the averting behavior models in that it is based on substitutability among market and nonmarket inputs into the production of health.

Hedonic property value models capture the value of differences in amenity and disamenity levels across residential locations, but they might also measure other environmental effects caused by the same environmental agent. For example, air pollution may reduce visibility and cause adverse health effects and damages to household materials, all of which would be reflected in property prices. Hedonic wage models can be used to value differences in the average levels of amenities or disamenities among cities or regions.

Human exploitation of natural and managed ecosystems in agriculture, commercial fisheries, and forestry is largely managed through market mechanisms. Thus the values of changes in these ecological services will be reflected in changes in prices and incomes. The cost function and production function models described in Chapter 9 provide the basis for estimating the resulting changes in producers' and consumers' surpluses. However, these models are also applicable where changes in other environmental services affect the productivity and costs of firms. Examples include the effects of water quality on processing costs and the effects of air pollution on manufacturing costs and on the costs of repair and maintenance of structures.

Finally, humans use both natural and managed ecosystems for various forms of recreation. The standard travel cost and random utility models are available to exploit information on the implicit price of access and on the cost of traveling to a site to estimate the values of sites and changes in their characteristics.

Table 14-1. Models for Estimating the Values of Environmental Service Flows

Type of environmental service flow	*Type of model*	*Described in (chapter)*
Human health		
Mortality risk	Averting behavior	10
	Contingent behavior	6
	Contingent valuation	6
	Hedonic wages	10, 12
	Stated choice	6
Chronic morbidity	Averting behavior	10
	Contingent behavior	6
	Contingent valuation	6
	Stated choice	6
Acute morbidity	Averting behavior	10
	Contingent behavior	6
	Contingent valuation	6
	Health production function	10
	Stated choice	6
Other direct impacts on humans: amenities, visibility, noise	Averting behavior	4
	Contingent behavior	6
	Contingent valuation	6
	Hedonic property values	11
	Hedonic wages, interurban	12
	Stated choice	6
Economic productivity of ecological systems	Changes in producers' and consumers' surpluses	9
Other ecological services: recreation	Contingent behavior	6
	Contingent valuation	6
	Random utility model	13
	Stated choice	6
	Travel cost model with weak complementarity	13
Effects on nonliving systems, such as materials	Averting behavior	4
	Changes in producers' and consumers' surpluses	9
	Contingent behavior	6
	Contingent valuation	6
	Hedonic property values	11
	Stated choice	6
Nonuse values such as ecological stability and biodiversity	Contingent valuation	6
	Stated choice	6

References

Adamowicz, W., J. Louviere, and M. Williams. 1994. Combining Revealed and Stated Preference Methods for Valuing Environmental Amenities. *Journal of Environmental Economics and Management* 26(3): 271–292.

Barbier, Edward B. 1994. Valuing Environmental Functions: Tropical Wetlands. *Land Economics* 70(2): 155–173.

———. 2000. Valuing the Environment as Input: Review of Applications to Mangrove-Fishery Linkages. *Ecological Economics* 35(1): 47–61.

Bell, Frederick W. 1989. *Application of Wetland Evaluation Theory to Florida Fisheries.* Tallahassee, FL: The Florida Sea Grant College.

Brookshire, David S., and Helen R. Neill. 1992. Benefits Transfer: Conceptual and Empirical Issues. *Water Resources Research* 28(3): 651–655.

Cameron, Trudy A. 1992. Combining Contingent Valuation and Travel Cost Data for the Valuation of Nonmarket Goods. *Land Economics* 68(3): 302–317.

Chichilnisky, Graciela, and Geoffrey Heal. 1998. Economic Returns from the Biosphere. *Nature* 391: 629–630.

Christensen, Norman L., Jr., and Jerry F. Franklin. 1997. Ecosystem Function and Ecosystem Management. In *Ecosystem Function and Human Activities: Reconciling Economics and Ecology,* edited by R. David Simpson and Norman L. Christensen, Jr. New York: Chapman and Hall.

Cooper, Joseph, and John Loomis. 1992. Sensitivity of Willingness-to-Pay Estimates to Bid Design in Dichotomous Choice Contingent Valuation Models. *Land Economics* 68(2): 211–224.

Costanza, Robert, Ralph d'Arge, Rudolph de Groot, Stephen Farber, and others. 1997. The Value of the World's Ecosystem Services and Natural Capital. *Nature* 387: 253–260.

Crocker, Thomas D., and John Tschirhart. 1992. Ecosystems, Externalities, and Economics. *Environmental and Resource Economics* 2(6): 551–567.

Cropper, Maureen L., Leland B. Deck, Nalin Kishor, and Kenneth E. McConnell. 1993. Valuing Product Attributes Using Single Market Data: A Comparison of Hedonic and Discrete Choice Approaches. *Review of Economics and Statistics* 75(2): 225–232.

Cropper, Maureen L., Leland B. Deck, and Kenneth E. McConnell. 1988. On the Choice of Functional Form for Hedonic Price Functions. *Review of Economics and Statistics* 70(4): 668–675.

Daily, Gretchen C. 1997. Introduction: What Are Ecosystem Services? In *Nature's Services: Societal Dependence on Natural Ecosystems,* edited by Gretchen C. Daily. Washington, DC: Island Press.

Daily, Gretchen C., Susan Alexander, Paul R. Ehrlich, and others 1997. Ecosystem Services: Benefits Supplied to Human Societies by Natural Ecosystems *Issues in Ecology* No. 2; (http://www.esa.org/sbi/issue2.pdf (accessed October 24, 2002).

Desvousges, William H., F. Reed Johnson, and H. Spencer Banzhaf. 1998. *Environmental Policy Analysis with Limited Information: Principles and Applications of the Transfer Method.* Cheltenham, U.K.: Edward Elgar.

Earnhart, Dietrich. 2001. Combining Revealed and Stated Preference Methods To Value Environmental Amenities at Residential Locations. *Land Economics* 77(1): 12–29.

———. 2002. Combining Revealed and Stated Data To Examine Housing Decisions Using Discrete Choice Analysis. *Journal of Urban Economics* 51(1): 143–169.

Eiswerth, Mark E., and W. Douglass Shaw. 1997. Adjusting Benefits Transfer Values for Inflation. *Water Resources Research* 33(10): 2381–2385.

Flores, Nicholas E., and Richard T. Carson. 1997. The Relationship between the Income Elasticities of Demand and Willingness To Pay. *Journal of Environmental Economics and Management* 33(3): 287–295.

Freeman, A. Myrick, III. 1979. *The Benefits of Environmental Improvement: Theory and Practice.* Baltimore, MD: The Johns Hopkins University Press for Resources for the Future.

Gosselink, James C., Eugene P. Odum, and R.M. Pope. 1974. *The Value of the Tidal Marsh*, Baton Rouge, LA: Department of Marine Sciences, Louisiana State University.

Huisman, Jef, and Franz J. Weissing. 2001. Fundamental Unpredictability in Multispecies Competition. *American Naturalist* 157(5): 488–494.

Kaoru, Yoshiaki, and V. Kerry Smith. 1990. *Black Mayonnaise and Marine Recreation: Methodological Issues in Valuing and Cleanup.* RFF discussion paper QE91–02. Washington, DC: Resources for the Future.

Kling, Catherine L. 1997. The Gains from Combining Travel Cost and Contingent Valuation Data To Value Nonmarket Goods. *Land Economics* 73(3): 428–439.

Levin, Simon A., and Stephen W. Pacala. Forthcoming. Ecosystem Dynamics. In *Handbook of Environmental Economics*, edited by Karl-Göran Mäler and Jeffery R. Vincent. Amsterdam: North-Holland.

Likens, G. 1992. *An Ecosystem Approach: Its Use and Abuse.* Excellence in Ecology, Book 3. Oldendorf/Luhe, Germany: Ecology Institute.

Lynne, Gary D., Patricia Conroy, and Frederick J. Prochaska. 1981. Economic Valuation of Marsh Areas for Marine Production Processes. *Journal of Environmental Economics and Management* 8(2): 175–186.

McConnell, Kenneth E., Quinn Weninger, and Ivar E. Strand. 1999. Joint Estimation of Contingent Valuation and Truncated Recreational Demands. In *Valuing Recreation and the Environment: Revealed Preference Methods in Theory and Practice*, edited by Joseph A. Herriges and Catherine L. Kling. Cheltenham, U.K.: Edward Elgar.

Mrozeck, Janusz R., and Laura O. Taylor. 2002. What Determines the Value of Life? A Meta-Analysis. *Journal of Policy Analysis and Management* 21(2): 269–286.

Rosenberger, Randall S., and John B. Loomis. 2003. Benefits Transfer. In *A Primer on Nonmarket Valuation*, edited by Patricia A. Champ, Kevin J. Boyle, and Thomas C. Brown. Dordrecht, Netherlands: Kluwer Academic Press.

Shabman, Leonard A., and Sandra S. Batie. 1978. The Economic Value of Coastal Wetlands: A Critique. *Coastal Zone Management Journal* 4(3): 231–237.

Shrestha, Ram K., and John B. Loomis. 2001. Testing a Meta-Analysis Model for Benefit Transfer in International Outdoor Recreation. *Ecological Economics* 39(1): 67–83.

Smith, V. Kerry, and Yoshiaki Kaoru. 1990. Signals or Noise? Explaining the Variation in Recreation Benefit Estimates. *American Journal of Agricultural Economics* 72(2): 419–433.

Smith, V. Kerry, George van Houtven, and Subhrendu K. Pattanayak. 2002. Benefit Transfer via Preference Calibration: "Prudential Algebra" for Policy. *Land Economics* 78(1): 132–152.

U.S. EPA (U.S. Environmental Protection Agency). 2000. *Guidelines for Preparing Economic Analyses.* Washington, DC: U.S. Environmental Protection Agency.

Whigham, Dennis F. 1997. Ecosystem Functions and Ecosystem Values. In *Ecosystem Function and Human Activities: Reconciling Economics and Ecology,* edited by R. David Simpson and Norman L. Christensen, Jr. New York: Chapman and Hall.

Name Index

Subject Index

About the Author

A. Myrick Freeman III has been on the faculty at Bowdoin College since receiving his Ph.D. from the University of Washington in 1965. Currently a research professor of economics, he has served as chair of the economics department and director of the environmental studies program. He also has held appointments as College Visiting Professor at the University of Washington, Robert M. La Follette Distinguished Visiting Professor at the University of Wisconsin–Madison, and senior fellow at Resources for the Future.

Freeman's principal research interests are applied welfare economics, benefit–cost analysis, and risk management as it relates to issues in environmental and resource management. Much of his work has been devoted to developing models and techniques for estimating the welfare effects of environmental changes—for example, the benefits of controlling pollution and the damage to natural resources as a result of the release of chemicals.

Freeman has authored or coauthored eight books, including *The Economics of Environmental Policy* (with Robert Haveman and Allen Kneese); *The Benefits of Environmental Improvement: Theory and Practice; Air and Water Pollution Control: A Benefit–Cost Assessment;* and, most recently, *The Economic Approach to Environmental Policy,* a collection of his essays. He has been a member of the Board on Toxicology and Environmental Health Hazards of the National Academy of Sciences. He has served as a member of the Advisory Council on Clean Air Compliance Analysis, the Clean Air Science Advisory Committee, and the Environmental Economics Advisory Committee of the U.S. Environmental Protection Agency.